AGRICULTURE ISSUES AND POLICIES SERIES

SOYBEAN AND WHEAT CROPS: GROWTH, FERTILIZATION, AND YIELD

AGRICULTURE ISSUES AND POLICIES SERIES

Agriculture Issues & Policies, Volume I
Alexander Berk (Editor)
2001. ISBN 1-56072-947-3

Agricultural Conservation
Anthony G. Hargis (Editor)
2009. ISBN 978-1-60692-273-6

Hired Farmworkers: Profile and Labor Issues
Rea S. Berube (Editor)
2009. ISBN 978-1-60741-232-8

Environmental Services and Agriculture
Karl T. Poston (Editor)
2009 ISBN: 978-1-60741-053-9

Weeds: Management, Economic Impacts and Biology
Rudolph V. Kingely (Editor)
2009 ISBN 978-1-60741-010-2

Effects of Liberalizing World Agricultural Trade
Henrik J. Ehrstrom (Editor)
2009 ISBN: 978-1-60741-198-7

Economic Impacts of Foreign-Source Animal Disease
Jace R. Corder (Editor)
2009 ISBN: 978-1-60741-601-2

Agricultural Wastes
Geoffrey S. Ashworth and Pablo Azevedo (Editors)
2009 ISBN: 978-1-60741-305-9

Soybean and Wheat Crops: Growth, Fertilization, and Yield
Samuel Davies and George Evans
2009 ISBN: 978-1-60741-173-4

AGRICULTURE ISSUES AND POLICIES SERIES

SOYBEAN AND WHEAT CROPS: GROWTH, FERTILIZATION, AND YIELD

SAMUEL DAVIES
AND
GEORGE EVANS
EDITORS

Nova Science Publishers, Inc.
New York

LIBRARY OF CONGRESS CATALOGING-IN-PUBLICATION DATA

Davies, Samuel, 1962-
Soybean and wheat crops : growth, fertilization, and yield / Samuel Davies and George Evans.
p. cm.
Includes bibliographical references and index.
ISBN 978-1-60741-173-4 (hardcover : alk. paper)
1. Soybean. 2. Wheat. I. Evans, George, 1961- II. Title.
SB205.S7D38 2009
633.3'4--dc22

2009010243

Published by Nova Science Publishers, Inc. ✢ New York

CONTENTS

PREFACE

The soybean is a species of legume native to East Asia. It is an annual plant that has been used in China for 5,000 years as a food and a component of drugs. Soy contains significant amounts of all the essential amino acids for humans, and so is a good source of protein. Soybeans are the primary ingredient in many processed foods, including dairy product substitutes and are an important global crop, providing oil and protein. On the other hand, wheat is a worldwide cultivated grass from the Middle East. Globally, after maize, wheat is the second most-produced food among the cereal crops. It is used to make flour, and for fermentation to make beer, alcohol or biofuel. This book addresses a wide variety of issues in the production and use of these two important crops. Among those included are pest infestation, quality of food produced for human as well as animal consumption, genetically modified plants and ways to increase productivity.

Chapter 1 - Seed coats represent the first tissue contacted by bruchids on host or non-host species suggesting its participation in the evolutionary adaptation of bruchids to legume seeds. On the cowpea (*Vigna unguiculata*) host seeds the oviposition and egg hatching phases of *Callosobruchus maculatus* are completed in about 6 days, eclosion occurs within the seed, and adult beetles emerge some 25-30 days after oviposition. Before the larva reaches the cotyledons, where it completes its life cycle, it is necessary to cross the seed coat, what may represent a critical event when infesting non-host seeds, because of physical and toxicity characteristics of this tissue. In the present chapter the authors present data on the influence of seed coat from several soybean (*Glycine max*) cultivars over the *C. maculatus* larvae capacity to penetrate, develop and survive on these seeds. Seed coat effects were evaluated by exposing the insects to different systems: natural soybeans; cowpea-based artificial seeds supplemented by soybean seed coat flour; or artificial cowpea seeds covered with natural soybean seed coats. Natural soybeans reduced both female oviposition, ranging from 100% (Tucunaré cultivar) to 35% (UFUS 2005) and larval eclosion (82.5% - Conquista cultivar - to 25% - commercial line). Major effects of natural soybeans were observed in respect to adult emergency, since no adult has emerged from any cultivar up to 40 days after oviposition. There were no positive correlations between thickness, pigmentation or surface texture of cultivars' seed coats and the larval ability of crossing this tissue. A delay of up to 116 % in the time for the larvae to cross the seed coats was observed. Some laid eggs showed abnormalities and others were completely withered. The surviving larvae that crossed the tissue, in the artificial soybean seed coat-cowpea covered system, reached 34 % of the mass of a normal larva. The incorporation of seed coat flour into artificial seeds revealed that the

UFV 20 Florestal was the most toxic cultivar (WD_{50} [dose that reduced larval weight to 50%] = 1.5%). Lowest levels of toxicity were observed for the UFUS 2005, Conquista, UFUS 2003 and Elite cultivars (WD_{50} varying from 10.5 to 12%). LD_{50} (doses that reduced the surviving larvae number to 50%) were also variable, ranging from 1% to 14% among the cultivars. Despite all variations, soybean seed coats were highly restrictive to the bruchid suggesting that the tissue plays an important role for evolutionary discrimination of legumes by this bruchid.

Chapter 2 - Soybean was grown in a lysimeters filled with loam and clay soils and was irrigated with water having three different levels of salinity (fresh water, and saline waters with 15 and 30 meq Cl/l).

During the soybean crop cycle, soil salinity was determined from the salt balance. Leaf-water potential, stomatal conductance and actual evapotranspiration were used as the water-stress indicators. Growth was measured through leaf area and dry matter and, finally, the yield and its components were determined. The water use efficiency was also calculated.

Without salt stress (treatments irrigated with fresh water), the effect of soil texture on the water relationship, productivity and water use efficiency of the soybean was not demonstrated.

With salt stress, all the parameters, in both types of soil, were coherent, indicating systematic differences between the saline treatments and the control treatments (treatments irrigated with fresh water).

Soil texture affects the soybean response to soil salinity. The saline treatments in the loam soil caused the values of the water stress indicators, of growth, of yield and of water use efficiency to be higher than the highest values observed for the same treatments on the clay soil.

The analysis of the relationship between relative yield and soil salinity indicates clearly that soybean shows a higher salt tolerance if it is cultivated in loam soil.

Chapter 3 - In recent years, though, expansion of soybean croplands has been increasingly important in the agricultural or production in many parts of the world. There are a lot attempts to set this cultivar in the modeling focus, from different points of view (microclimate, irrigation, crop, land surface, climate cahnge, etc.). However, regardless the model is used, the interaction of surface and subsurface runoff and soil moisture, the simulation of total evaporation (or latent heat) are always highly ranked in the modeling hierarchy.

This chapter deals with the simulation of the water balance components of the soybean canopy using a surface scheme. In that sense the authors used the hydrological module in the Land-Air Parameterization Scheme (LAPS) developed at Faculty of Agriculture, Department for Field and Vegetable Crops, University of Novi Sad (Serbia). It is designed as a software package that can be run as part of an environmental model or as a stand-alone one. The LAPS includes modeling the interaction of the land surface and the atmosphere, under processes divided into three sections: subsurface thermal and hydraulic processes, bare soil transfer processes and canopy transfer processes. They are: interaction of radiation with vegetation, evaporation from bare soil, evapotranspiration including transpiration and evaporation of intercepted water and dew, conduction of soil water through the vegetation layer, vertical water movement in the soil, surface and subsurface runoff, heat conduction in the soil and momentum transport within and above the vegetation. The scheme has seven prognostic variables: three temperature variables (foliage, soil surface and deep soil), one interception storage variable, and three soil moisture storage variables. For the upper boundary conditions the following forcing variables are used: air temperature, water vapor pressure, wind speed, short wave and long wave radiation and precipitation at a reference level within the atmospheric boundary layer. The hydrological module is designed as a three-layer model, which is used to describe the vertical transfer of water in the soil. The LAPS uses the morphological and physiological characteristics of the plant community for deriving the coefficients and resistances that govern all the fluxes between the surface and atmosphere.

In order to simulate partitioning of the soybean canopy water into water balance components, during short and long period (day, growing season), several simulation were performed. The corresponding forcing, morphological, physiological and soil data as well as the observations were inserted from data sets comprising different agroecological soybean regions: Paragaminas (Brasil), Marchfeld (Austria) and Caumont (France).

Chapter 4 - During the past 30 years, breeding programs have developed many soybean varieties for their adaptation to different geographical areas and for improving seed characteristics: increasing protein and oil concentrations, improving protein quality, reducing antinutritional compounds, etc. The differentiation among the increasing number of soybean cultivars is not an easy task since many of them are genetically very close. Traditional methodologies for the identification of soybean cultivars were based on phenotypic characters from the leaf, stem, and seed. Since many different soybean cultivars are indistinguishable based on these features, other methodologies have raised as alternatives for cultivar characterization.

The characterization of soybean cultivars through the analysis of proteins has been reviewed in this work. Special emphasis was made on the use of electrophoretic and chromatographic techniques. The discussion of the results obtained by our research team in relation with the differentiation of 91 soybean varieties through their protein profiles obtained by a rapid chromatographic methodology will also be included.

Chapter 5 - Soybean (*Glycine max* (L) Merr) production in the smallholder farming areas of southern Africa is constrained by soil acidity and nutrient deficiency among other factors. A study was conducted to determine the performance of four soybean cultivars commonly grown in Southern Africa, in acid soil, their response to liming, and N and P fertilisation. Soybean was grown over two cropping seasons at a research station and in a sub-humid smallholder farming area in north-eastern Zimbabwe. Liming increased the number of nodules and nodule dry matter yield (NDMY) in both cropping seasons in all the four soybean cultivars tested but the differences were only significant in the second cropping season (nodule number, $p=0.004$; NDMY, $p=0.025$). In both seasons liming increased grain yield (season 1, $p=0.046$; season 2, $p=0.023$) but cultivar differences were not different. Addition of P fertiliser increased P uptake, grain and stover yield and liming further enhanced both P uptake and grain and stover yield. Addition of 30 kg N ha^{-1} as ammonium nitrate significantly ($p<0.05$) reduced nodules numbers and grain yield compared to treatments where N had not been applied. It was concluded that soybean productivity in acid soil prevalent in humid and sub-humid areas of SSA can be effectively increased through liming and P fertilisation. Application of mineral N not supported by soil testing reduces the effectiveness of biological N fixation thereby depriving farmers of a cheaper source of N where soil N is relatively high.

Chapter 6 - Production of healthy food in sufficient amounts is, among the others, hindered by the activity of various plant pathogens, of which fungi have an especially negative impact. *Fusarium* species produce a broad spectrum of toxins including fumonisins, trichothecenes of the A- and B-type, and zearalenone. When contaminated plants are used as food and feed the toxins involved exhibit a lot of harmful effects on humans and animals. The occurrence of *Fusarium* toxins in cereals has been registered all over the world. However, in contrast to corn and wheat, no special attention has been paid to the study of *Fusarium* toxins in soybean. Because of that the aim of this study was to gain an insight into the presence of deoxynivalenol (DON) in soybean grown in Serbia, based on the analysis of 42 soybean and soybean meal samples collected during 2004–2007, as well as to compare the obtained results with data pertaining to a number of countries. Samples were analyzed by liquid chromatography on ODS Hypersil column with DAD detector. The DON content was above the limit of quantification (0.040 mg/kg) in 16.7%, with an average content in positive samples of 0.248 mg/kg (concentration range 0.10–0.45 mg DON/kg). However, none of the samples contained DON above the advisory level of 500 μg/kg of DON, passed by the European Union, which is not related strictly to soybean but to cereal products as consumed and other cereal products at retail stage.

Chapter 7 - The yield of Roundup Ready® soybean plants infested by larvae of *Dectes texanus* LeConte was compared to that of uninfested plants obtained from the same fields in two successive years in west-central Kansas. Plants were significantly larger in all respects in 2006 than in 2005, due largely to a lower plant population, but there was no reduction in either pod number or total seed weight as a consequence of larval boring in either year. Infested plants had marginally greater total seed weights than uninfested plants in 2005, an effect attributed to females preferring larger plants for oviposition when average plant size was small. Mean stem diameters were not notably different between infested and uninfested plants as observed in certain earlier studies, suggesting that *D. texanus* populations may now be exploiting smaller size classes of soybean plants to a greater extent. Although our results suggest no impact of larval boring on yield, the possibility of some yield impact under different growing conditions cannot be ruled out. However, any yield losses arising from *D. texanus* larval boring are likely small compared to those arising from pre-harvest lodging that results when mature larvae girdle plants prior to overwintering.

Chapter 8 - Fusarium Head Blight (FHB), also known as scab, is a devastating disease of enormous economic importance throughout the world that attacks all classes of barley and wheat. Every year, all of the most important cereal producers in the world are affected by this disease. *Fusarium graminearum* and *Fusarium culmorum* are considered the most pathogenic and widespread agents of FHB in wheat. Both *F. graminearum* and *F. culmorum* produce deoxynivalenol (DON, also known as vomitoxin), a mycotoxin of the trichotecenes group, one of the most widespread mycotoxins in cereals, which can be synthesised in the field in small grain cereals and in maize. This mycotoxin has a great impact on the health of animals and humans, due to their cytotoxic activity and immunosuppressive effects. As a consequence, FHB and DON contamination are responsible for serious direct and indirect economic losses. Economic losses can be attributed to yeld loss due to fungal disease, reduced technological and nutritional quality of grain, reduced crop value due to mycotoxin contamination, reduced animal productivity, human health costs. Prevention, monitoring, sampling, chemical analysis, litigation, mitigation, and research costs also need to be taken into account. Moreover, FHB and DON limits, which have been established in many parts of the world for animal feeding and human consumption, have made the worldwide cereal market much more selective.

F. graminearum and F. culmorum inoculum is usually abundant and its production, dispersal and deposition are influenced by weather conditions. Warm, humid weather, frequent rainfall and heavy dew favour spore germination. Temperature and water activity of the host are the fundamental factors that affect fungal growth inside grain. DON synthesis is mainly influenced by temperature and water ativity and its production seems to play a decisive role on the aggressiveness of *F. graminearum* and *F. culmorum*: it has been suggested that in the presence of DON, plant defence mechanisms are not triggered fast enough, thereby leading to an increased aggressiveness of the pathogen.

The accumulation of DON in the grain is intimately related to the development of the disease in the field. Prevention strategies through pre-harvest agronomic management can achieve the quality and safety standards required by the market and international regulations. Among the agricultural practices, FHB development and DON contamination in wheat grains are mainly affected by tillage, crop rotation, varieties and fungicide application. None of these strategies on their own are able to significantly reduce the impact of this disease. Nevertheless, careful planning of all the different management decisions canl lead to the designing of crop management systems able to reduce FHB and mycotoxin incidence and severity. In other words, it is necessary for producers to make use of the available wheat varieties that are known to be more resistant and to apply all of the good agricultural practices tuned to a determined cultivation area.

Useful tools for prevention are represented by FHB and DON predictive models: the use of a model to predict the outcome of a disease is desirable to enhance and trigger management opportunities with the aim of reaching high technological, nutritional and productivity quality and safety of the production.

Chapter 9 - In the 1980s, Scottish grain distillers began to change from imported maize to home-grown wheat as their preferred adjunct, requiring an annual intake in excess of 0.5 million tonnes. For ease of processing, soft-milling varieties were required, and low grain protein contents were desirable. However, as this was a localised market, requiring a small proportion of the UK wheat harvest, it received little attention from wheat breeders until European interest developed in the use of wheat-based fuel ethanol as a partial petrol replacement. A greater number of varieties with potentially high alcohol yields are now being entered into national trials, but breeders face problems in early generation selection for alcohol yield, as rapid testing procedures are still being developed, including the use of Near Infra-Red Spectroscopy (NIR). Research to locate genetic factors responsible for alcohol yield, on wheat chromosomes, is also at an early stage, although this should facilitate the use of DNA-based selection systems in future breeding programmes. Changes in the quantity and timing of nitrogen fertiliser may also be necessary, as grain nitrogen content has a significant and negative effect on alcohol yield and reduced inputs are also desirable to enhance the energy balance associated with fuel ethanol production. However, these have to be achieved without a deleterious effect on grain yield. As a number of current varieties have good alcohol yield potential, but may have agronomic weaknesses, an alternative approach for cultivation is in the form of varietal mixtures. Complex mixtures, i.e. those with four or more components, have also been shown to increase grain yield and to reduce the spread of disease and thus the need for prophylactic spraying of fungicides. Mixtures are also likely to provide greater stability, across sites and seasons, for both yield and quality.

Chapter 10 - Genetic improvement of yield potential has always been an important objective in China. Averaged annual genetic gain in grain yield ranged from 13.96 kg/ha/year to 72.11 kg/ha/year or from 0.31% to 1.23% annually in different wheat regions. The genetic improvement in grain yield was primarily attributed to increased grain weight per spike, reduced plant height, and increased harvest index. Three dwarfing genes and 1B/1R translocation have been successfully used in wheat production. *Rht-D1b* (45.5%) and *Rht 8* (46.8%) were more frequent, followed by *Rht-B1b* (24.5%). The frequencies of *Rht-B1b* and *Rht-D1b* increased, from 8.6% to 32.2% and 36.2% to 53.4%, respectively, whereas the frequency of *Rht8* has remained constant over time, when compared with cultivars released before and after 1990. From the late 1970s to the early 1990s, wheat breeding in autumn-sown wheat regions focused on the utilization of the 1B/1R translocation. The dominant *Vrn-D1* allele showed the highest frequency in Chinese wheat cultivars (37.8%), followed by the dominant *Vrn-A1*, *Vrn-B1*, and *Vrn-B3* alleles. All cultivars released in the Northern Winter Wheat Zone were winter type. Winter (53.0%), spring (36.1%) and early heading (10.9%) cultivars were grown in the Yellow and Huai River Valleys Facultative Wheat Zone. Most of the spring genotypes from this zone carried only the dominant *Vrn-D1* allele, which was also predominant (64.1%) in the Middle and Lower Yangtze Valleys Autumn-Sown Spring Wheat Zone and Southwestern Autumn-Sown Spring Wheat Zone. The average frequency of the photoperiod-insensitive *Ppd-D1a* allele was 66.0%, with the frequencies of 38.6% and 90.6% in landraces and improved cultivars, respectively. Therefore, in addition to utilization of dwarfing genes and the 1B/1R translocation, dominant vernalization and photoperiod genes for early maturity have also contributed to yield improvement of Chinese wheat. The future challenge of wheat breeding is to continually raise grain yield, or to both maintain the genetic gain in grain yield and improve grain quality, without increasing inputs for the wheat based double cropping system.

Chapter 11 - Wheat is the second most important cereal crop in Bangladesh. A unique feature of wheat in Bangladesh is 100% adoption of modern varieties. The present chapter provides an account of the growth performance of wheat in Bangladesh over the past four decades. The chapter then examines the productivity performance of the wheat producers as well as its determinants at the farm-level using a survey data of 293 households collected from three wheat growing regions in 2004. Results reveal that the area under wheat increased six folds from only 132,000 ha in 1971 to 832,000 ha in 2000 but then declined sharply to 479,050 ha in 2006. Consequently, total production and yield grew at an annual rate of 6.9% and 1.9%, respectively. The actual yield level increased from 0.9 t/ha to 1.5 t/ha over this 36 year period. Farm-level result reveals that the environmental production conditions within which the farmers operate significantly affect productivity as well as technical efficiency of wheat production, an issue commonly ignored in the existing literature. Wheat productivity is significantly lower in low lying areas and poor soils. Productivity is also significantly affected by a delay in sowing. Technical efficiency of wheat production in Bangladesh is estimated at 83%, implying that production can be increased by 20% [{(100-83)/83}*100] through reallocation of resources alone. Analysis of the determinants of technical efficiency reveals that a host of managerial and socio-economic factors significantly affect performance of wheat producers. Farmers' education, access to agricultural information, training and use of mechanical power significantly improves technical efficiency, whereas a delay in sowing and fertilization, and poor sourcing of seeds (i.e., from local market and/or neighbours) significantly reduces efficiency. Large farms are more efficient relative to small and medium sized farms.

Geography does matter. Productivity of wheat is significantly lower in Jamalpur region. Policy implications include, soil fertility improvement through soil conservation and crop rotation, improvement in managerial practices through extension services and adoption of modern technologies, promotion of education and training targeted to farmers, strengthening the research-extension link, and development of new varictics that have higher yield potential and are also suitable for marginal areas.

Chapter 12 - Losses in biomass production and protein caused by stress factors such as drought, waterlogging, salinity, or heavy-metal contamination are accompanied by structural damages to cellular membranes of crop plants, to alterations in enzyme activities, and to cytoplasmic dehydration. Water deficit leads to the closure of stomata, to reduced uptake of aerial CO_2, and thus to the inhibition of photosynthesis and the excessive formation of reactive oxygen species in chloroplasts and other cell compartments. Stress-activated phospholipases and esterases catalyze the separation of ethanolamine (EA), choline, and their phosphate esters, as well as fatty acids and serine from the phospholipid molecule of cell membranes. The membrane protein component is subject to proteolysis. The liberated serine is a precursor for the formation of further EA and choline. In a research project extending over 3 decades, the membrane lipid moieties, ethanolamine and choline were applied at 1.5 kg/ha as a foliar spray to potted cereal plants and to cultures in more than 150 field trials. External application of the biogenic stress metabolites was expected to initiate the plants' resistance and tolerance mechanisms by pretending a stress situation. Actually, the treatment confined yield losses in drought-stressed wheat, rye, and barley by stabilizing water household, photosynthesis, and protein production. Relative to untreated, drought-stressed plants, increases in biomass (5-20 %) and protein content (5-7 %) were recorded in 14 to 58 % of the experiments which had been conducted at 17 Experimental Agrostations of different soil conditions but little climatic extremes. Plant responses were significant under drought conditions and on poor soils. They were negligible in the absence of abiotic stress. The major short- and long-term stress responses were followed on the biochemical and ultrastructural level. Under the temperate climatic conditions of Central Europe, the ecotoxicologically unobjectionable treatment has not yet become part of daily agricultural practice. In the light of the world-wide food crisis, an application of stress control agents could help stabilize crop production in semi-arid and saline regions and during temporary rainfall deficits. The treatment should also revalorize the economic role of the multitude of traditional cereal cultivars in use which are adapted to the local soil and climate conditions.

Chapter 13 - The composition and feeding value of wheat is usually more variable than that of other cereals. Protein levels in wheat, for example, can vary from 10-18%. But wheat provides 10% less metabolisable energy than corn, mainly due to the presence of soluble non-starch polysaccharides. These non-starch polysaccharides cause increased digesta viscosity in the gut, leading to reduced nutrient digestibility and metabolisable energy values, especially in young birds. Wide variation in the metabolizable energy of wheat is a major concern to the poultry industry. Currently, exogenous enzymes, containing xylanase activity, are routinely used to mitigate the adverse effects of non-starch polysaccharides and to minimize the variation in the performance of poultry fed wheat-based diets. With the use of enzyme supplementation, wheat can be used without any limitation in poultry diet. Feeding of whole wheat grains to poultry has also generated interest in recent years as a mean reducing processing and production costs, and improving gut health.

In: Soybean and Wheat Crops
Editors: S. Davies, G. Evans
ISBN: 978-1-60741-173-4

Chapter 1

INFLUENCE OF THE SOYBEAN SEED COAT UPON SEED INFESTATION AND DEVELOPMENT OF THE INSECT *CALLOSOBRUCHUS MACULATUS*

Antonia Elenir A. Oliveira, Kátia V. S. Fernandes*, Amanda J. Souza and Patrícia O. Santos

Laboratório de Química e Função de Proteínas e Peptídeos LQFPP, Centro de Biociências e Biotecnologia - CBB, Universidade Estadual do Norte Fluminense Darcy Ribeiro - UENF, Campos dos Goytacazes - RJ, Brazil

ABSTRACT

Seed coats represent the first tissue contacted by bruchids on host or non-host species suggesting its participation in the evolutionary adaptation of bruchids to legume seeds. On the cowpea (*Vigna unguiculata*) host seeds the oviposition and egg hatching phases of *Callosobruchus maculatus* are completed in about 6 days, eclosion occurs within the seed, and adult beetles emerge some 25-30 days after oviposition. Before the larva reaches the cotyledons, where it completes its life cycle, it is necessary to cross the seed coat, what may represent a critical event when infesting non-host seeds, because of physical and toxicity characteristics of this tissue. In the present chapter we present data on the influence of seed coat from several soybean (*Glycine max*) cultivars over the *C. maculatus* larvae capacity to penetrate, develop and survive on these seeds. Seed coat effects were evaluated by exposing the insects to different systems: natural soybeans; cowpea-based artificial seeds supplemented by soybean seed coat flour; or artificial cowpea seeds covered with natural soybean seed coats. Natural soybeans reduced both female oviposition, ranging from 100% (Tucunaré cultivar) to 35% (UFUS 2005) and larval eclosion (82.5% - Conquista cultivar - to 25% - commercial line). Major effects of natural soybeans were observed in respect to adult emergency, since no adult has

* Corresponding author: Kátia Valevski Sales Fernandes. Universidade Estadual do Norte Fluminense; Av. Alberto Lamego, 2000 – Horto; Campos dos Goytacazes – RJ, Brasil. CEP: 28013-602. cowpkat@uenf.br ; Fax: 55 22 27261520.

emerged from any cultivar up to 40 days after oviposition. There were no positive correlations between thickness, pigmentation or surface texture of cultivars' seed coats and the larval ability of crossing this tissue. A delay of up to 116 % in the time for the larvae to cross the seed coats was observed. Some laid eggs showed abnormalities and others were completely withered. The surviving larvae that crossed the tissue, in the artificial soybean seed coat-cowpea covered system, reached 34 % of the mass of a normal larva. The incorporation of seed coat flour into artificial seeds revealed that the UFV 20 Florestal was the most toxic cultivar (WD_{50} [dose that reduced larval weight to 50%] = 1.5%). Lowest levels of toxicity were observed for the UFUS 2005, Conquista, UFUS 2003 and Elite cultivars (WD_{50} varying from 10.5 to 12%). LD_{50} (doses that reduced the surviving larvae number to 50%) were also variable, ranging from 1% to 14% among the cultivars. Despite all variations, soybean seed coats were highly restrictive to the bruchid suggesting that the tissue plays an important role for evolutionary discrimination of legumes by this bruchid.

Keywords: bruchid, cowpea, legume discrimination, seed coat, seed defense, soybean

INTRODUCTION

The seed coat consists of several layers of specialized maternal cell types that provide an important interface between the embryo and the external environment during embryogenesis, dormancy and germination (Haughn & Chaudhury, 2005). In certain species, seed coat is sometimes represented by a rudimentary testa, where the first external covering is the pericarp, derived from the ovary wall. In the Fabaceae, the seed coat originates from the two-ovule integuments (De Souza & Marcos-Filho, 2001). During seed ontogeny, the outer integument gives rise to several distinct layers transforming itself into the testa, while in many species, the inner integument disappears (Esau, 1977; Miller et al. 1999). Especially among *Glycine* and *Phaseolus* species, the seed coat is a true testa. Special features described for *Glycine* seed coats comprise: two waxy layers instead a single outermost cuticle layer (Ragus, 1987); small openings denominated pores, or pits, (or antipits [Ma et al. 2004]) unevenly distributed throughout the testa surface (La Scala Jr. et al. 1999); and the adherence of the membranous inner endocarp epidermis of the pod wall to the seed coat surface (Gijzen et al. 1999). The vast majority of paper work published on soybean seed coat relates to water uptake issues (Calero et al. 1981; Yaklich et al. 1984: Harris, 1987; Koizumi et al. 2008).

Among the major seed coat functions are: preservation of the integrity of seed parts, regulation of aqueous and gaseous exchanges between the embryo and the external environment, dispersion process of some seeds and in the protection of the embryo against mechanical damage and attacks of pests and pathogens (Zeng et al. 2004).

Insect injury to leguminous seeds poses a serious problem for agriculture and food processing. Despite the abundance of defensive chemicals, such as lectins, proteinase inhibitors and secondary metabolites (tannins, alkaloids, cyanogenic glucosides) found in these seeds (Carlini & Grossi-de-Sá, 2002), several members of the seed-eating Bruchidae family are major pests of cultivated legumes, developing within and consuming seed tissues, which would be utilized for human consumption (Southgate, 1979), without suffering any severe damages. Among these insect species, *Callosobruchus maculatus* (F.) is of considerable importance as a cosmopolitan pest of stored cowpea *Vigna unguiculata* (L.) and to a lesser extent of other stored legumes like mung bean (*Vigna radiata*), adzuki beans (*Vigna angularis*) and pigeon pea (*Cajanus cajan*) (Jackai & Daoust, 1986). More recently, adaptation to a novel host legume, *Cicer arietinum,* has also been reported (Fricke & Arnqvist, 2007). The female and male couple and the females begin oviposition within 1 hour. *C. maculatus* females cement their eggs onto the seed surface. Oviposition is completed in about 8 days, and adults die about 10-12 days after emergence. Eggs hatch after 3–5 days. Before the larva arrives to the cotyledons, where it will complete the life cycle, it is necessary that the larvae cross the seed coat, which represents a critical event (Credland, 1987). Larval development and pupation take place inside the seed, adults emerge from the seed some 25-30 days after oviposition, and adult emergence leads to a reduction in seed quality, poor field performance and lower acceptability in the food industry (Caswell, 1968; Southgate, 1979; Credland, 1987). The larvae of this species feed and develop exclusively on the seed of legumes and the adults do not require food or water and spend their limited lifespan mating and laying eggs on beans (Beck & Blumer, 2007).

Seed defense against insect attacks may involve a diverse set of contributors, going from a tissue to a molecular level of action. Approaches on the seed coat level contribution are not numerous. One of these studies analyzed the perforation capacity of *C. maculatus* larvae in seeds of 73 species and showed that the seed coat prevented the penetration of larvae in 69.5% of the tested seeds. However, there was no direct relationship between hardness and / or thickness of the seed coat with the larval abilities for crossing or not this tissue (Janzen, 1977), since the work showed a high mortality of larvae during the perforation of the very thin seed coats of *Erythrina berteroana* and *Ormosia venezolana* seeds (Janzen, 1977). Research on infestation of *Phaseolus lunatus* by *Acanthoscelides obtectus* larvae showed that these were unable to penetrate the seeds and the reasons for that also pointed to not physical aspects of the tissue, since *P. lunatus* seed coat is not as hard as others from different species of *Phaseolus* (Thiéry, 1984). Thiéry *et al.* (1994) showed that the survival of *A. obtectus* depends on the ability of the first instar larvae to pierce the seed coat of *P. vulgaris* and only 57 % of the larvae successfully penetrated the seeds. The seed coat of fourteen *Vicia faba* genotypes acted like a barrier for the two bruchid species, *Callosobruchus chinensis* (L.) and *Callosobruchus maculatus* (F.). Only 45-58% of the neonate larvae were able to perforate the seed coat and thus reached the cotyledons; the other larvae began to perforate the seed coat and died while trying to penetrate it (Boughdad et al. 1986; Desroches et al., 1995).

Previous studies have shown the presence of vicilin-like 7S storage globulin, normally thought to be expressed only in the embryo, in the seed coat of some legume seeds. Oliveira et al. (1999) isolated, from *Canavalia ensiformis* seed coat, a protein that was seen to be homologous to canavalin (vicilin-like protein from *Canavalia* seed cotyledons) by N-terminal amino acid sequence determination. This protein interfered with the normal development of the cowpea weevil *C. maculatus,* reducing the mass and number of surviving larvae. The presence of phaseolin (a vicilin-like protein) peptides in the seed coat of the legume *Phaseolus lunatus* L. (lima bean) was demonstrated by N-terminal amino acid sequencing. Utilizing an artificial seed system assay it was showed that phaseolin was detrimental to *C. maculatus* with a LD_{50} of 3.5%. The level of phaseolin in the seed coat (16.7% of the total protein) was found to be sufficient to deter this bruchid larval development (Moraes et al. 2000). Silva et al. (2004) have confirmed that the seeds of the common bean (*Phaseolus vulgaris* L.) do not support development of the same bruchid. Analysis of the *P. vulgaris* seed coat suggested that neither thickness nor the levels of compounds such as tannic acid, tannins, or HCN were important for the seed resistance. On the other hand, the authors showed that phaseolin is detected in the seed coat by Western blotting and N-terminal amino acid sequencing and that this protein was detrimental to the development of *C. maculatus*. Studies have suggested that the resistance of some *Vigna unguiculata* cultivars seeds to *C. maculatus* attack is due to variant forms of vicilins which are refractory to hydrolysis by midgut proteinases, what possibly leads to low dietary amino acid supply to the larvae (Xavier-Filho et al. 1991; Sales el al. 1992; Macedo et al. 1993). The mechanism of action of variant vicilins seems to be also linked to their chitin-binding power (Sales et al. 1996; Firmino et al. 1996; Yunes el al. 1998). This mechanism may be similar to the one attributed to the action of chitin-binding proteins (N-acetylglucosamine-specific lectins, chitinases, hevein and antimicrobial peptides) which are involved in defense mechanisms of plants against insects and pathogens (Chrispeels & Raikhel, 1991).

A water-soluble polysaccharide from the Jack bean [*Canavalia ensiformis* (L) DC] seed coat that was shown to be highly detrimental to larval development of the cowpea weevil *C. maculatus* was also isolated (Oliveira et al. 2001). Determination of the composition and structure of this polysaccharide showed that it is a galactorhamnan with an M_w of 883.0, containing 92 % rhamnose and 8 % galactose. The polymer is formed by a main chain of rhamnose ($1\rightarrow2$) substituted at *O*-4 by galactose non-reducing end units. Immunolocalisation by light and electron microscopy showed that this polysaccharide is localized in the innermost cell layer of the seed coat and also in the cytoplasmic space of cotyledonary tissues.

Soybean seed coat proteins have been previously identified, among these a 41 kDa peroxidase enzyme (Buttery & Buzzel, 1968; Gijzen, 1997), a 32 kDa class I chitinase (Gijzen et al. 2001), a 21 kDa trypsin inhibitor (Kunitz, 1945; Koide & Ikenaka, 1973) and an 8 kDa hydrophobic protein (Gijzen et al. 1999a); however the functions of these proteins are not completely understood. Soybean seed coat peroxidases (SBPs) were shown to be very stable at high temperatures, extremes of pH and in organic solvents (Nissum et al. 2001). The mature protein showed higher than 70% amino acid sequence identity to peroxidases from other legumes recruited in various defense response processes (Henriksen et al. 2001). Examples of such plant defense relation with peroxidases are found in Hammerschmidt et al. (1982) and El-Turk et al. (1996). Hammerschmidt et al. (1982) showed the association of enhanced peroxidase activity with induced systemic resistance of cucumber to *Colletotrichum lagenarium*. El-Turk et al. (1996) reported the nucleotide sequences of four pathogen-induced alfalfa peroxidase-encoding cDNAs. Gijzen et al. (2001) showed that a 32 kDa soybean seed coat protein presented an N-terminal cysteine-rich hevein domain typically found in class I chitinases and in other chitin-binding proteins. The protein was abundant in soluble extracts from soybean seed coats. We have recently shown that a soybean seed coat protein fraction was able to inhibit the growth of *Fusarium lateritium* and *Fusarium oxysporum* phytopathogenic fungi. The antifungal fraction revealed the presence of peroxidase, vicilin and of a 24 kDa protein, homologous to acid phosphatases. Germination experiments revealed that both acid phosphatase and peroxidase were exuded during seed imbibition what might be an indicative of a protective role for these proteins during seed germination (Santos et al. 2008).

The expression of the *C. maculatus* detrimental compounds in the seed coats of the non-host seeds may have been important for the evolutionary discrimination of legume seeds by this bruchid. Observations done by our group have shown that a great percentage of the *C.*

maculatus larvae that die in non-host *Canavalia ensiformis* (Oliveira et al., 1999), *Phaseolus lunatus* (Moraes et al., 2000) and *Phaseolus vulgaris* (Silva et al., 2004) seeds does not cross the seed coat and that vicilin-like proteins isolated from these seed coats were toxic to this insect larvae.

Although an enormous number of toxic proteins and peptides found in seed cotyledons have been related to the resistance of some seeds against insects (Macedo et al. 1993; Gomes el al. 1996; Carlini et al. 1997; Carlini & Grossi-de-Sá, 2002; Moura et al. 2007), almost no work has considered the toxicity of the seed coat as an important factor for such resistance, despite the knowledge that this tissue is the first natural barrier encountered by pests of stored seeds. The overall strategy has been proposed by Miller et al. (1999), who proposes that tissues of developing soybean seed coats may be targeted for modification of tissue expression aiming to influence properties of mature seeds. Additionally, a large amount of hydrophobic proteins which may prevent pathogen attachment and penetration to the seed coat and act as deterrent or toxin to other organisms, such as insects, have been located at the special adherence sites formed by the merging of pod wall tissues and soybean seed coats. Considering this aspect, Gijdzen et al. (1999b) suggested that it may be possible to alter seed coat characteristics through manipulation of gene expression in the ovary wall. Although seeds have been the subject of extensive studies for many years, their seed coats are just beginning to be examined from the perspective of biochemistry, molecular genetics and control of development. Within the seed coat are a number of unique tissues that undergo differentiation to serve specific functions in the seed. A large number of genes are known to be specifically expressed within the seed coat tissues; however, very few of them are understood functionally. The seed coat synthesizes a wide range of novel compounds that may serve the plant in diverse ways, including defense and control of development. The use of seed coat biotechnology to enhance seed quality and yield or to generate novel components has not been exploited, largely because of lack of knowledge of the genetic systems that govern seed coat development and composition as well as the function of these compounds (Moïse et al. 2005).

In the work presented here, we summarize some of the results obtained by our research group when studying the specific relationship among the soybean seed and the bruchid *C. maculatus*, with major emphasis on the influence of the soybean seed coat tissue upon the disruption of seed infestation by the insect.

MATERIAL AND METHODS

Seeds

Commercial soybean (*Glycine max)* and cowpea (*Vigna unguiculata* cv. Fradinho) seeds were bought from local markets at Campos dos Goytacazes, RJ, Brazil.

Non-commercial soybean cultivars (cv.) - Conquista, UFV-20 Florestal, UFUS 2003, UFUS 2005, Tucunaré and Elite - were obtained from the Departamento de Fitotecnia of the Universidade Federal de Viçosa (UFV), Viçosa-MG, Brazil.

Insect

Callosobruchus maculatus (Coleoptera: Bruchidae) was obtained from a colony maintained in the Laboratório de Química e Função de Proteínas e Peptídeos from the Centro de Biociências e Biotecnologia, Universidade Estadual do Norte Fluminense, Campos dos Goytacazes, RJ, Brazil. The insects were reared on cowpea seeds (cv. Fradinho) at 28 °C, 60-80 % r.h., and a photoperiod of L12:D12.

Seed Coat Thickness, Texture and Coloration

The seeds of cowpea (cv. Fradinho) and soybean (commercial seeds and Conquista, UFV-20 Florestal, UFUS 2003, UFUS 2005, Tucunaré, Elite cultivars) were de-coated and the thickness of 3 randomly chosen seed coat fragments were measured by an electronic digital Marathon micrometer (0-25 mm measuring range). The seed coats pigmentation and texture were analyzed through observation under a stereoscopic microscope coupled to a digital CCD video camera. The results were analyzed through Student's *t* test, and significant differences were determined at $P < 0.05$ (Bridge & Sawilowsky, 1999).

Natural Soybean Seeds Experiments

Sixty natural soybean seeds from commercial line or from the Conquista, UFV-20, Florestal, UFUS 2003, UFUS 2005, Tucunaré and Elite cultivars, placed into glass flasks, were exposed to ten *C. maculatus* females (of 3-days-old) during 24 h, inside a BOD incubator chamber, at 28 °C and 70% r.h. After this time, the females were removed and the eggs laid on the seeds were counted. After 8 and 12 days of incubation at 28 °C, 70% r.h., the larval eclosion was observed and registered and the seeds were incubated for other additional 28 days, in the same previous conditions. Adult emergence was evaluated after the whole 40 days-period. The larval development was accompanied throughout the period, by daily observations with a stereoscopic microscope coupled to a digital CCD video camera, since the posture day (first day after oviposition) until the complete perforation of the seed coat or until larval death. The necessary time (days) for the surviving larvae to perforate the seed coat completely was also observed. Control experiments using cowpea host seeds were performed at the same above described conditions. The results were analyzed through Student's *t* test, and significant differences were determined at $P < 0.05$ (Bridge and Sawilowsky, 1999).

Artificial Cowpea Seeds Covered with Natural Soybean Seed Coats Experiments

Seed coat pieces were separated from cotyledons and the effectiveness of these tissues as barriers against the penetration of the C. maculatus larvae was studied by using an artificial seed system (Macedo et al. 1993), modified by us to include a covering piece made of natural soybean seed coat (Figure 6A'). Artificial seeds (final mass of 400 mg) were made by using a finely ground decorticated cowpea seed meal, from an insect susceptible *Vigna unguiculata* genotype (cv. Fradinho). 50 mg of the flour of *V. unguiculata* was placed into a cylindrical brass mould and over that powder a piece of natural seed coat, with the external part of the seed coat facing down, was laid. Later, 400 mg of *V. unguiculata* flour were added over the seed coat piece. The artificial seed was made by pressing the mixture with the help of a hand press, and after removal of the seed from the mould, the *V. unguiculata* flour excess (50 mg) was removed from the outer surface to expose the piece of seed coat (Figure 6A'). The parts of the artificial seed that were not covered by the seed coat fragment were protected with a plastic film and the seeds were presented to fertilized females in glass vials. Three females (3-days-old) were transferred to each vial (two seeds per vial) and kept inside it for 24 h at 28 °C, at 70% r.h. After this time, the females and the excess of eggs were removed; three eggs laid on the piece of seed coat were left per seed. After 8 days of incubation at 28 °C, at 70% r.h., the larval eclosion was observed and recorded. After completion of 20 incubation days, larvae that crossed the seed coat or died inside the eggs were counted. The seeds were opened and the weights of all larvae were taken and compared with larvae grown in control artificial cowpea seeds. Experiments were run in triplicate (total of 9 eggs). These 9 eggs laid on seed coats were considered as 100 % when calculating the percentages of larval eclosion, larva that crossed the seed coat and larvae that died inside the eggs. Control seeds consisted of artificial cowpea seeds covered by a cowpea seed coat fragment, manufactured identically as above described.

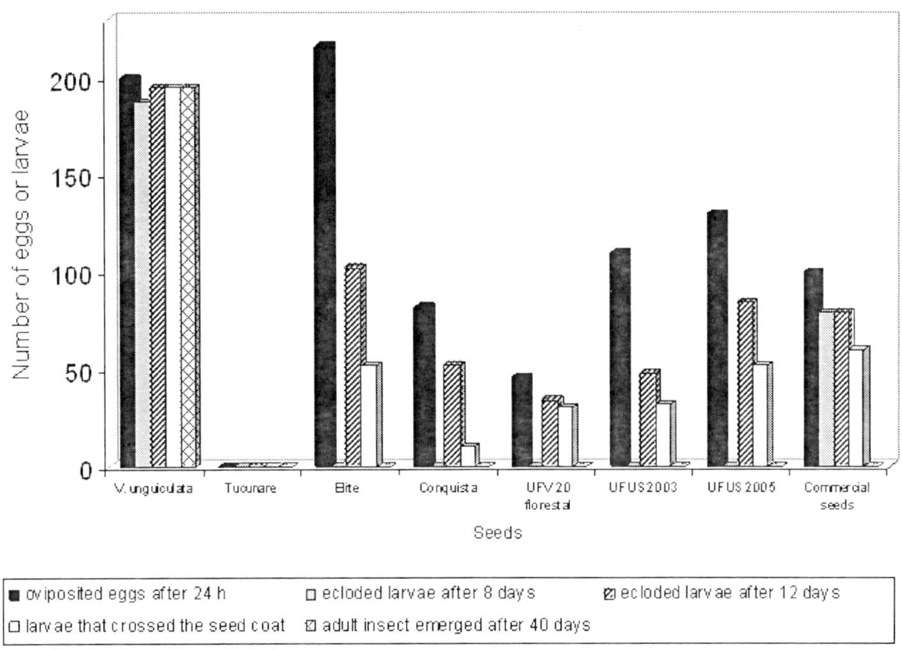

Figure 1. Performance and survival of *Callosobruchus maculatus* on natural host *Vigna unguiculata* seeds (positive control) and *Glycine max* seeds (cvs. Tucunaré, Elite, Conquista, UFV-20 Florestal, UFUS 2003, UFUS 2005 and commercial soybeans). Data refer to the infestation of sixty seeds from each species by ten *C. maculatus* females.

Artificial Cowpea Seeds Added by Soybean Seed Flour Experiments

Soybean seed coats were separated from cotyledons and ground to a flour. To test the potential deleterious effects of the seed coat flours on larval development, we employed an artificial seed system (Macedo et al. 1993). The seed coat flours were added to the cowpea meal in concentrations of 2, 8 and 16 %. The seeds were exposed to 3-days-old fertilized females during 24 h, at 28 °C and 70% r.h. After this time, the females were removed and three eggs were left per seed after removing the excess of eggs laid on the seeds. After 20 days, infested seeds were opened and the weight and the number of larvae were taken. Control artificial seeds were exclusively composed by *V. unguiculata* (cv. Fradinho) seed meal. Experiments consisted of 2 seeds per assay and were run in triplicate (total of 6 seeds and 18 eggs per tested dose). Dose-response curves were drawn utilizing the number and the average weight of the surviving larvae found for each tested dose, comparing with larvae grown in the control artificial seeds. The results were analyzed through Student's *t* test, and significant differences were determined at $P < 0.05$ (Bridge & Sawilowsky, 1999). The dose-response curves values were used to calculate the dose that reduced larval weight to 50 % (WD_{50}), and the number of surviving larvae to 50 % (LD_{50}) as well the lethal dose (LD).

RESULTS

Performance and Survival of *Callosobruchus maculatus* on Natural Soybean Seeds

The survival of *C. maculatus* on natural soybean seeds was evaluated and compared with the insect development in the control *V. unguiculata* (cv. Fradinho) seeds (Figure 1). Our results showed that the female oviposition, larval eclosion and adult emergence were drastically reduced in some soybean seeds when compared with the host seed. In *V. unguiculata* seeds, the females laid a total of 200 eggs and this number was similar to that of eggs laid on the Elite soybean cultivar (216 eggs). To other soybean cultivars, the reduction in female oviposition varied from 35 % in UFUS 2005 (130 eggs) to 77% in UFV 20 Florestal (46 eggs) (Figure 1). In the Tucunaré cultivar, oviposition was not observed, in three independent experiments. The larval eclosion, the complete perforation of the seed coat and adult emergency were also quite affected. At the eighth day after oviposition, 187 larvae had ecloded in the host seed and 79 in commercial soybeans, but no larval eclosion was observed in the other soybean cultivars. In the twelfth day after oviposition, eclosions of 162 larvae in the Elite cultivar, 52 in the Conquista, 34 in the UFV 20 Florestal, 48 in the UFUS 2003 and 84 larvae in the UFUS 2005 were observed (Figure 1). Twenty-five days after oviposition, adult emergency from the host seed is already observed (data not shown) and 40 days after oviposition, 195 insects of the 200 eggs laid on host seeds had emerged. In contrast to these data, no emergency of insects adult was observed from soybean seeds at any time interval. Forty days after oviposition, the seeds were opened and the amount of larvae that crossed the seed coats was analyzed. The number of larvae that crossed completely the seed coat varied vastly from one seed cultivar to the other (Figure 1). In the Elite cultivar only 52 larvae crossed the seed coat, which represent 32 % of the eggs laid on this seed coat. The amount of larvae that crossed the seed coat of the Conquista (10), UFV 20 Florestal (30), UFUS 2003 (32), UFUS 2005 (52) and commercial soybean (60) represents 17.5, 88, 66, 61.9 and 75% of the eggs laid on each of these seed coats, respectively (Figure 1). These results showed reductions up to 82.5% (Conquista cultivar) in the amount of larvae that reaches the seed cotyledons, when compared to the bruchid performance onto the host cowpeas. All the larvae that crossed the seed coat were found dead in the cotyledons surface (Figure 5L).

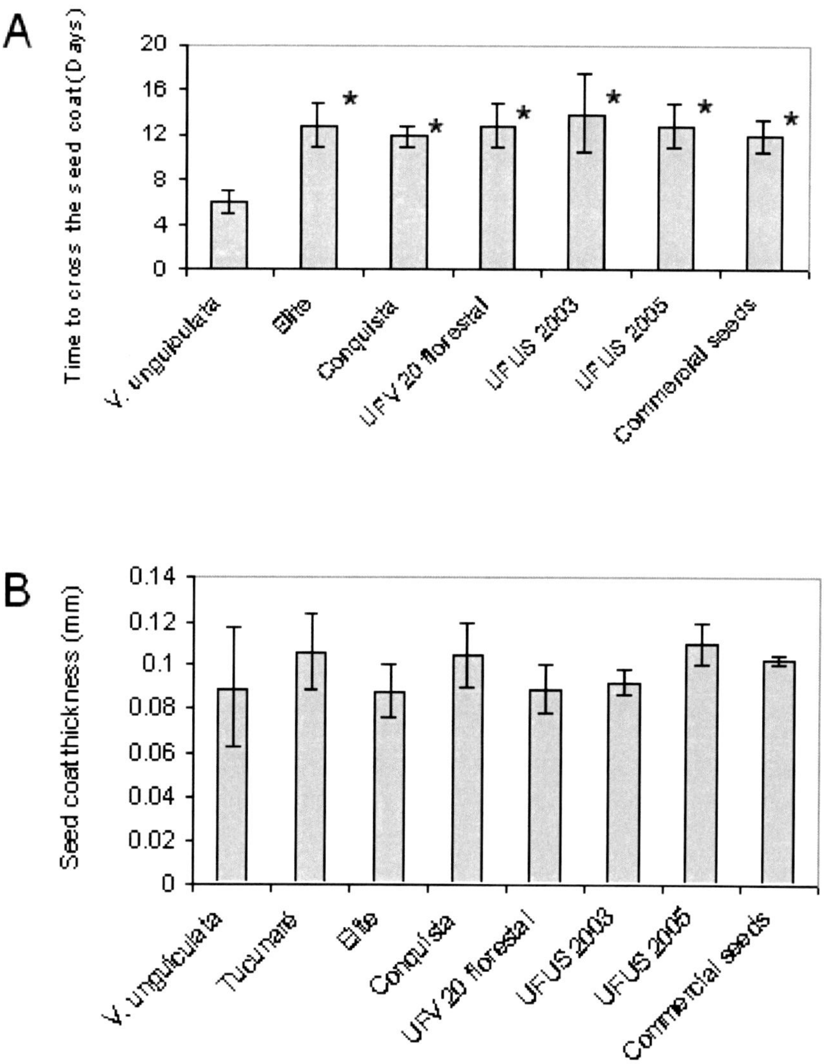

Figure 2. A - Time (days) consumed by the surviving larvae to perforate completely the soybean seed coats. The data represent the mean of twelve larvae per experiment. B - Seed coat thickness (cm). Measuremens were done in triplicate and the data are the mean of these results.

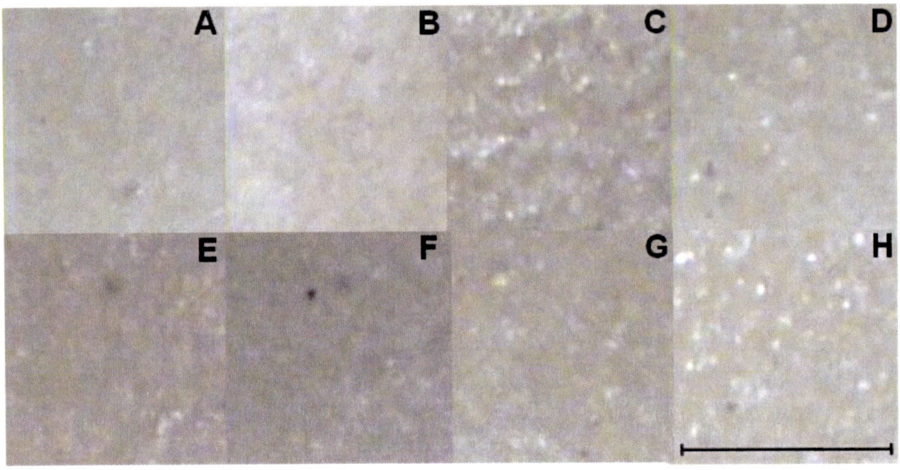

Figure 3. Seed coat surfaces observed under a stereoscopic microscope coupled to a digital CCD video camera. A - *Vigna unguiculata* (cv. Fradinho); B – cv. Elite; C - cv. Conquista; D - Commercial soybeans; E - cv. UFV 20 Florestal; F - cv. UFUS 2003; G - cv. UFUS 2005 and H - cv. Tucunaré. Bar = 0.5 mm.

Influence of the Seed Coat on the Larval Ability to Perforate the Seed Coat

The time necessary for the surviving larvae to perforate completely the seed coat was evaluated. We observed that in the host *V. unguiculata* seeds (positive control), at the sixth day after the oviposition the larvae already crossed completely the seed coat and reached the cotyledons, while for soybean seeds this time varied from 12 to 13 days (Figure 2A). This result shows an increase of up to 116 % in the necessary time for the larva to perforate the seed coat.

When the thickness of the seed coat was evaluated, we noticed that was no significant variation in the thickness of the studied seed coats. The result indicates that there is no relationship between the thickness and the ability of the larva for crossing this tissue (Figure 2B).

The textures and pigmentation of the seed coat surfaces were also analyzed and we observed that the seed coat of Conquista (Figure 3C), commercial soybean (Figure 3D), UFUS 2005 (Figure 3G) and Tucunaré (Figure 3H) soybean cultivars are slightly more rough. However there was no direct correlation between this characteristic and the performance of the eggs or ecloded larvae on the seed coats. The pigmentation of the seed coats did not differ in the studied seeds (Figure 3).

Influence of the Seed Coat on the Development of Neonate Larvae

Observations of the development of *C. maculatus* neonate larvae on natural seeds of *V. unguiculata* showed that, at the first day after oviposition, the egg content is clear, transparent (Figure 4A); at the fifth day, the larva is apparently formed (Figure 4B); in the inferior surface of the egg of the fifth day it becomes visible the perforation through which the larva digs the surface of the seed towards the cotyledons (Figure 4C); at the sixty day, we can observe flour inside the egg giving a complete whitish aspect to the egg (Figure 4D), the larva is partially inside of the cotyledon (Figure 4E), the seed coat is completely perforated (Figure 4F) as it is also perforated the cotyledon (Figure 4G).

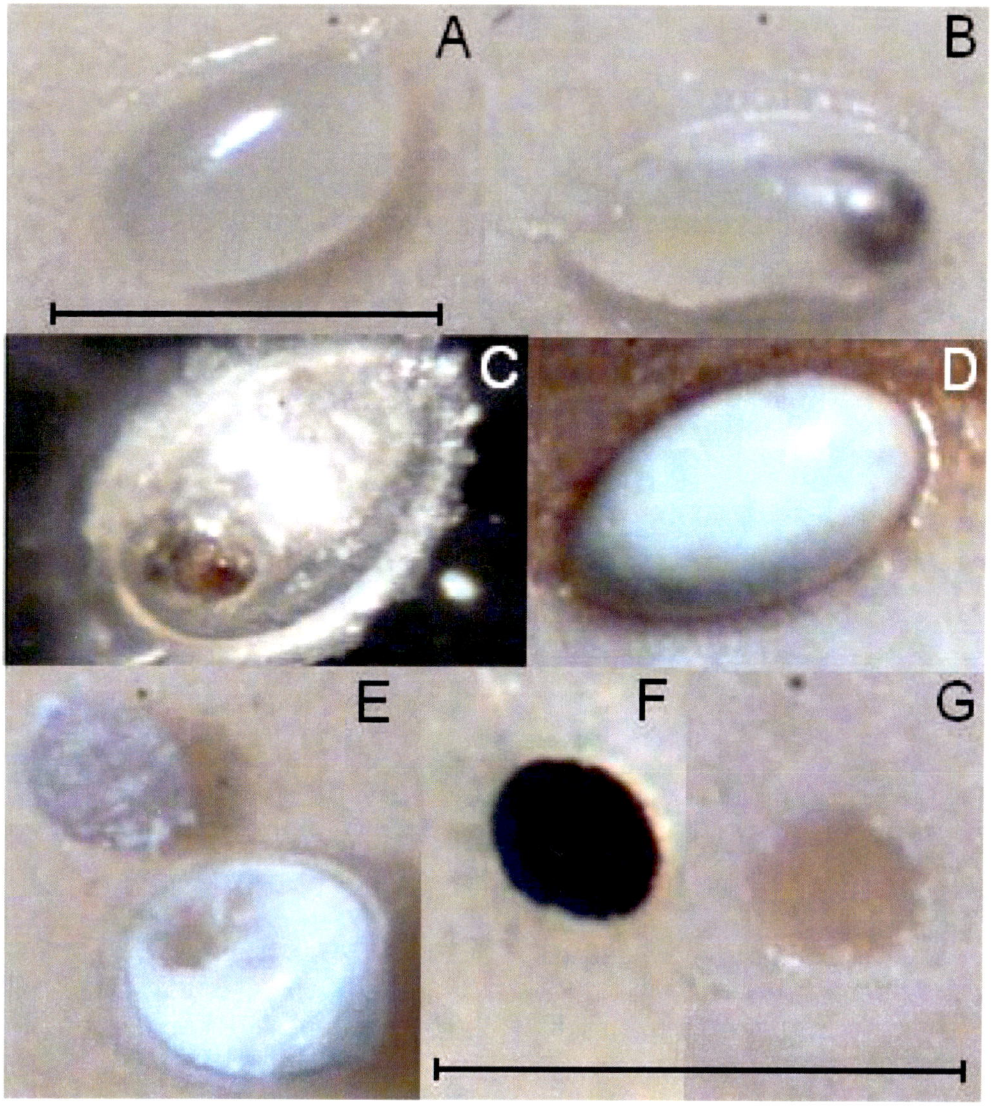

Figure 4. Development of *Callosobruchus maculatus* larvae on natural *Vigna unguiculata* (cv. EPACE 10) seeds from the first day of oviposition to the sixth day of development. A - Egg at the first day; B - Egg at the fifth day; C - Inferior surface of the egg (fifth day); D - Egg at the sixty day; E - Larva penetrating the seed (sixty day); F - Hole across the seed coat (sixty day); G - Hole on the cotyledon surface (sixty day). Bars= 0.5 mm. The bar inside panel A is equal for panels B-E.

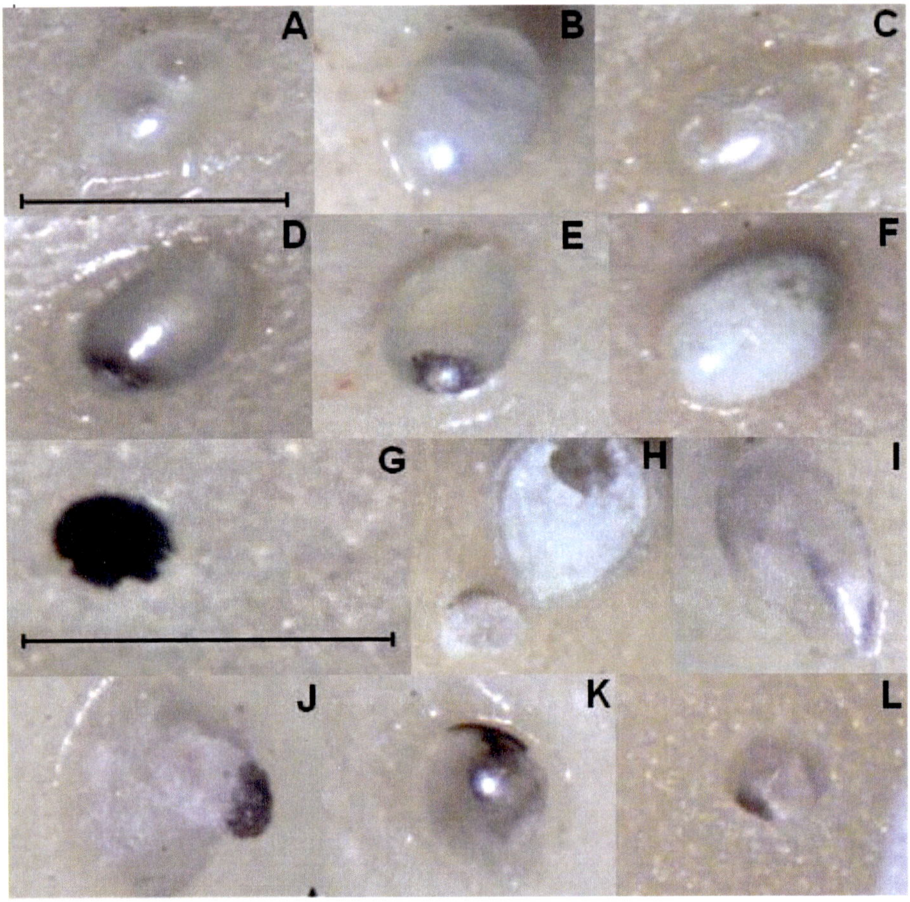

Figure 5. Development of *Callosobruchus maculatus* larvae on natural soybean seeds from the first day of oviposition to the fortieth day of development. A - Egg at the first day (commercial soybeans); B - Egg at the second day (commercial soybeans); C - Egg at the fourth day (commercial soybeans); D - Egg at the sixth day (commercial soybeans); E - Egg at the eighth day (commercial soybeans); F: Egg at the tenth day (cv. UFUS 2003); G - Hole across the seed coat of cv. UFUS 2005 (twelfth day); H - Larva penetrating the cv. Elite (twelfth day); I - Holed egg at the twelfth day (cv. Conquista); J - Larva at the twelfth day (cv. Elite). K - Dead larva at the fortieth day (cv. UFUS 2005); L - Dead larva inside the cotyledon at the fortieth day (cv. Conquista). Bars = 0.5 mm. The bar inside panel A is equal for panels B-F and H-L.

Influence of the Seed Coat on the Development of Neonate Larvae

Observations of the development of *C. maculatus* neonate larvae on natural seeds of *V. unguiculata* showed that, at the first day after oviposition, the egg content is clear, transparent (Figure 4A); at the fifth day, the larva is apparently formed (Figure 4B); in the inferior surface of the egg of the fifth day it becomes visible the perforation through which the larva digs the surface of the seed towards the cotyledons (Figure 4C); at the sixty day, we can observe flour inside the egg giving a complete whitish aspect to the egg (Figure 4D), the larva is partially inside of the cotyledon (Figure 4E), the seed coat is completely perforated (Figure 4F) as it is also perforated the cotyledon (Figure 4G).

The neonate larvae development on natural soybean seed coats was similar in all studied seeds. The eggs showed anomalies already visible after the first day of oviposition, consisting of bubbles in the egg surface (Figure 5A) and alteration in the content appearance (Figure 5B). Other eggs were completely withered at the fourth day after oviposition (Figure 5C). Some neonate surviving larvae were apparently formed inside the egg at the sixth day (Figure 5D) and at the eighth day (Figure 5E). At the tenth day after oviposition, flour was seen inside some eggs (Figure 5F) indicating that the larva began to dig the seed cotyledon. Only at the twelfth day after oviposition, the complete perforation of the seed coat and the initial perforation of the cotyledon were observed (Figure 5G). In this stage the larva was partially inside the cotyledon (Figure 5H). Although the normal behavior of the larvae is to open holes in the inferior surface of the egg in order to penetrate the seed cotyledon, an inverse behavior was observed for some larvae laid on the soybean seed coats. Some eggs showed a hole in the superior surface after the twelfth day after oviposition (Figure 5I), through which some larvae came out (Figure 5J). Up to the fortieth day after oviposition, no emergency of adult insects was observed; all larvae were dead inside the eggs (Figure 5K) or in the surface of the cotyledons (Figure 5L).

Influence of the Seed Coats on Larval Mass

Artificial seeds covered with natural seed coat (Figure 6) confirmed the negative interference of the soybean seed coats in the *C. maculatus* larval development. In these experiments, a total of 9 eggs was considered as 100 % of oviposition. All artificial seeds covered with soybean seed coats reduced the number of oviposited eggs, ecloded larvae and larvae that perforated completely the seed coat (Figure 6A). The percentage of larvae that died inside the eggs varied from 25 % for the larvae oviposited onto commercial and UFV 20 Florestal soybean seed coats to 100 % for those onto Elite, Conquista, Tucunaré and UFUS 2003 cultivars. About 93 % of the larvae which survived in experiments with commercial soybean seed coat succeed in perforating completely the seed coat; however, after 20 days of development, these larvae had only about 23.8 % (3.34 mg) of the mass of a control larva which developed in artificial seeds covered with *V. unguiculata* natural seed coat (14 mg) (Figure 6B).

Toxicity of the Seed Coat Flour to Larvae

The toxicity experiments showed that flour from all seed coats interfered with the normal development of the *C. maculatus* larvae, reducing considerably the weight and survival of the 20 days-old larvae. The toxicity levels varied from cultivar to cultivar and were dose-dependent.

The results corresponding to the incorporation of the seed coat flour are shown in figure 7. The seed coat of UFV 20 Florestal cultivar was highly toxic and no larva survived in artificial seeds containing 2% of this flour. In artificial seeds containing 16.0 % of UFUS 2003 and UFUS 2005 seed coat flours no surviving larva was found after 20 days. In respect to the negative impact on the larvae weight, a similar toxicity was observed to Elite, Conquista, Tucunaré and commercial soybeans when incorporated at a 16 % level, and reductions as high as 88% could be observed (Figure 7A and Figure 8). Although some larvae have survived in the seeds containing 16 % of the soybean seed coat flours, the number of surviving larvae was drastically reduced for about 30 % (in artificial seeds containing Conquista, Tucunaré and commercial soybean seed coat flours) or 10% in artificial seeds containing seed coat flour from Elite cultivar (Figure 7B).

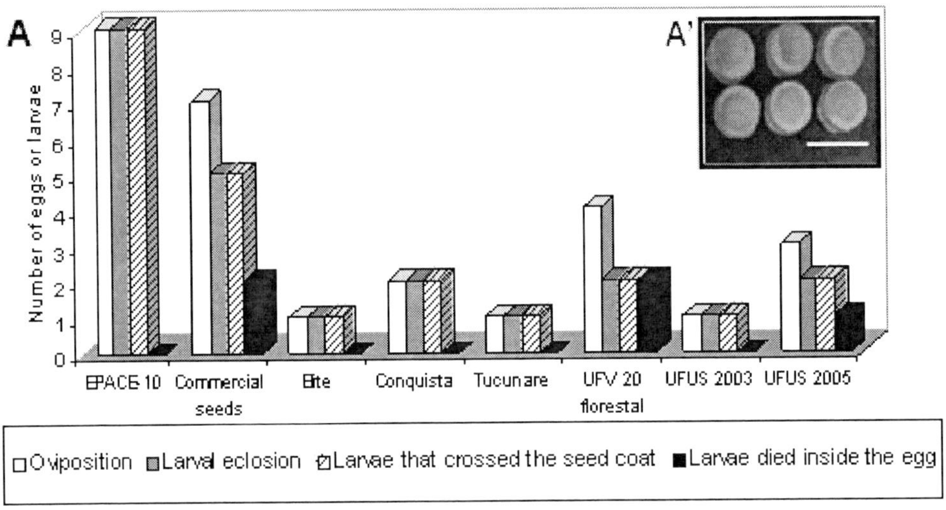

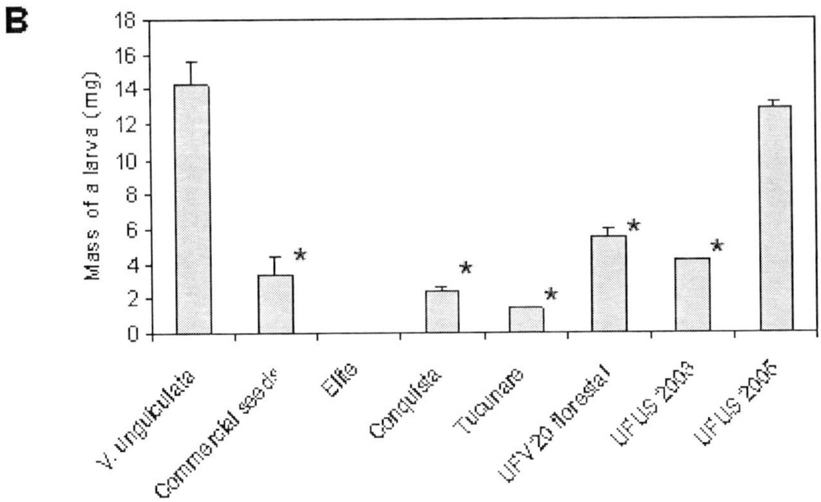

Figure 6. A - Oviposition and performance of *Callosobruchus maculatus* larvae on artificial seeds covered with natural seed coats of cowpea (control seeds) and soybeans from commercial, Elite, Conquista, Tucunaré, UFV Florestal, UFUS 2005 and UFUS 2005 cultivars. A' – Picure of artificial seeds covered with natural seed coats. B - Mean mass of *Callosobruchus maculatus* larva which survived on artificial seeds covered with natural seed coats. Experiments were done in triplicate and the data shown are the average of these results. Bar = 1 cm.

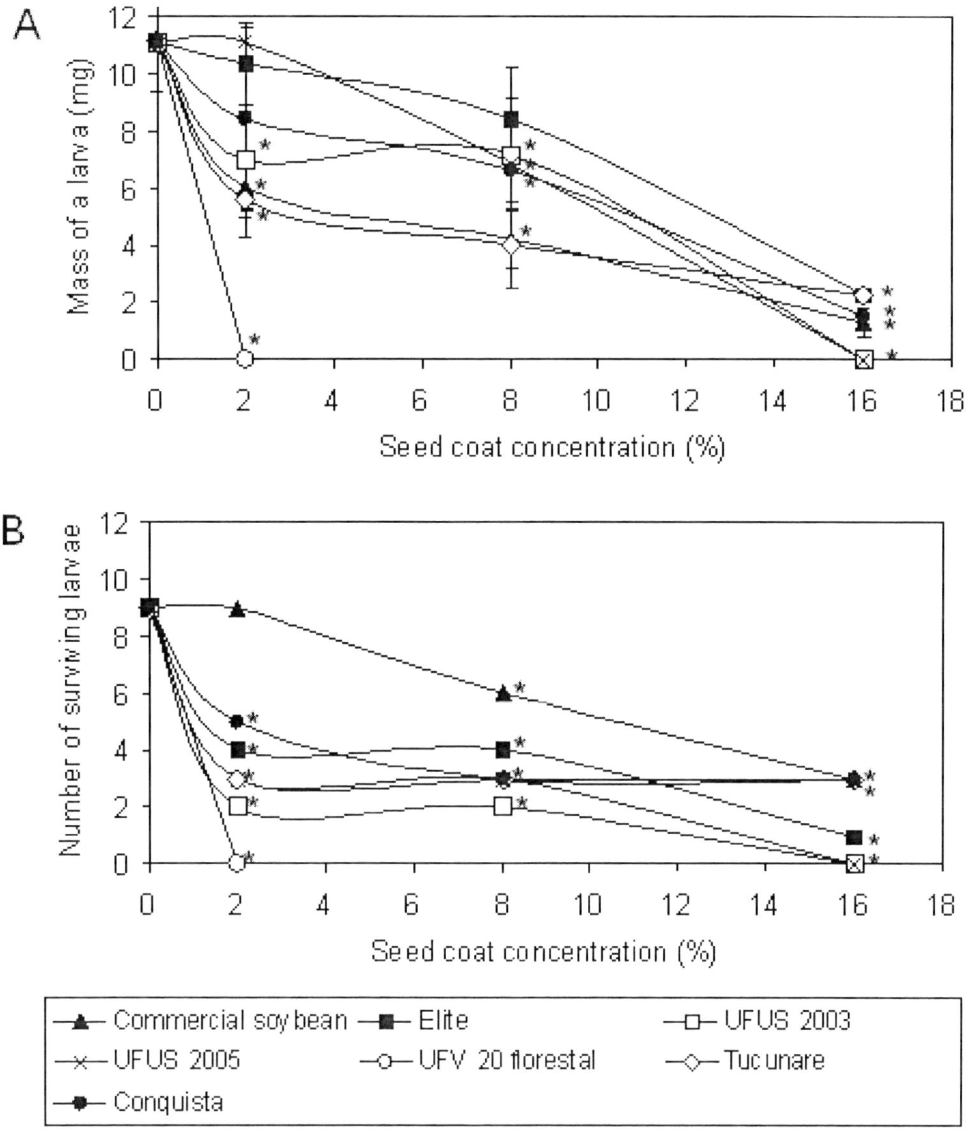

Figure 7. Performance (A) and survival (B) of *Callosobruchus maculatus* on artificial seeds containing different concentrations (0, 2, 8 and 16%) of soybean seed coats from commercial, Elite, Conquista, Tucunaré, UFV Florestal, UFUS 2003 and UFUS 2005 cultivars. Experiments were done in triplicate and the data shown are the average of these results.

The values of WD_{50} (dose that reduced larval weight to 50%) and LD_{50} (dose that reduced the number of surviving larvae to 50%) were calculated to all soybean seed coat flours and the results are shown in figure 9. The smallest values of WD_{50} (Figure 9A) were found for the seed coat of UFV 20 Florestal (1.5 %), Tucunaré (3.0 %) and soybean commercial seeds (3.3%). Thus, these seed coats were the most efficient in the reduction of larval mass (Figure 9A). The seed coat flours of UFUS 2005, Conquista, UFUS 2003 and Elite cultivars showed the highest values of WD_{50} (10.5, 11, 11 and 12%, respectively) (Figure 9A); even though, these were highly toxic to the bruchid when compared to control host seed coats (Figure 8). Lowest LD_{50} values were found for the seed coats of UFV 20 Florestal, UFUS 2005 and UFUS 2003 (1.0, 1.0 and 1.3 %, respectively) (Figure 9B). Although it has been observed a great variation in the values of WD_{50} and LD_{50}, all soybean seed coats may be considered as highly toxic to *C. maculatus*, since during the insect development in natural seeds the larvae firstly enter in contact with this tissue, being exposed and relying on a diet 100% based on this tissue compounds.

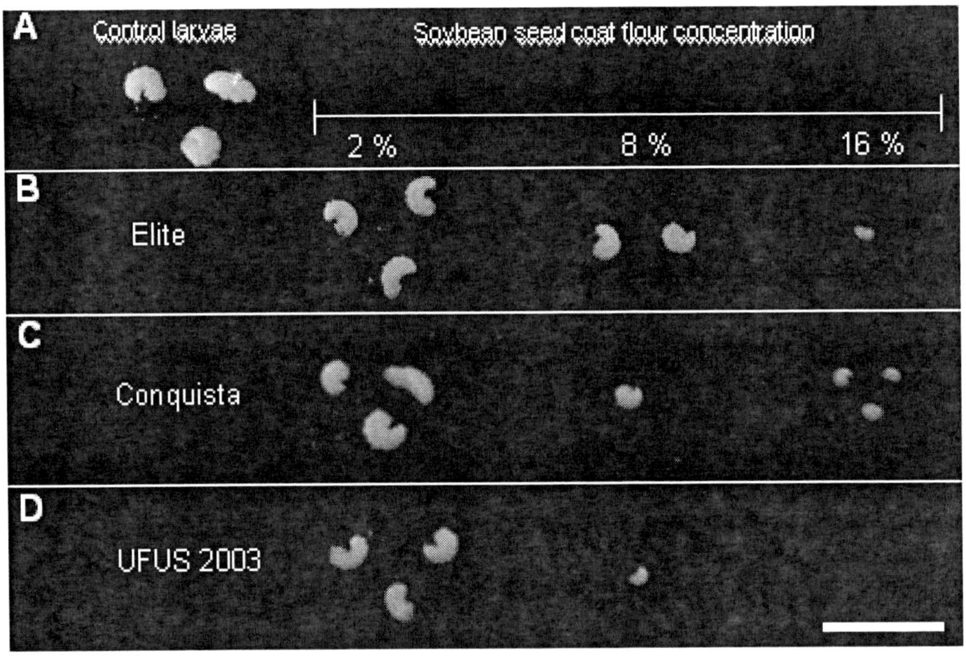

Figure 8. Picture of developing *Callosobruchus maculatus* control larvae and larvae grown on artificial seeds containing different concentrations (2, 8 and 16%) of soybean seed coats from Elite, Conquista and UFUS 2003 cultivars.

Conclusion

The soybean seed coat prevented the penetration of up to 82.5% of the *Callosobruchus maculatus* larvae. There was not direct relationship between thickness, color and/or texture of the seed coat from several soybean cultivars with the ability of the larva to cross or not the seed coats of these seeds. A high retard in the larval perforation of the soybean seed coat and abnormalities in neonate larvae were observed during perforation attempts. Other larvae began to perforate the seed coat and died while trying to penetrate the cotyledons. The reasons for this are probably of non-physical nature, since the seed coat flours were very toxic to insect larvae reducing the mass and the number of surviving larvae reared on artificial seeds. We believe that the expression of the *C. maculatus* detrimental compounds in the seed coats of the non-host soybean seeds reinforces the idea that the seed coat plays an important role for the evolutionary discrimination of legume seeds by this bruchid.

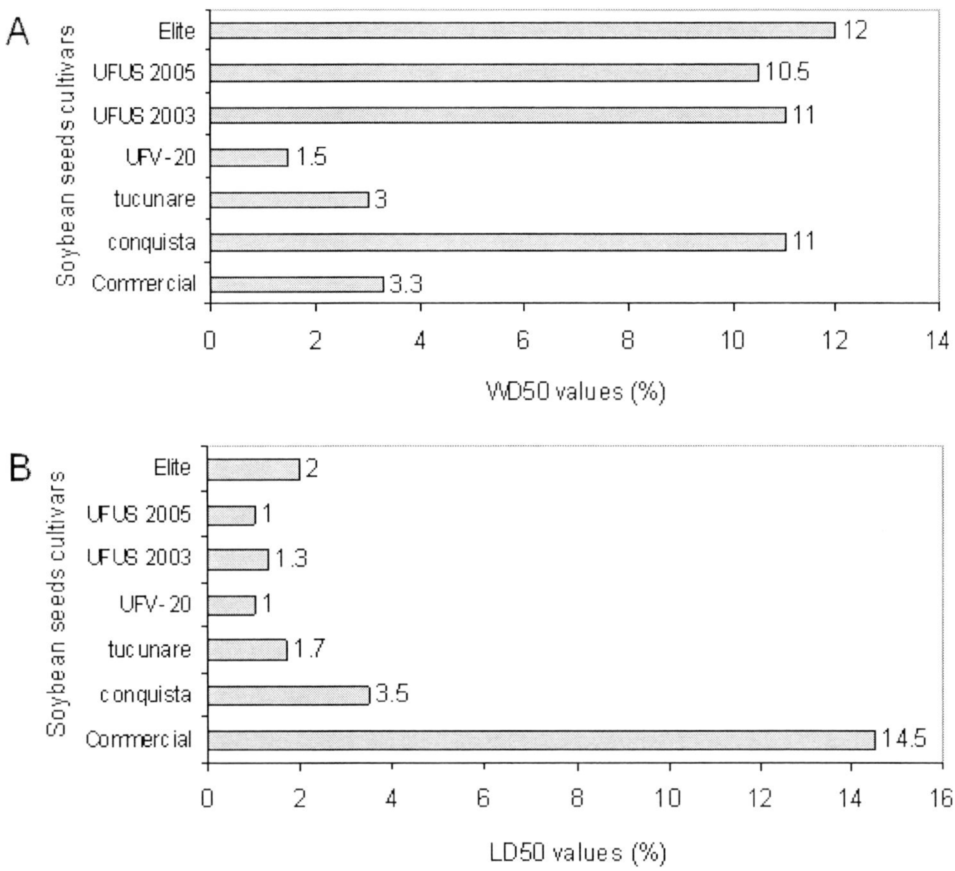

Figure 9. Toxicity of the soybean seed coat flour from commercial, Elite, Conquista, Tucunaré, UFV Florestal, UFUS 2005 and UFUS 2005 cultivars. A - WD_{50} (lethal dose that reduced larval weight to 50%); B - LD_{50} (lethal dose that reduced the number of surviving larvae to 50%).

ACKNOWLEDGMENTS

This work was supported by grants from the Brazilian agencies FAPERJ, CNPq, and Universidade Estadual do Norte Fluminense Darcy Ribeiro. We would like to thank Dr. Roberto Ferreira da Silva from CCTA-UENF and Departamento de Fitotecnia da Universidade Federal de Viçosa (UFV) for the soybean seeds from the Conquista, UFV-20 Florestal, UFUS 2003, UFUS 2005, Tucunaré and Elite cultivars.

REFERENCES

Beck, C. W. & Blumer, L. S. (2007) A Handbook on Bean Beetles, *Callosobruchus maculatus*. URL: www.beanbeetles.org.

Boughdad A.; Gillon Y. & Gagnepain C. (1986). Influence du tegument des graines mures de *Vicia faba* sur le developpement larvaire de *Callosobruchus maculatus*. *Entomologia Experimentalis et Applicata* 42, 219-223.

Bridge, P. D. & Sawilowsky, S. S. (1999). Increasing physicians' awareness of the impact of statistics on research outcomes: comparative power of the T-test and Wilcoxon rank-sum test in small samples applied research. *Journal of Clinical Epidemiology*. 52, 229-235.

Buttery, B. R. & Buzzel, R. I. (1968). Peroxidase activity in the seeds of soybean varieties. *Crop Science*. 8, 722-725.

Calero, E.; Wes, S. H. & Hinson, K. (1981). Water absorption of soybean seed and associated causal factors. *Crop Science*. 21, 926-933.

Carlini, C. R. & Grossi-de-Sá, M. F. (2002). Plant toxic proteins with insecticidal properties. A review on their potentialities as bioinsecticides. *Toxicon*. 40, 1515–1539.

Carlini, C. R.; Oliveira, A. E. A.; Azambuja, P.; Xavier-Filho, J. & Wells, M. A. (1997). Biological effects of canatoxin in different insect models: evidence for a proteolytic activation of the toxin by insect cathepsin-like enzymes. *Journal of Economic Entomology*. 90, 340-348.

Caswell, G. H. (1968). The storage of cowpea in the northern states of Nigeria. *Proceedings of the Agricultural Society of Nigeria* 5, 4–6.

Chrispeels, M. J. & Raikhel, N. V. (1991). Lectin, lectin genes and their role in plant defense. *Plant Cell*. 3, 1-9.

Credland, P. F. (1987). Effects of host change on the fecundity and development of an unusual strain of *Callosobruchus maculatus* (F.) (Coleoptera: Bruchidae). *Journal of Stored Products Research*, 23: 91–98.

De Souza, F. H. D. & Marcos-Filho, J. (2001). The seed coat as a modulator of seed-environment relationships in Fabaceae. *Revista Brasileira de Botânica*. 24, 365-375.

Desroches, P.; El Shazly, E.; Mandon, N.; Duc, G. & Huignard, J. (1995). Development of *Callosobruchus chinensis* (L.) and *Callosobruchus maculatus* (F.) (Coleoptera: Bruchidae) in seeds of *Vicia faba* L. differing in their tannin, vicine and convicine contents. *Journal of Stored Products Research*. 31, 83-89.

El-Turk, J.; Asemota, O.; Leymarie, J.; Sallaud, C.: Mesnage, S.; Brenda, C.; Buffard, D.; Kondorosi, A. & Esnault, R. (1996). Nucleotide sequences of four pathogen-induced alfalfa peroxidase-encoding cDNAs. *Gene*. 170, 213-216.

Esau, K. (1977). Anatomy of seed plants. 2nd ed. John Wiley, New York. 550 p.

Fricke, C. & Arnqvist, G. (2007) Rapid adaptation to a novel host in a seed beetle (*Callosobruchus maculatus*): the role of sexual selection. *Evolution*. 61, 440-454.

Firmino, F.; Fernandes, K. V. S.; Sales, M. P.; Gomes, V. M.; Miranda, M. R. A.; Domingues, S. J. S. & Xavier-Filho, J. (1996). Cowpea (*Vigna unguiculata*) vicilins associate with putative chitinous structures in midgut and feces of the bruchid beetles *Callosobruchus maculatus* and *Zabrotes subfasciatus*. *Brazilian Journal of Medical and Biological Research*. 29, 749-756.

Gijzen, M. (1997). A deletion mutation at the *ep* locus causes low seed coat peroxidase activity in soybean. *The Plant Journal*. 12, 991-998.

Gijzen, M.; Kuflu, K.; Qutob, D. & Chernys, J. T. (2001). A class I chitinase from soybean seed coat. *Journal of Experimental Botany*. 52, 2283-2289.

Gijzen, M.; Miller, S. S.; Bowman, L. A.; Batchelor, A. K.; Boutilier, K. & Miki, B. L. A. (1999a). Localization of peroxidase mRNAs in soybean seeds by in situ hybridization. *Plant Molecular Biology*. 41, 57-63.

Gijzen, M.; Miller, S. S.; Kuflu, K.; Buzzell, R. L. & Miki, B. L. A. (1999b). Hydrophobic protein synthesized in pod endocarp adheres to the seed surface. *Plant Physiology*. 120, 951-959.

Gomes, V. M.; Oliveira, A. E. A. & Xavier-Filho, J. (1996). A chitinase and a β-1,3-glucanase isolated from the seeds of cowpea (*Vigna unguiculata* L. (Walp.)) inhibit growth of fungi and insect pests of the seed. *Journal of the Science of Food and Agriculture*. 72, 86-90.

Hammerschmidt, R.; Nuckles, E. M. & Kuc, J. (1982). Association of enhanced peroxidase activity with induced systemic resistance of cucumber to *Colletotrichum lagenarium*. *Physiologia Plantarum*. 20, 73–82.

Harris, W. H. (1987). Comparative ultrastructure of developing seed coats of 'hard-seeded' and 'soft-seeded' varieties of soybean, *Glycine max* (L.) Merril. *Botanical Gazette*. 148, 324-331.

Haughn, G. & Chaudhury, A. (2005). Genetic analysis of seed coat development in *Arabidopsis*. *TRENDS in Plant Science*. 10, 472-477.

Henriksen, A.; Mirza, O.; Indiani, C.; Teilum, K.; Smulevich, G.; Welinder, K.G. & Gajhede, M. (2001). Structure of soybean seed coat peroxidase: a plant peroxidase with unusual stability and haem-apoprotein interactions. *Protein Science*. 10, 108-115.

Jackai, L. E. N. & Daoust, R. A. (1986). Insect pests of cowpea. *Annual Review of Entomology*. 31, 95–119.

Janzen, D. H. (1977). How southern cowpea weevil larvae (Bruchidae: *Callosobruchus maculatus*) die on non host seeds. *Ecology*. 58, 921-927.

Koide, T. & Ikenaka, T. (1973). Studies on soybean trypsin inhibitors. Amino acid sequence of the carboxyl-terminal region and the complete amino acid sequence of the soybean trypsin inhibitor. *European Journal of Biochemistry*. 32, 417-431.

Koizumi, M.; Kikuchi, K.; Isobe, S.; Ishida, N. Naito, S. & Kano, H. (2008). Role of the seed coat in imbibing soybean seeds observed by micro-magnetic resonance imaging. *Annals of Botany*. 102, 343-352.

Kunitz, M. (1945). Crystallization of a trypsin inhibitor from soybean. *Science*. 101, 668-669.

La Scala JR., N.; Florentino, A. O. & Carvalho, N. M. (1999). Differences and similarities in the pore size distribution of soybean seed coats. *Seed Science and Technology*. 27, 365-369.

Ma, F.; Teterson, C. A. & Gijzen, M. (2004). Reassessment of the pits and antipits in soybean seeds. *Canadian Journal of Botany*. 82, 654-662.

Macedo, M. L. R.; Andrade, L. B. S.; Moraes R. A. & Xavier-Filho, J. (1993). Vicilin variants and the resistance of cowpea (*Vigna unguiculata*) seeds to the cowpea weevil (*Callosobruchus maculatus*). *Comparative Biochemistry and Physiology*. 105, 89-94.

Miller, S. S.; Bowman, L. A.; Gijzen, M. & Miki, B. L. A. (1999) Early development of the seed coat of soybean. *Annals of Botany*. 84, 297-304.

Moïse, J. A.; Han, S.; Gudynaitę-Savitch, L.; Johnson, D. A. & Mikica, B. L. A. (2005). Seed coats: Structure, development, composition, and biotechnology. *In vitro Cellular and Developmental Biology - Plant*. 41: 620-644.

Moraes, R.; Sales, M. P.; Pinto, M. S. P.; Silva L. B.; Oliveira A. E. A.; Machado, O. L.T.; Fernandes K.V.S. & Xavier-Filho, J. (2000). Lima bean (*Phaseolus lunatus*) seed coat phaseolin is detrimental to the cowpea weevil (*Callosobruchus maculatus*). *Brazilian Journal of Medical and Biological Research*. 33, 191-198.

Moura, F. T.; Oliveira, A. S.; Macedo, L. L.; Vianna, A. L.; Andrade, L. B.; Martins-Miranda, A. S.; Oliveira, J. T.; Santos, E. A. & Sales, M. P. (2007). Effects of a chitin-binding vicilin from *Enterolobium contortisiliquum* seeds on bean bruchid pests (*Callosobruchus maculatus* and *Zabrotes subfasciatus*) and phytopathogenic fungi (*Fusarium solani* and *Colletrichum lindemuntianum*). *Journal of Agricultural and Food Chemistry*. 55, 260-266.

Nissum, M.; Schiodt, C. B. & Welinder, K. G. (2001). Reactions of soybean peroxidase and hydrogen peroxide pH 2.4– 12.0, and veratryl alcohol at pH 2.4. *Biochimica and Biophysica Acta*. 545, 339-348.

Oliveira, A. E. A.; Sales, M. P.; Machado, O. L. T.; Fernandes, K. V. S. & Xavier-Filho, J. (1999). The toxicity of Jack bean (*Canavalia ensiformis*) cotyledon and seed coat proteins to the cowpea weevil (*Callosobruchus maculatus*). *Entomologia Experimentalis et Applicata*. 92, 249-255.

Oliveira, A. E. A.; Sassaki, G. L.; Iacomini, M.; Cunha, M.; Fernandes, K. V. S. & Xavier-Filho, J. (2001). Isolation and characterization of a galactorhamnan polysaccharide from the seed coat of *Canavalia ensiformis* that is toxic to the cowpea weevil *Callosobruchus maculatus*. *Entomologia Experimentalis et Applicata*. 101, p. 225-231.

Ragus, L. N. (1987) Role of water absorbing capacity in soybean germination and seedling vigour. *Seed Science and Technology*. 15, 285-296.

Sales, M. P.; Fernandes, K. V. S.; Gomes, V. M. & Xavier-Filho, J. (1996). Chitin-binding proteins from cowpea (*Vigna unguiculata*) seeds. *Brazilian Journal of Medical and Biological Research*. 29, 319-326.

Sales, M.P.; Macedo, M. R. L. & Xavier-Filho, J. (1992). Digestibility of cowpea (*Vigna unguiculata*) vicilins by pepsin, papain and bruchid midgut proteinases. *Comparative Biochemistry and Physiology*. 103, 945- 950.

Santos, P. O.; Santos, I. S.; Gomes, V. M.; Machado, O. L. T.; Fernandes, K. V. S., Xavier-Filho, J. & Oliveira, A. E. A. (2008). In vitro evaluation of antifungal activity of soybean (*Glycine max*) seed coat proteins. *Journal of Stored Products Research*. 44, 310– 315.

Silva, L. B.; Sales, M. P.; Oliveira A. E. A.; Machado O. L. T., Fernandes K. V. S. & Xavier-Filho, J. (2004). The seed coat of *Phaseolus vulgaris* interferes with the development of the cowpea weevil (*Callosobruchus maculatus* (F) (Coleoptera: Bruchidea)]. *Anais da Academia Brasileira de Ciências*. 76, 57-65.

Southgate, B. J. (1979). Biology of the Bruchidae. *Annual Review of Entomology*. 24, 449–473.

Thiéry D. (1984). Hardness of some Fabaceous seed coats in relation to larval penetration by *Acanthoscelides obtectus* (Say) (Coleoptera: Bruchidae). *Journal of Stored Products Research*. 20, 177-181.

Thiéry, D.; Jarry, M. & Pouzat, J. (1994). To penetrate or not penetrate? A behavioral choice by bean beetle frist-instar larvae in response to *Phaseolus vulgaris* seed surface quality. *Journal of Chemical Ecology*. 20, 1867-1875.

Xavier-Filho, J. (1991). The resistance of seeds of cowpea (*Vigna unguiculata*) to the cowpea weevil (*Callosobruchus maculatus*). *Memórias do Instituto Oswaldo Cruz*. 86, 75-77.

Yaklich, R. W.; Vigil, E. L. & Wergin, W. (1984). Scanning electron microscopy of soybean seed coat. *Scanning Electron Microscopy*. 2, 531-544.

Yunes, A. N. A.; Andrade, M. T.; Sales, M. P.; Moraes, R. A.; Fernandes, K. V. S.; Gomes, V. M. & Xavier-Filho, J. (1998). Legume seed vicilins (7S storage proteins) interfere with the development of the cowpea weevil [*Callosobruchus maculatus* (F.)]. *Journal of the Science of Food and Agriculture*. 76, 111-116.

Zeng, C.-L.; Wang, J.-B.; Liu, A.-H. & Wu, X.-M. (2004). Seed coa microsculpturing changes during seed development in diploid and amphidiploid *Brassica* species. *Annals of Botany*. 93, 555-566.

In: Soybean and Wheat Crops
Editors: S. Davies, G. Evans
ISBN: 978-1-60741-173-4

Chapter 2

EFFECTS OF SOIL TEXTURE AND SOIL SALINITY ON THE PLANT WATER RELATIONSHIP, GROWTH, YIELD AND WATER USE EFFICIENCY OF THE SOYBEAN CROP

***N. Katerji*[a], *M. Mastrorilli*[b], *F. Lahmer*[c], *A. Hamdy*[c]**
[a]INRA, Unité de Recherche Environnement et Grandes Cultures, 78850 Thiverval-Grignon, France
[b]CRA, Research Unit for Cropping Systems in Dry Environments, 70125 Bari, Italy
[c]CIHEAM, Mediterranean Agronomic Institute, 70010 Valenzano, Italy

ABSTRACT

Soybean was grown in a lysimeters filled with loam and clay soils and was irrigated with water having three different levels of salinity (fresh water, and saline waters with 15 and 30 meq Cl/l).

During the soybean crop cycle, soil salinity was determined from the salt balance. Leaf-water potential, stomatal conductance and actual evapotranspiration were used as the water-stress indicators. Growth was measured through leaf area and dry matter and, finally, the yield and its components were determined. The water use efficiency was also calculated.

Without salt stress (treatments irrigated with fresh water), the effect of soil texture on the water relationship, productivity and water use efficiency of the soybean was not demonstrated.

With salt stress, all the parameters, in both types of soil, were coherent, indicating systematic differences between the saline treatments and the control treatments (treatments irrigated with fresh water).

Soil texture affects the soybean response to soil salinity. The saline treatments in the loam soil caused the values of the water stress indicators, of growth, of yield and of water use efficiency to be higher than the highest values observed for the same treatments on the clay soil.

The analysis of the relationship between relative yield and soil salinity indicates clearly that soybean shows a higher salt tolerance if it is cultivated in loam soil.

Key-words: nitrogen balance; salt balance; evapotranspiration; abiotic stress; pre-dawn leaf water potential

I- INTRODUCTION

According to the classification made by Ayers and Westcott (1985), soybean is considered to be a moderately salt tolerant species. Katerji et al. (2000) classify soybean among those species sensitive to salinity, like all grain legumes (Maas and Hoffman, 1977). Since various varieties of soybean were considered in the studies carried out to classify the species. These nuances in classification can probably be explained by the large range of varietal response to soil salinity demonstrated by the soybean (Velagaleti and Schweitzer, 1993).

The effect of salinity on bacterial activity in nitrogen fixation is one of the hypotheses that may explain grain legumes' sensitivity to salt (Pessarakli et al., 1989). In the case of soybean, various authors (Bernstein and Ogata, 1966; Wilson, 1970; Tu, 1981) have noted a decrease in the number and the dry weight of nodules at increasing levels of salinity. However, nitrogen fixation as a percentage of the total nitrogen uptake varies considerably, from 25% to 85% for soybean (Patterson and LaRue, 1983, Wagner and Zapata, 1982; Zapata et al., 1987). According to Patterson and LaRue (1983), the large range of these estimations may be explained by differences among varieties. This is probably one of the causes of the variability in tolerance to salinity previously found in soybean by Velagaleti and Schweitzer (1993).

Generally, environmental factors (climate, soil texture) can be considered among the causes that vary crop tolerance to soil salinity, but in literature these factors have not been studied in depth. More is known about the role of climate (temperature, air water deficit, and evaporative demand) thanks to the review by Shalhavet (1994). The role of climate has also been demonstrated by Katerji et al. (2000) with studies on the broad-bean. These authors have noticed that the same variety of leguminous crop can be either tolerant or sensitive to soil salinity, depending on the level of the evaporative demand during the crop cycle. Additionally, the role of soil texture as a cause of variability in tolerance to salinity has not yet been studied in the literature. However for leguminous crop and for the soybean in particular, it is admitted that the physical soil properties interact with the symbiotic nitrogen fixation (Bouniols et al., 1990).

The objective of this paper is to analyse the tolerance of soybean cultivated in clay and loam soils, characterised by a wide range of salinity, so as to determine the possible role of soil texture in salt tolerance. The analysis carried out was based on observations at court term on the level of the plant water relationship, at mean term on the level of growth in leaf surface and dry matter, and at long term on the level of yield, its components, seasonal evapotranspiration and water use efficiency. The analysis was also based on observations relative to the nitrogen balance carried out on soybean by van Hoorn et al. (2001), with the same experimental set-up.

The three questions asked in this study are:

- In the absence of salt stress, is soybean sensitive to the soil texture?
- In the presence of salt stress, is soybean tolerance to soil salinity modified by the soil texture?
- If so, how can the change in tolerance to salinity be interpreted in relationship to the soil texture?

II- Materials and Methods

The study was conducted in Bari (southern Italy). The climate is characterised by hot dry summers, with maximum air temperature sometimes higher than 40 C° and minimum relative air humidity often less than 20%.

1- Crop

Soybean (Glycine max, Talon variety) was grown from 18 July 1995 until 16 October 1995. The crop was sown at a density of 45 seeds per lysimeter, reduced to 25 plants at the two-leaf stage and 18 plants at harvest time because of successive samplings. Fertilising was done at the rate of 250 kg P_2O_5/ha at 23 D.A.S.(Days After Sowing) before the second irrigation and at the rate of 50 kg N/ha, divided over three dates at 27, 48 and 65 d.a.s, always two days after irrigation.

The main phenological stages observed during the crop cycle were: plant emergence (12 D.A.S.), the beginning of anthesis (40 D.A.S.), the beginning of pod formation (52 D.A.S.) and harvest (90 D.A.S.).

2- Set –Up

The set-up consisted in 30 cylindrical lysimeters of reinforced fibreglass with an internal diameter of 1.2 m and a depth of 1.2 m. A layer of coarse sand and gravel, 0.10 m thick, was overlaid by a repacked soil profile of 1 m. At the bottom of the lysimeter, a pipe serving as a drainage outlet connected the lysimeter to a drainage reservoir. A 15 lysimeter block was filled with loam, and a second block with clay. Table 1 presents the physical properties of the soil after the lysimeters were filled.

The set-up was covered at a height of 4 m by a sheet of transparent plastic. This sheeting excluded any rain, but also attenuated solar radiation by up to 10 %.

The lysimeters were irrigated with three different qualities of water: the control treatment with fresh water (C) containing 3.7 meq Cl/l and an electrical conductivity (EC) of 1.1 dS/m, and two saline treatments (15 and 30) containing 15 and 30 meq Cl/l and an EC of 2.3 and 3.6 dS/m respectively, obtained by adding equivalent amounts of NaCl and $CaCl_2$ to fresh water. For each water quality, five lysimeters were available. Table 2 presents the chemical composition of the waters.

Table 1. Soil properties

soil	particle size in percentage of mineral parts			$CaCO_3$ (%)	% water (v/v)		bulk density (kg/dm^3)
	<2 μm	2-50μm	>50μm		pF2.0	pF4.2	
loam	19	49	32	25	36.3	20.4	1.45
clay	47	37	16	5	42	24	1.45

Table 2. Chemical composition of irrigation water (meq/l), electrical conductivity (dS/m), and Sodium Adsorption Ratio (SAR)

Water quality	Ca^{2+}	Mg^{2+}	Na^+	K^+	Cl^-	HCO_3^-	SO_4^{2-}	EC(dS/m)	SAR
C (fresh water)	6.2	3.1	2.3	0.4	3.7	7.3	0.6	0.9	1.1
15 (15 meq Cl/l)	10.8	3.1	8.7	0.4	15	0.8	0.8	2.3	3.3
30 (30 meq Cl/l)	16.7	3.4	16.2	0.4	30	0.7	0.7	3.6	5.1

Just before sowing, 10 litres of fresh water were applied to all treatments to obtain a sufficient emergence. Once the different qualities of water were applied, at each irrigation surplus of water was added to provide a leaching fraction of about 0.2.

The evapotranspiration between the irrigation interval was calculated as the difference between the amounts of irrigation and drainage water. Soil moisture sampling in the past had shown almost the same moisture content after each irrigation, corresponding to the field capacity.

To determine the soil salinity, the average chloride concentration of the soil water was calculated from the balance of irrigation and drainage water (Katerji et al., 1992) and converted into EC of the soil water by the equation $\ln EC = 0.824 \ln Cl - 1.42$, established for this type of irrigation water and soil (van Hoorn et al., 1993). This EC value of soil water was divided by two for the conversion into ECe. Owing to leaching at each water application, the soil salinity remained almost constant or increased slightly from the start to the end of the plant life cycle. The salt balance has the advantage of covering the whole soil volume of the lysimeter, as compared to the point measurements of soil or soil water sampling.

3- Water Stress of the Plant

The parameters used to characterise the water stress of the plant were the leaf-water potential and stomatal conductance. The leaf-water potential was determined with a pressure chamber (Scholander et al., 1965) on one leaf per lysimeter (five leaves per treatment), taken from the upper part of the canopy to avoid senescent leaves. The stomatal conductance was measured with a diffusion porometer on the upper leaf surface of two leaves per lysimeter (ten leaves per treatment).

These measurements were carried out at two time periods. During the entire vegetative cycle, the predawn leaf-water potential was regularly determined at sunrise. Throughout the day, on particular days, the leaf-water potential and stomatal conductance were simultaneously measured each hour.

4- Growth and Yield

Leaf area, plant height and dry matter were determined at the successive phenological stages on two plants per lysimeter.

At harvest, the yield of grain and straw and the yield components (number of pods per plant, number of grains per pod, 1000-grain weight) were measured from all plants per lysimeter.

5 - Statistical Analysis

For the statistical analysis, each of the 5 lysimeters was treated as a replicate. In order to analyse the effect of texture, comparisons were made between the two control treatments irrigated with fresh water in clay and in loam soil. In order to analyse the effect of the salinity, the treatments C (control), 15 and 30 were compared separately in loam and in clay soils.

Each comparison between the two soil types, or among the three water qualities, affecting plant water-status, crop growth, evapotranspiration and yield was statistically analysed with the general linear model (SAS, 1999). Treatments were further compared by calculating the least significant difference (LSD; $P < 0.05$).

III- Results and Discussion

1- Soil Salinity

Table 3 shows the values of ECe, determined, after each irrigation, with the Salt Balance method. The values of ECe observed in the three treatments on the loam soil were slightly superior to the same treatments on the clay soil. However the differences (0.4 - 0.7 dS/m) between the mean values observed during the whole crop cycle are not higher than 10%.

Table 3. Values of ECe (dS/m) determined during the soybean crop cycle observed on loam and clay soils.

Soil		loam			*clay*	
D.A.S.	C	15	30	C	15	30
11	0.8	4.08	6.96	0.8	3.71	6.29
25	0.8	4.11	6.92	0.8	3.73	6.25
44	0.8	4.14	6.96	0.8	3.85	6.30
61	0.8	4.22	7.03	0.8	3.81	6.29
91	0.8	4.39	7.14	0.8	3.83	6.27
mean	*0.8*	*4.19*	*7.00*	*0.8*	*3.79*	*6.28*

2- Effect of Soil Texture on Plant Water Relationships and the Productivity of the Treatments Irrigated With Fresh Water.

Figure 1 presents the values of pre-dawn leaf-water potential (fig. 1a), leaf surface (fig. 1b), and dry matter (fig. 1c).observed during the vegetative cycle on the control treatments in loam and clay soil

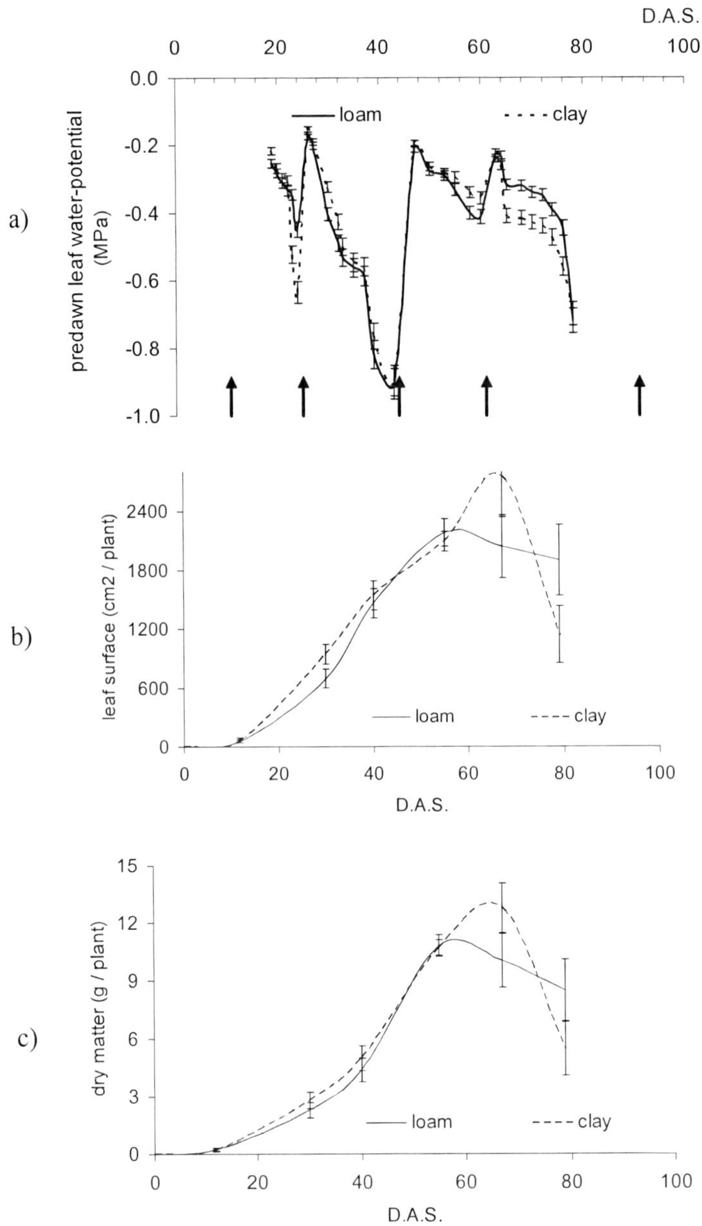

Figure 1. Values of: pre-dawn leaf-water potential (a), leaf surface/plant (b), and dry matter/plant (c) observed during the soybean crop cycle on the control treatments in loam and clay soils. Irrigation timing is indicated by the arrows.

Table 4 shows data for the same treatments regarding yield (Y), its components, and seasonal evapotranspiration (ET). This table also includes the data on water use efficiency calculated as the Y/ET relationship.

Table 4. Yield, its components, seasonal evapotranspiration (ET) and water use efficiency (WUE) observed on the control treatments in loam and clay soils.

soil	*loam*		*clay*	
plant/m^2	18	a	18	a
n° pod/plant	44	a	41	a
n° grain/pod	2.6	b	2.7	a
1000-grain weight (g)	164	a	161	a
grain yield (g/ m^2)	334	a	311	a
straw (g/m^2)	261	a	257	a
seasonal ET (mm)	410	a	430	a
WUE (kg/m^3)	0.81	a	0.72	a

The predawn leaf-water potential shows an increase, after each irrigation, and then decreases during the irrigation interval. Differences in pre-dawn leaf-water potential between the two soil types can be observed, but these differences do not show a clear trend. Soil texture did not show a clear effect on the predawn leaf-water potential.

According to Mastrorilli et al. (1993), the soybean stomatal conductance is not affected until the pre-dawn leaf-water potential exceeds the threshold of -0.4MPa. The two control treatments had, therefore, optimal water conditions during the vegetative cycle, except for the 13 days between 30 and 43 DAS.

Between the emergence and the 55 DAS, the observed values per plant of leaf surface and dry matter are perfectly identical in loam and clay soils. After that, the leaf surface and the dry matter observed did not show a clear trend. Soil texture did not show a clear effect on growth in leaf surface and dry matter either.

The values measured for loam and clay soils for yield in straw and grain are not significantly different. The same can be observed, in general, for the yield components measured for the two soil types.

In loam and clay soils, the measured seasonal evapotranspiration (ET) and the values calculated for water use efficiency are not significantly different.

Van Hoorn et al. (2001) determined the nitrogen balance for control treatments in loam and clay soils. This determination later allowed for the estimation of the quantity of organic nitrogen (the sum of the nitrogen fixation and the nitrogen derived by the transformation of organic matter). These quantities are equivalent to 23.5 and 21.2 g/m^2 in loam and clay soil respectively. These values are very similar and represent 69% and 66% of the nitrogen absorbed by the plants cultivated in loam and clay soil, respectively.

3.-Effect of the Soil Texture on Plant Water Relationships and Productivity of the Treatments Irrigated With Saline Waters

3-1- Effect on the Plant Water Relationship

Figure 2 shows the values of pre-dawn leaf-water potential observed in the loam and clay soils for the control treatments (irrigated with fresh water) and saline treatments (irrigated with waters having 15 and 30 meq Cl/l).

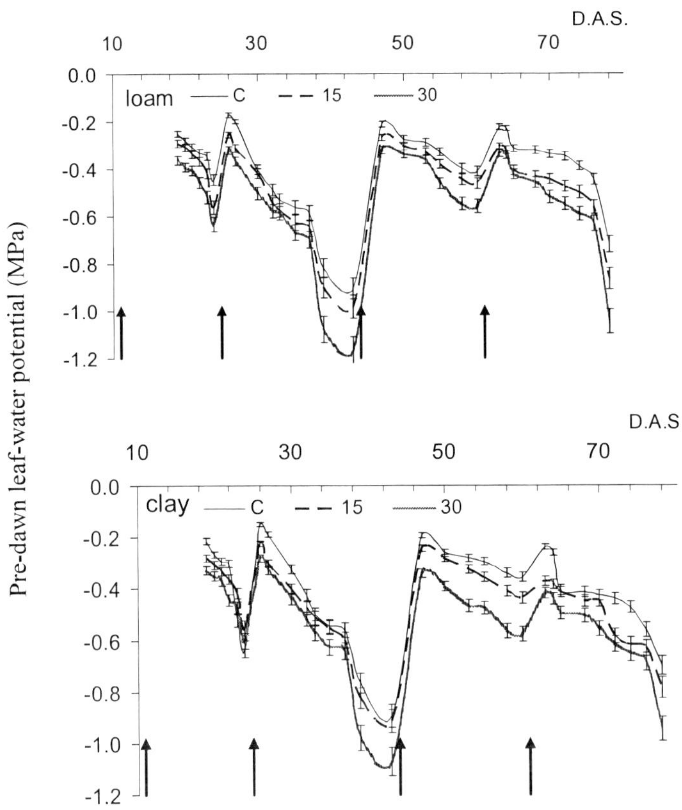

Figure 2. Pre-dawn leaf-water potential observed during the soybean crop cycle on the loam and clay soils for the three water-quality treatments (control, C, 15 and 30 meq/l). Irrigation timing is indicated by the arrows.

The variations in pre-dawn leaf-water potential observed in saline treatments are the same as those observed in control treatments. However, the observed values differed significantly for the three water qualities. In loam and clay soils, the 15 meq/l treatments had an intermediate position between the control and the 30 meq Cl/l treatments. The largest difference between the three treatments was always observed just before irrigation. Soil texture did not significantly affect the pre-dawn leaf-water potential.

Figures 3 and 4 show the hourly values of leaf-water potential and stomatal conductance for the six treatments during the daytime on day 64 D.A.S., just after irrigation at a high predawn-leaf-water potential and day 76 D.A.S. at a lower predawn leaf-water potential.

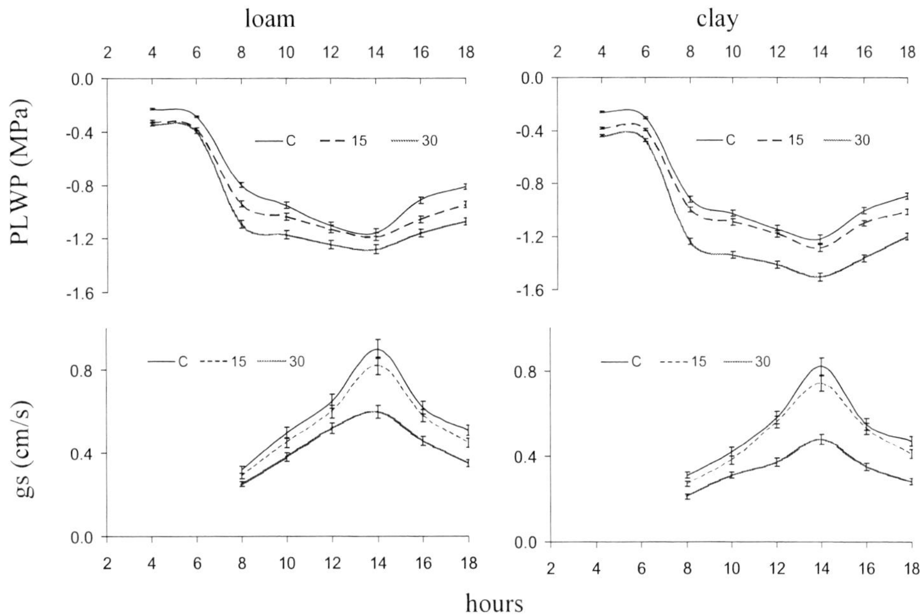

Figure 3. Pre-dawn leaf-water potential (PLWP) and stomatal conductance (gs) observed during the daytime (day 64 D.A.S) on loam and clay soils for three water-quality treatments (control C, 15 and 30 meq/l).

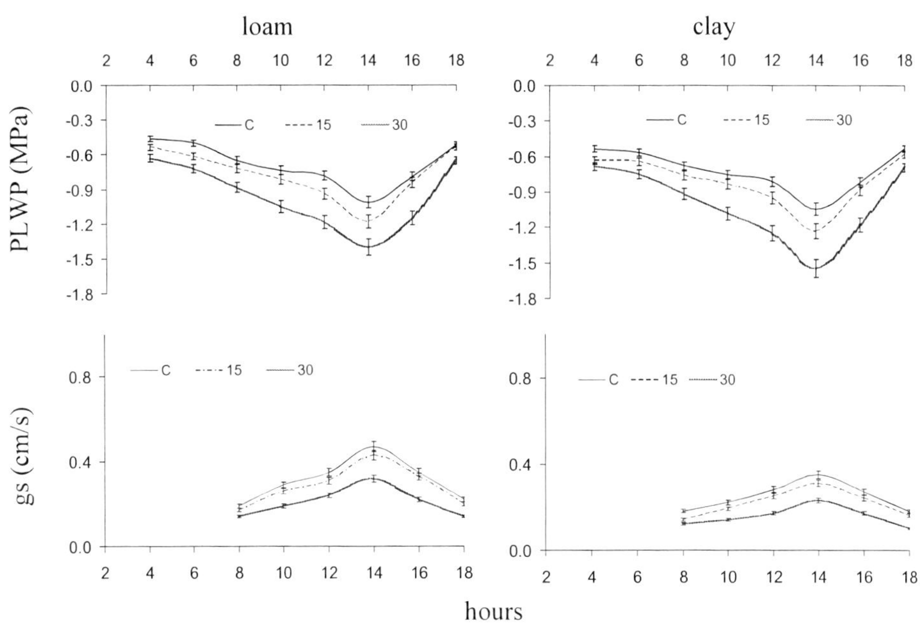

Figure 4. Pre-dawn leaf-water potential (PLWP) and stomatal conductance (gs) measured during the daytime (day 76 D.A.S) on loam and clay soils for three water-quality treatments (control C, 15 and 30 meq/l).

The leaf-water potential decreased after dawn, attained a minimum around 14h, and increased again afterwards. The stomatal conductance increased during the morning, attained a maximum value around noon and then decreased to reach its minimum value again around 18h. All treatments displayed the same development in leaf-water potential and stomatal conductance. However, it is possible to observe:

- The effect of salinity on the hourly values of leaf-water potential and of stomatal conductance. The higher the salinity, the lower the leaf-water potential and stomatal conductance.
- The effect of the soil texture on the hourly values of leaf-water potential and of stomatal conductance. Clay soil showed, in all treatments, lower values than loam soil.
- The differences observed are consistently significant in the case of the more saline treatments (30 meq Cl/l).

3-2 Effect on Growth

Figures 5 and 6 show the values per plant of leaf surface and the above-ground dry matter observed for all treatments on the loam and clay soils.

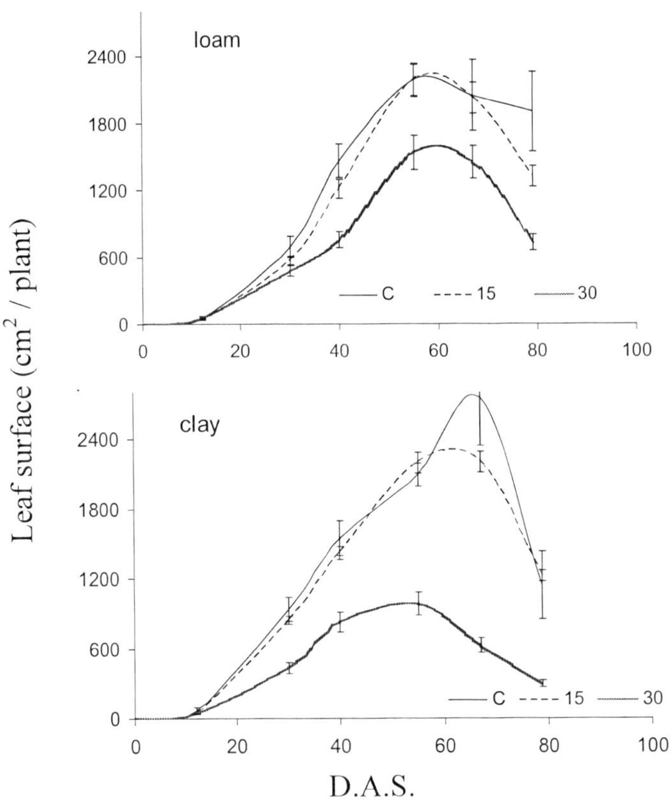

Figure 5. Values of leaf surface /plant observed during the soybean crop cycle on the loam and clay soils for three water-quality treatments (control C, 15 and 30 meq/l).

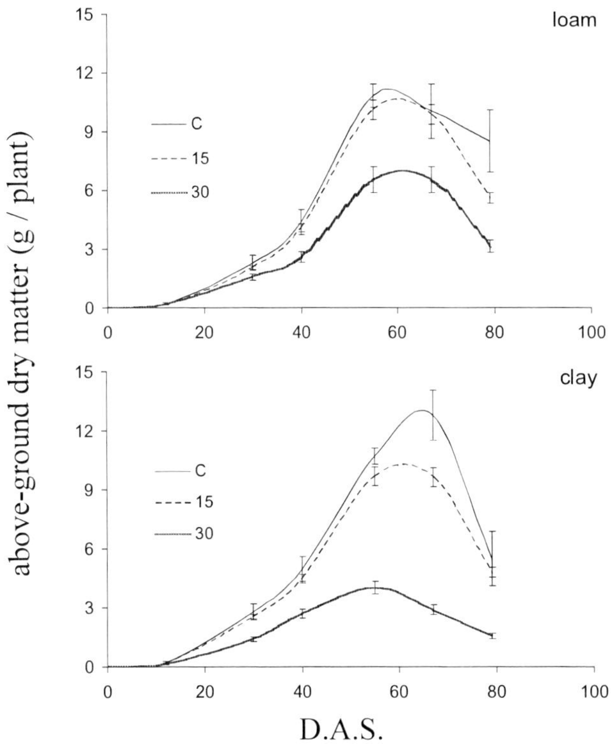

Figure 6. Values of a bove-ground dry matter / plant observed during the soybean crop cycle on the loam and clay soils for three water-quality treatments (control C, 15 and 30 meq/l)..

For loam and clay soils, the control treatments and the 15 meq/l treatments show the same growth in leaf surface and in dry matter before the maximal growth stage. However, it is possible to observe values significantly lower for the 15 meq /l treatments during the maximal growth stage.

The salt and texture effects are more evident for the treatments with the highest saline concentrations:

- The treatments with the highest saline concentration (treatments irrigated with 30 meq/l) show an increase in leaf surface and in dry matter that is significantly lower than their respective control groups. This is true for the entire length of the crop cycle.
- The highest values for leaf surface and dry matter observed for the treatments with the highest saline concentration cultivated in the clay soil are 37% lower on average than those observed for the same treatment cultivated in loam soil.

3.3- Effect on Yield and Water Use Efficiency

Table 5 presents observations regarding productivity, its components and the seasonal evapotranspiration obtained on the loam and clay soils for all treatments. Also shown on the table are the data calculated for water use efficiency, treatment by treatment.

Table 5. Yield, yield components, seasonal evapotranspiration (ET), and water use efficiency (WUE) observed on loam and clay soils for three irrigation water-qualities (control C, 15 and 30 meq/l).

soil	*loam*						*clay*					
water quality	*C*		*15*		*30*		*C*		*15*		*30*	
plant/m^2	18	a	18	a	18	a	18	a	18	a	18	a
n° pod/plant	44	a	43	a	32	b	41	a	34	b	25	c
n° grain/pod	2.6	a	2.5	b	2.5	b	2.7	a	2.5	ab	2.4	b
1000-grain weight (g)	164	a	154	a	130	b	161	a	148	b	100	c
grain yield (g/m^2)	334	a	294	a	180	b	311	a	221	b	106	c
straw (g/m^2)	261	a	269	a	169	b	257	a	253	a	251	a
seasonal ET (mm)	410	a	376	a	306	b	430	a	361	b	300	c
WUE (kg/m^3)	0.81	a	0.78	a	0.59	b	0.72	a	0.61	b	0.35	c

The yields of straw and grain are reduced in loam and clay soils with the increase of soil salinity. However, in the case of loam soil the differences in yield between the control treatment and the 15 meq/l treatment are not significant. Additionally, these two treatments do not present differences in the level of water use efficiency. Moreover, the reduction in straw and grain yield observed for the 30 meq Cl/l treatment is significant.

In the case of the clay soil, the reductions in yield (straw and grain) and in water use efficiency observed, on the one hand, between the control and the 15 meq/l treatments and, on the other hand, between the 15 meq/l and the 30 meq Cl/l treatments were significant.

The reductions observed for the grain yield of the treatments irrigated with the saline waters are due to the simultaneous reductions of three parameters: the number of pods/plant, the number of grains/pod and the 1000-grain weight.

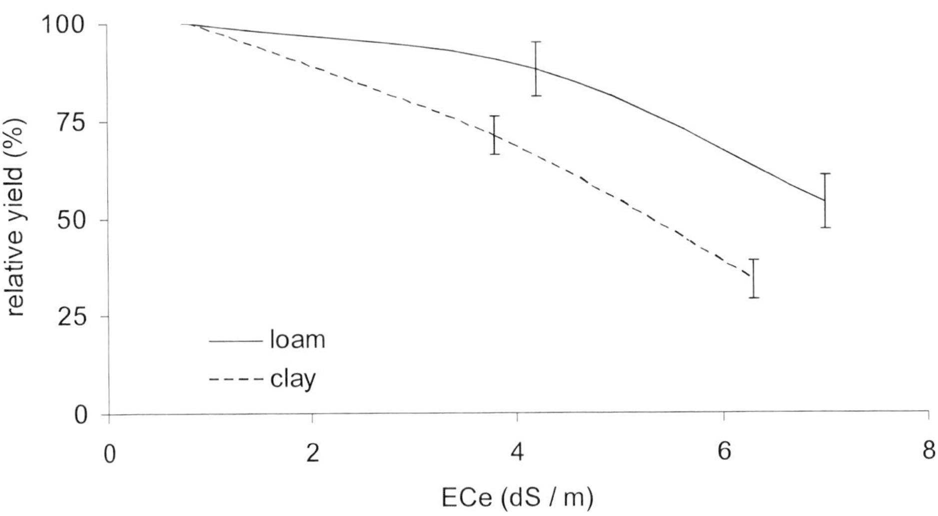

Figure 7. Relation between relative yield of soybean and soil salinity, observed on loam and clay soil.

Figure 7 illustrates the relationship between the reduction in relative grain yield (relationship between the yield observed for the saline treatment and the yield observed for the control treatment) and the soil salinity (ECe). With the same level of soil salinity, the relative yield observed in the clay soil is significantly lower than that observed in the loam soil.

According to the observations of Van Hoorn et al. (2001), salinity and soil texture did not affect the nitrogen content (mg N/ dry matter) of soybean. Moreover, the quantity of organic nitrogen varies with the soil salinity and texture, according to the dry matter accumulation. These quantities are 20 and 9.4 g/m^2 respectively for the 15 and 30 meq Cl/l treatments in loam soil. In the case of clay soil, these quantities are inferior to those of the same treatments in loam soil: 28% and 64% for the 15 and 30 meq Cl/l treatments, respectively. Therefore, the organic nitrogen in loam soil ensures between 67% (for the 15 meq/l treatment) and 50% (for the 30 meq Cl/l treatment) of the plants' nitrogen uptake. Instead, in the case of the same treatments in clay soil, the organic nitrogen fixation ensures only 59% and 26% of the plants' nitrogen uptake.

IV- Discussion and Conclusion

The observations obtained in this study provide the answers to the three questions posed in the introduction to this chapter.

In the absence of salt stress, soil texture does not have a significant influence on the plant water relationships, growth and yield for soybean. It can simply be noted that the plant water status, yield and water use efficiency have a slight advantage in loam soil. Differently from other species like sugar-beet, potato and sunflower (see the review by Katerji and Mastrorilli, 2009), soybean is not strongly influenced by soil texture.

In the presence of salt stress, soybean's tolerance to salinity changes depending on the soil texture. The same soybean cultivar tolerated growing in loam soil better than in clay soil (see table 5 and Fig. 7).

Two hypotheses can be formulated to explain soybean's greater tolerance of salinity in loam soil:

1. Soybean is a species that has high nitrogen requirements (Bouniols et al., 1990). The quantities of nitrogen provided by fertilisation in this study were lower than the plants' requirements (van Hoorn et al., 2001). The higher organic nitrogen in loam soil allows the plant to produce a greater yield and better tolerate the soil salinity. Instead, the lesser nitrogen fixation in clay soil limits the growth and productivity of the plant. The greater nitrogen fixation in loam soil can be explained by two possible causes:

- ✓ The texture of clay having a negative effect on nitrogen fixation. This effect is not demonstrated in a comparison between the two control treatments. Actually, according to the estimates by van Hoorn et al. (2001), the quantities of nitrogen fixed by the two control treatments are very similar.

✓ The effect of salinity on nitrogen fixation being stronger in the case of clay soil. The measurements of soil salinity do not support this hypothesis. Instead, on the contrary, they are slightly higher in the saline treatments in loam soil than in the same treatments in clay soil (tab. 3).

The previous discussion, therefore, leads us to reject this first hypothesis.

2. The unfavourable effect of the clay texture, barely noticeable in the control treatment, is accentuated in the presence of salt stress, associated with a water shortage during a short period of the vegetative cycle. The water status of the plants cultivated in clay soils is much less favourable than of the same plants cultivated in loam soil (Figs. 3 and 4). As a consequence, it is possible to conclude that the water uptake is easier for the soybean cultivated in loam soil. There are various reasons: the depth of rooting was superior in loam than in clay soils (Smith and Harris, 1981) and the hydraulic conductivity in clay soil is inferior to that in loam soil (Castrignanò et al., 1998). As a result, growth (Figs 5 and 6) and the appearance of the reproductive organs are reduced in clay soil (tab. 5). Differently from plants cultivated in loam soil, the nitrogen requirements of plants cultivated in clay soil are inferior, therefore, ensured largely by the quantity of mineral nitrogen provided by fertilisation. Moreover, the organic nitrogen fixation is inhibited until the mineral nitrogen in the soil is reduced (Alston and Graham, 1982). The lower fixation observed in clay soil is not the cause of the lower productivity of the soybean observed in clay soil, but, rather, the cause of lower nitrogen requirements by plants cultivated in this type of soil. For this reason, van Hoorn et al. (2001) did not observe any effect from texture on soybean's nitrogen content.

As can be seen, this last hypothesis seems to be more in line with the experimental observations.

The preceding results call attention to the role of soil texture as an environmental factor capable of making significant changes in crop tolerance to salinity, in particular of soybean. This role, often ignored in the literature, should be taken into consideration when considering management strategies for saline soils. When talking about soybean, the level of soil salinity can not continue to be the only criteria taken into consideration. Soil texture also plays a fundamental role.

References

Alston, A.M., Graham, R.D., 1982. The influence of soil nitrogen status and previous crop on nitrogen fixation (acetyline reduction) in barrel modec, Medicago trunculata Gaertn. Aust. *J. Soil Sci.* 27, 462–469.

Ayers, R.S., Westcot, D.W., 1985. *Water quality for agriculture.* FAO Irrigation and Drainage Paper 29 rev. 1, Rome, 174 pp.

Bernstein, L., Ogata, G., 1966. Effects of salinity on nodulation, nitrogen fixation, and growth of soybean and alfalfa. *Agron. J.* 58, 201–203.

Bouniols A, Tancogne M, Merrien, A, Blanchet R, 1990.L'alimentation azotée des légumineuses à graines dans l'agriculture Française actuelle : exemple du soja. CR. *Acad. FR.* 76, 109-115.

Castrignanò, A., Katerji, N., Mastrorilli, M., 1998. Modeling crop response to soil salinity: review and proposal of a new approach. *Ecol. Model.* 111, 107–120.

Katerji, N., van Hoorn, J.W., Hamdy, A., Bouzid, N., El-Sayed Mahrous, S., Mastrorilli, M., 1992. Effect of salinity on water stress, growth and yield of broadbeans. *Agric. Water* Manage. 21, 107–117.

Katerji, N., van Horn, J.W., Hamdy, A., Mastrorilli, M., 2000. Salt tolerance classification of crops according to soil salinity and to water stress day index. Agric. *Water Manage*., 43, 99-109.

Katerji, N., Mastrorilli, M., 2009. The effect of soil texture on the water use efficiency of irrigated crops: Results of a multi-year experiment carried out in the Mediterranean region. *Eur. J. Agron*., 30, 95-100.

Maas, E.V., Hoffman, G.J., 1977. Crop salt tolerance, current assessment. J. Irrig. Drain. *Div. ASCE* 103, 115–134.

Mastrorilli, M., Losavio, N., Rana, G., Katerji, N., 1993. Comparison of water stress indicators for soybean. *Acta Horticulturae*, 335, 359-364.

Patterson, T.G., La Rue, T.A., 1983. Nitrogen fixation of soybeans: seasonal and cultivar effects, and comparison of estimates. *Crop Sci.* 23, 488–492.

Pessarakli, M., Huber, J.T., Tucker, T.C., 1989. Protein synthesis in grean beans under salt stress with two nitrogen sources. *J. Plant Nut.* 12, 1261–1377.

Shalhevet, J., 1994. Using water of marginal quality for crop production: major issues. Agric.*Water Manage*. 25, 233–269.

Scholander, P.J., Hammel, H.T., Bradstreet, E.D., Hemmingsen, E.A., 1965. Sap pressure in vascular plants. *Sci.* N.Y. 148, 339–346.

Smith, R.C.G., Harris, H.C., 1981. Environmental resources and restraints to agricultural production in a Mediterranean-type environment. *Plant and soil*, 58,1-3, 31-57.

Tu, J.C., 1981. Effect of salinity on rhizobium–root-hair interaction, nodulation and growth of soybean. *Can. J. Plant Sci.* 61, 231–239.

Van Hoorn, J.W., Katerji, N., Hamdy, A., Mastrorilli, M., 1993. Effect of saline water on soil salinity and on water stress, growth and yield of wheat and potatoes. Agric. *Water Manage.* 23, 247–265.

Van Hoorn, J.W., Katerji, N., Hamdy, A., Mastrorilli, M., 2001. Effect of salinity on yield and nitrogen uptake of four grain legumes and on biological nitrogen contribution from the soil. Agric. *Water Manage*. 51, 87–98.

Velagaleti, R., Schweitzer, S.M., 1993. General effects of salt stress on growth and symbiotic nitrogen fixation in soybean. In: Pessarakli, M. (Ed.), *Plant and Crop Stress*, pp. 461–471.

Wagner, G.H., Zapata, F., 1982. Field evaluation of reference crops in the study of nitrogen fixation by legumes using isotope techniques. *Agron*. J. 74, 607–612.

Wilson J.R., 1970. Response to salinity in glycine: VI. Some of soybean. *Science* 215, 1631-1632.

Zapata, F., Danso, S.K.A., Hardarson, G., Fried, M., 1987. Time course of nitrogen fixation in field-grown soybean using nitrogen-15 methodology. *Agron*. J. 79, 172–176.

In: Soybean and Wheat Crops
Editors: S. Davies, G. Evans

ISBN: 978-1-60741-173-4

Chapter 3

MODELING THE WATER BALANCE COMPONENTS OF THE SOYBEAN CANOPY BY SOIL-VEGETATION-ATMOSPHERE TRANSFER MODEL

D. T. Mihailović [*] ***and B. Lalić***
Faculty of Agriculture, Department for Field and Vegetable Crops,
University of Novi Sad, 21000 Novi Sad, Serbia

ABSTRACT

In recent years, though, expansion of soybean croplands has been increasingly important in the agricultural or production in many parts of the world. There are a lot attempts to set this cultivar in the modeling focus, from different points of view (microclimate, irrigation, crop, land surface, climate cahnge, etc.). However, regardless the model is used, the interaction of surface and subsurface runoff and soil moisture, the simulation of total evaporation (or latent heat) are always highly ranked in the modeling hierarchy.

This chapter deals with the simulation of the water balance components of the soybean canopy using a surface scheme. In that sense we used the hydrological module in the Land-Air Parameterization Scheme (LAPS) developed at Faculty of Agriculture, Department for Field and Vegetable Crops, University of Novi Sad (Serbia). It is designed as a software package that can be run as part of an environmental model or as a stand-alone one. The LAPS includes modeling the interaction of the land surface and the atmosphere, under processes divided into three sections: subsurface thermal and hydraulic processes, bare soil transfer processes and canopy transfer processes. They are: interaction of radiation with vegetation, evaporation from bare soil, evapotranspiration including transpiration and evaporation of intercepted water and dew, conduction of soil water through the vegetation layer, vertical water movement in the soil, surface and subsurface runoff, heat conduction in the soil and momentum transport within and above

[*] Corresponding author: D.T. Mihailovic; E-mail address: guto@polj.ns.ac.yu; Phone: +381 21 4853203; Fax: +381216350552

the vegetation. The scheme has seven prognostic variables: three temperature variables (foliage, soil surface and deep soil), one interception storage variable, and three soil moisture storage variables. For the upper boundary conditions the following forcing variables are used: air temperature, water vapor pressure, wind speed, short wave and long wave radiation and precipitation at a reference level within the atmospheric boundary layer. The hydrological module is designed as a three-layer model, which is used to describe the vertical transfer of water in the soil. The LAPS uses the morphological and physiological characteristics of the plant community for deriving the coefficients and resistances that govern all the fluxes between the surface and atmosphere.

In order to simulate partitioning of the soybean canopy water into water balance components, during short and long period (day, growing season), several simulation were performed. The corresponding forcing, morphological, physiological and soil data as well as the observations were inserted from data sets comprising different agroecological soybean regions: Paragaminas (Brasil), Marchfeld (Austria) and Caumont (France).

1. Introduction

1.1. Soybean-water Relationship

Irrigation applications for soybean (*Glycine max*), or high amounts of rainfall that occur during vegetative growth, are normally not beneficial unless soil water levels are extremely low. In fact, excessive water during the early growth stages may stimulate vegetative growth without leading to a yield increase. If lodging occurs, vegetative stage irrigation can substantially depress yields.

The total water used (evapotranspiration usually denoted as ET) by a fully irrigated soybean crop may vary from 508 mm to 660 mm during the growing season (Figure 1). About 65 percent of total water use occurs during the Beginning Flower to Beginning Seed Enlargement reproductive stages. In an average season, the water-use rate will reach a peak of about 8.1 mm per day during the late flowering and early pod development stages, but during the mid to late reproductive stages it may average 6.4 mm per day. However, the actual ET amount in any given year will vary daily. Note that a well-watered soybean crop may transpire up to 12.8 mm of water on a hot, windy day in late-July or August. The most important times for soybean plants to have adequate water are during pod development (Beginning Pod Elongation – End of Pod Elongation) and seed fill (Beginning Seed Enlargement – End of Seed Enlargement). These are the stages when water stress can lead to a significant decrease in yield. Irrigation also may be required prior to these stages on sandy soils (insufficient water-holding capacity) or during very low rainfall years on medium- and fine-textured soils. However, if water is applied during flowering, it is important to follow with adequate water during seed fill. Irrigation at flowering typically increases the number of seeds produced per plant, but any subsequent water stress will reduce the size of those seeds such that the yield response to an irrigation at flowering may be no more or even less than not irrigating at flowering. Irrigation at flowering also may increase the potential for a white mold infection due to the humid in-canopy conditions after the irrigation. Though soybean roots can reach depths of 1.52 m to 1.82 m, the largest concentration of roots and the majority of soil water extraction occur in the top 0.61 m to 0.91 m of the soil profile. Irrigation water should thus be limited to a 0.91 m penetration depth to avoid leaching nitrates into the local aquifer [1-4].

Figure 1. Soybean field at Rimski Sancevi (Serbia) [5].

1.2. Modeling Soybean Water Balance

The soybean water balance and its components can be considered and modeled for different purposes. For example, CROPGRO is a process-oriented model that simulates a crop carbon balance, a crop and soil water balance, and a crop and soil nitrogen balance. State variables are the amounts, masses and numbers of tissues, and rate variables are the rates of inputs transformations and losses from state variables [6,7]. This model as a stand alone, in modified versions, can be used for (i) determining the productivity levels, potential yields, yield gaps [8], (ii) simulating soybean growth in controlled environments [9], (iii) simulating phasic temperature and photo period control for soybean [10], etc.

Water balance components and related hydrological variables of the soybean field can be simulated by a Soil-Vegetation-Atmosphere Transfer (SVAT) model that is forced by the products of RegCM (Regional Climate Model) for both present and future (warm-up) conditions. This is important for the projection of the climate change impact on the regional hydrological cycle that is a basis to start the discussion on the vulnerability of agricultural production system and to see the performance and usefulness of the RegCM products. While changes in atmospheric composition are expected to exert an increasing radiative forcing of climate change leading to further warming of global mean temperatures and shifts in precipitation patterns, these are not the only climatic processes which may influence crop and also soybean production. Climate models used to drive crop models may, therefore, need to consider changes in the land surface, either as imposed boundary conditions or as feedbacks from an interactive climate–vegetation model. Crops may also respond directly to changes in atmospheric composition, such as the concentrations of carbon dioxide (CO_2), ozone (O_3) and compounds of sulphur and nitrogen, so crop models should consider these processes as well as climate change. Changes in these, and the responses of the crops, may be intimately linked with meteorological processes so crop and climate models should consider synergies between climate and atmospheric chemistry. Some crop responses may occur at scales too small to significantly influence meteorology [11,5,12], so may not need to be included as feedbacks within climate models. However, the volume of data required to drive the appropriate crop models may be very large, especially if short-time-scale variability is important. Implementation of crop models within climate models would minimize the need to transfer large quantities of data between separate modeling systems. It should also be noted that crop responses to climate change may interact with other impacts of climate change, such as hydrological changes. For example, the availability of water for irrigation may be affected by changes in runoff as a direct consequence of climate change, and may also be affected by climate-related changes in demand for water for other uses. It is, therefore, necessary to consider the interactions between the responses of several impacts sectors to climate change. Overall, there is a strong case for a much closer coupling between models of climate, crops and hydrology, but this in itself poses challenges arising from issues of scale and errors in the models. A strategy is proposed whereby the pursuit of a fully coupled climate–chemistry–crop–hydrology model is paralleled by continued use of separate climate and SVAT but with a focus on consistency between the models [13-17]

Crop models and SVAT models have been designed for analyzing the interactions between plant canopy processes and the environment. They give priceless information for production and yield monitoring, management of water resources, assessment of water requirements and more recently for carbon cycle studies in relation with climate research. The use of such models over large areas is limited by our ability to provide them with the required input information. On one hand, it is almost impossible to obtain requisite plant and soil characteristics directly from networks of ground observations. On the other hand, remote sensing techniques can provide information on plant canopy processes that may be used for driving or constraining crop and SVAT models over large areas. These models may be operated without a systematic use of remote sensing data by intrinsically providing the means for interpolating energy and water fluxes or biomass production between remote sensing data acquisitions [18]. Compared to classical methods for mapping evapotranspiration, based on models such as SEBAL [19,20], the use of crop or SVAT models makes it possible to continuously monitor evapotranspiration along the whole crop cycle instead of estimating it for only snapshots derived from images.

2. The Structure of the Land-Air Parameterization Scheme (LAPS)

2.1. The Governing Equations and Approaches

The net radiation absorbed by the canopy and soil is assumed to be partitioned into sensible heat, latent heat, and storage terms, as

$$R_{nf} = \lambda E_f + H_f + C_f \frac{\partial T_f}{\partial t} \tag{1}$$

$$R_{ng} = \lambda E_g + H_g + C_g \frac{\partial T_g}{\partial t}, \tag{2}$$

where R_{nf}, R_{ng} are net radiation fluxes (W m^{-2}); λ is latent heat of vaporization (J kg^{-1}); λE_f, λE_g are evapotranspiration fluxes (W m^{-2}); H_f, H_g are sensible heat fluxes (W m^{-2}); C_f, C_g are heat capacities (J K^{-1} m^{-2}); and T_f, T_g are surface temperatures (K). The subscripts f and g refer to the upper-level canopy and soil, respectively (Figure 2). The fluxes of sensible and latent heat from the canopy and ground are represented by electrical analog models in which the fluxes are proportional to potential differences (in temperature or vapor pressure) and inversely proportional to resistances, which are equivalent to the inverse integrals of conductances over a specified length scale. For example, an aerodynamic resistance is calculated by integrating the inverse of a turbulent transfer coefficient between the reference points. The LAPS schematic diagram in Figure 2 [21,22] shows that these heat fluxes may be written as shown in Eqs. (3) through (6)

$$\lambda E_f = \frac{\rho c_p}{\gamma}\left[e_*(T_f) - e_a\right]\left(\frac{W_f}{\overline{r_b}} + \frac{1-W_f}{\overline{r_b} + \overline{r_c}}\right), \tag{3}$$

where ρ and c_p are the density and specific heat of air (kg m^{-3}, J kg^{-1} K^{-1}); γ is the psychrometric constant (mb K^{-1}); $e_*(T_f)$ is saturated vapor pressure at temperature T_f (mb); e_a is canopy air space vapor pressure (mb); W_f is canopy wetness fraction; $\overline{r_b}$ is bulk canopy boundary layer resistance (s m^{-1}); and $\overline{r_c}$ is bulk canopy stomatal resistance (s m^{-1}).

$$\lambda E_g = \frac{\rho c_p}{\gamma}\frac{\alpha_s e_*(T_g) - e_a}{r_{surf} + r_d}, \tag{4}$$

where α_s is a factor to correct for soil dryness [23], $e_*(T_g)$ is saturated vapor pressure at temperature T_g (mb), r_{surf} is soil surface resistance (s m^{-1}), and r_d is aerodynamic resistance between soil surface and canopy air space (s m^{-1}).

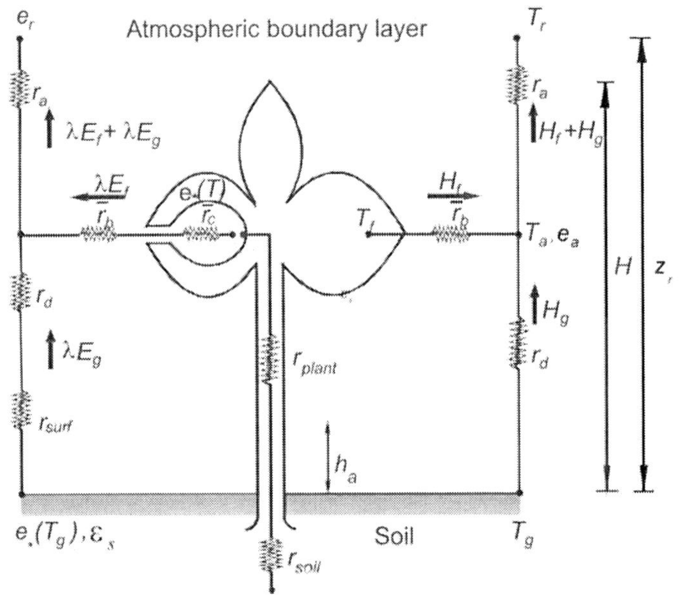

Figure 2. Schematic diagram of the Land-Air Parameterization Scheme (LAPS). The transfer pathways for latent and sensible heat fluxes are shown on the left- and right-hand sides of the diagram, respectively.

$$H_f = \frac{2(T_f - T_a)}{\bar{r}_b} \rho c_p \tag{5}$$

$$H_g = \frac{(T_g - T_a)}{r_d} \rho c_p , \tag{6}$$

where T_a is canopy air space temperature (K).

The latent and sensible heat fluxes from the soil and canopy combine to give the total surface fluxes, which are transferred from the canopy air space to the reference height, z_r and are given by

$$\lambda E_f + \lambda E_g = \frac{(e_a - e_r)}{r_a} \frac{\rho c_p}{\gamma} \tag{7}$$

$$H_f + H_g = \frac{(T_a - T_r)}{r_a} \rho c_p , \tag{8}$$

where e_r is vapor pressure at the reference height (mb), r_a is aerodynamic resistance (s m^{-1}), and T_r is air temperature at the reference height, z_r (K). Substituting (3) through (8) into (1) and (2) yields two differential equations for T_f and T_g. These are solved simultaneously using a backwards implicit method.

Derivation of the equation for wind profile inside a canopy in the "sandwich" approach Let us consider an element of the canopy volume having an area S and height H. The loss of air particles' momentum due to close contact with the plant leaves comes from the drag force arising on the leaf surface. This drag force, F_d, produces a shearing such that $d\tau/dz$, the vertical gradient of shear stress, τ, is equal to the drag force per volume V, i.e.,

$$\frac{d\tau}{dz} = F_d / V . \tag{9}$$

The drag force per leaf unit area, S_l, is parameterized to be proportional to the wind speed, u, i.e., the volumetric kinetic energy $1/2\rho u^2$ with the coefficient of proportionality C_d - the leaf drag coefficient. So,

$$\frac{F_d}{S_l} = \frac{1}{2} C_d \rho u^2 . \tag{10}$$

Note that S_l is the area of all leaves in the considered volume. Following the definition of leaf area index (LAI), we can write $LAI=S_l/2S$, since LAI is defined in terms of only one side of the leaf. Using $\tau = \rho K_s\, du/dz$ (where K_s is the turbulent transfer coefficient and z is the vertical coordinate), Eqs. (9)-(10), and keeping in mind that the volume occupied by plants is $S{\cdot}H$, after some manipulation we arrive at

$$\frac{d}{dz}\left(K_s \frac{du}{dz} \right) = \frac{C_d \overline{L_d}(H-h)}{H} u^2 . \tag{11}$$

where $\overline{L_d}$ is the area-averaged canopy density and h is the canopy bottom height (the height of the base of the canopy).

2.2. Deriving Aerodynamic Resistances for Calculating air Temperature Inside the Canopy

The canopy air space temperature, T_a, in land surface schemes can be determined diagnostically from the energy balance equation. This procedure comes from the equality of the sensible heat flux from the canopy to some reference level in the atmosphere, and the sum of the sensible heat fluxes from the ground and from the leaves to the canopy air volume [21,24], i.e.,

$$T_a = \frac{\dfrac{2T_f}{r_b} + \dfrac{T_g}{r_d} + \dfrac{T_r}{r_a}}{\dfrac{2}{r_b} + \dfrac{1}{r_d} + \dfrac{1}{r_a}} . \tag{12}$$

The aerodynamic resistance r_a between z_r and the water vapor and sensible heat source height, h_a, [21], can be defined as

$$r_a = \int_{h_a}^{H} \frac{1}{K_s} dz + \int_{H}^{z_r} \frac{1}{K_s} dz , \tag{13}$$

where K_s is the turbulent transfer coefficient (momentum/moisture/heat) inside and above the canopy in the intervals (h_a, H) and (H, z_r), respectively. The aerodynamic resistance in canopy air space, r_d, can be written in the form

$$r_d = \int_{z_g}^{H} \frac{1}{K_s} dz + \int_{h}^{h_a} \frac{1}{K_s} dz , \tag{14}$$

where z_g is the effective ground roughness length, while the area-averaged bulk boundary layer resistance, $\overline{r_b}$, has the form [21]

$$\frac{1}{\overline{r_b}} = \int_{h_a}^{H} \frac{\overline{L_d}\sqrt{u(z)}}{C_s P_s} dz , \tag{15}$$

where $\overline{L_d}$ is related to leaf area index LAI as $LAI = \overline{L_d}(H - h)$; $u(z)$ is the wind speed; C_s is the transfer coefficient [21]; and P_s the leaf shelter factor. Eqs. (12)-(15) can be modified to take into account the effects of nonneutrality. According to Sellers et al. (1986) [21], the position of the canopy source height, h_a, can be estimated by obtaining the center of gravity of the $1/\overline{r_b}$ integral. Thus,

$$\int_{h}^{h_a} \frac{\overline{L_d}}{r_b} dz = \int_{h_a}^{H} \frac{\overline{L_d}}{r_b} dz = \frac{1}{2} \int_{h}^{H} \frac{\overline{L_d}}{r_b} dz = \frac{1}{2\overline{r_b}} . \tag{16}$$

We may obtain h_a by successive estimation [21,23] until the foregoing equality is reached.

2.3. Calculating the Wind Profile Inside Tall Grass Canopies

We consider the canopy to be a block of constant-density porous material sandwiched between two heights, H and h [21,22]. The differential equation describing the wind profile within such a "sandwiched" canopy architecture can be written in the form of Eq. (11). To solve this equation, we have to know how K_s depends on parameters representing the canopy's aerodynamic and morphological features. K-theory is a commonly used method in modeling the turbulence within a plant canopy. Although its use may be physically unrealistic for this application, it yields reasonable results, so we shall use this method until suitable second-order closure models are developed and then applied to the problem. Recently, several papers have been published focusing on the closure problem, particularly within forest canopies [26-28]. However, because the traditional K-theory approach can be applied successfully within tall grass canopies [21], we shall stay with it.

A number of assumptions are offered about the variation of K_s within tall grass canopies [29,30]. Among those, we chose the approach where K_s is proportional to wind speed, u, i.e.,

$$K_s = \sigma u \,, \tag{17}$$

where the scaling length, σ, is an arbitrary, unknown constant. Combining Eqs. (11) and (17) produces an equation for the wind speed inside the canopy

$$\frac{d^2u^2}{dz^2} = \frac{2C_d \overline{L_d}(H-h)}{\sigma H}u^2 \,. \tag{18}$$

A particular solution of this equation can be found in a form that approximates the wind profile within the tall grass canopy fairly well [31]:

$$u(z) = u(H)\exp\left[-\frac{1}{2}\beta\left(1-\frac{z}{H}\right)\right] \tag{19}$$

where $u(H)$ is the wind speed at the canopy height and β is the extinction parameter, defined as

$$\beta^2 = \frac{2C_d \overline{L_d}(H-h)H}{\sigma} \,, \tag{20}$$

where σ still remains an unknown constant. Its value can be determined as a function of the morphological and aerodynamic characteristics of the underlying tall grass canopy in the following manner. We shall use the lower boundary condition at the canopy bottom $z = h$ in terms of the shear stress, τ, just above and below the indicated level. Therefore,

$$\tau\big|_h = \rho C_{dg} u^2\big|_h \,, \tag{21}$$

where C_{dg} is the leaf drag coefficient estimated from the size of the roughness elements of the ground [21], i.e.,

$$C_{dg} = \frac{k^2}{\left[\ln\dfrac{h}{z_g}\right]^2} \,, \tag{22}$$

where k is the von Karman constant, taken to be 0.41, and

$$\tau|_h = \rho K_s \frac{du}{dz}|_h \,. \tag{23}$$

Combining Eqs. (17), (19), (21), and (23), after some simple algebra we get an equation for the parameter σ of the form

$$2C_{dg}H = \beta\sigma \,. \tag{24}$$

Finally, substituting β from Eq. (20) in Eq. (24) and solving for the scaling length, σ, we reach

$$\sigma = \frac{2C_{dg}^2 H}{C_d \overline{L_d}(H-h)}, \tag{25}$$

which expresses σ through the morphological and aerodynamic parameters describing the stand canopy in the "sandwich" approach. Figure 3 depicts calculated values of σ as a function of the leaf drag coefficient, canopy density, canopy height, and canopy bottom height for three tall grass canopy heights. In these calculations we used Eqs. (22) and (25). It can be seen that lower values of tall grass canopy height and density $\left[C_d \overline{L}_d (H-h)\right]$ cause rapid growth of the scaling length. Such a plant canopy architecture includes more air space, so physically the turbulent transport of momentum, heat and water vapor more closely resembles the energy and momentum exchange above bare soil. In contrast, for values of $C_d \overline{L}_d (H-h)$ that are typical of tall grass canopy (0.5 to 0.7) according to Sellers et al. (1986), σ has a lower value. It means that the turbulent transfer coefficient, K_s, within a dense tall grass canopy, where eddies are small and dissipate quickly, tends to have lower values.

To our knowledge, the literature contains no information on how to reliably estimate canopy bottom height. The only information available is the h values of 1 m to 2.5 m reported for tall grass canopies by Dubov et al. (1978), Goudriaan (1977), Sellers (1986), Xue et al. (1991), Mihailovic and Kallos (1997), and Mihailovic et al. (2000) [32,33,21,34,22,35]. Using these data, we found a functional dependence of h on the canopy height, H, using a fifth-degree polynomial fitting procedure (Figure 4). This dependence can be helpful in land surface schemes using the "sandwich" approach to describe the vegetation layer. We checked the representativeness of this curve indirectly by calculating the zero plane displacement, d, and roughness length, z_0, for tall grass canopies of known morphological characteristics, using maize data from Wilson et al. (1982) and van Pul (1992) [36,37]. The derived z_0/H and d/H values of 0.08 and 0.71 for the Wilson et al. case and 0.051 and 0.76 for the van Pul case are within the range of values reported by Uchijima (1976) [38]: 0.05-0.15 for z_o/H and 0.53-0.8 for d/H.

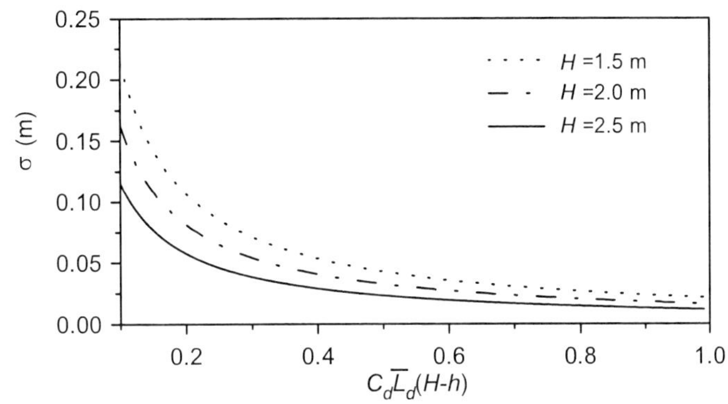

Figure 3. Calculated values of the scaling length, σ, as a function of the leaf drag coefficient (C_d), area-averaged canopy density ($\overline{L_d}$) canopy height (H), and canopy bottom height (h), for three tall grass canopy heights.

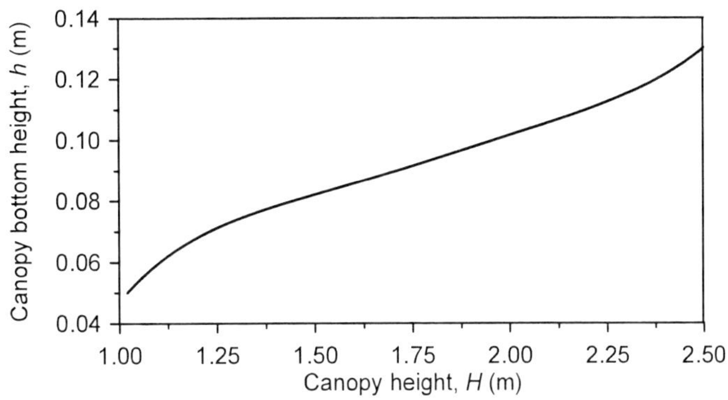

Figure 4. Calculated values of the canopy bottom height (h) as a function of the canopy height (H) for tall grass vegetation. The fitting curve is drawn using data from Dubov et al. (1978), Sellers et al. (1986), Mihailovic and Kallos (1997), Mihailovic et al. (2000) and Mihailovic et al. (2006) [30,31,19,35,36].

Between the canopy height and canopy bottom height, the wind profile attenuates exponentially according to Eq. (19), while beneath the canopy bottom height it follows a classical logarithmic profile of the form

$$u(z) = \frac{u(H)\exp\left[-\frac{1}{2}\beta\left(1-\frac{h}{H}\right)\right]}{\ln\frac{h}{z_g}}\ln\frac{z}{z_g}. \qquad (26)$$

Finally, since we know $u(z)$, the variation of wind speed with height inside the canopy, we can derive an expression for the canopy source height, h_a. Assuming that the canopy density is constant with height and combining Eqs. (15) and (16), we reach equality:

$$2\int_{h}^{h_a}\sqrt{u(z)}dz = \int_{h_a}^{H}\sqrt{u(z)}dz\,. \tag{27}$$

If we solve these integrals, keeping in mind that the functional form of $u(z)$ is given by Eq. (19), we get

$$h_a = H\left\{1+\frac{4}{\beta}\ln\frac{1+2\exp\left[-\frac{1}{4}\beta\left(1-\frac{h}{H}\right)\right]}{3}\right\}. \tag{28}$$

The temperature inside the canopy air space, T_a, is determined diagnostically using Eq. (12). In that formula the three aerodynamic resistances, r_a, r_b, and r_d, are calculated as follows:

$$r_a = \frac{1}{u_*}\left\{\frac{2kH}{\sigma\beta\ln\frac{H-d}{z_0}}\left[\exp\left[\frac{1}{2}\beta\left(1-\frac{h_a}{H}\right)\right]-1\right]+\frac{1}{k}\ln\frac{z_r-d}{H-d}\right\} \tag{29}$$

$$r_b = \frac{1}{\sqrt{u_*}}\frac{\beta C_s P_s\sqrt{k}}{4H\overline{L_d}\sqrt{\ln\frac{H-d}{z_o}}\left[1-\exp\left[-\frac{1}{4}\beta\left(1-\frac{h_a}{H}\right)\right]\right]} \tag{30}$$

$$r_d = \frac{1}{u_*}\left\{\frac{2kH}{\sigma\beta\ln\frac{H-d}{z_0}}\left[\exp\left[\frac{1}{2}\beta\left(1-\frac{h}{H}\right)\right]-\exp\left[\frac{1}{2}\beta\left(1-\frac{h_a}{H}\right)\right]-1\right]\right.$$

$$\left.+\frac{\exp\left[\frac{1}{2}\beta\left(1-\frac{h}{H}\right)\right]}{k\ln\frac{H-d}{z_o}}\ln^2\frac{h}{z_g}\right\} \tag{31}$$

The values for the leaf shelter factor, P_s, and transfer coefficient, C_s, used in the simulation are listed in Mihailovic and Kallos (1997) [22]. The effect of atmospheric nonneutrality in Eqs. (29)-(31) (i.e., their dependence on T_a and T_r) is included in u_* and accordingly in other calculations. In Eqs. (29)-(31) is supposed a logarithmic wind profile above a canopy.

In a sparse tall grass canopy (one in which the plant spacing is the order of the canopy height or larger), K_s is strongly affected by processes in the environmental space, including the plants and the space above the bare soil fraction. Therefore, K_s inside a sparse canopy, denoted K_s^s, can be represented by some combination of turbulent transfer coefficients in that space. If, as a working hypothesis, we assume a linear combination weighted by the fractional vegetation cover, σ_f (a measure of how sparse the tall grass is), then we can define K_s^s as

$$K_s^s = \sigma_f \sigma u + k u_* (1 - \sigma_f) z , \tag{32}$$

where u_* is the friction velocity above the bare soil fraction. In the case of dense vegetation (σ_f = 1), Eq. (32) reduces to Eq. (17). Otherwise, when σ_f = 0, Eq. (32) represents the turbulent transfer coefficient over bare soil. We can use Eq. (32) in calculating the wind speed inside a sparse tall grass canopy. For that purpose we have to slightly modify Eq. (11) to take into account σ_f

$$\frac{d}{dz}\left(K_s^s \frac{du}{dz} \right) = \sigma_f \frac{C_d \overline{L_d}(H-h)}{H} u^2 \tag{33}$$

Replacing the expression for K_s^s given by Eq. (32) in Eq. (33) produces

$$\frac{d}{dz}\left\{ \left[\sigma_f \sigma u + k u_* (1-\sigma_f) z \right] \frac{du}{dz} \right\} = \sigma \frac{C_d \overline{L_d}(H-h)}{H} u^2 . \tag{34}$$

After differentiation and grouping, the terms we reach are

$$a(u,z) \frac{d^2 u}{dz^2} + b(z) \frac{du}{dz} + c \left(\frac{du}{dz} \right)^2 = g u^2 , \tag{35}$$

where

$$a(u,z) = \sigma_f \sigma u + k u_* (1-\sigma_f) z \qquad c = \sigma_f \sigma$$

$$b(z) = k u_* (1-\sigma_f) z \qquad g = \sigma_f \frac{C_d \overline{L_d}(H-h)}{H}$$

To get the wind profile inside the sparse canopy, we can solve Eq. (35) numerically, using the fourth-order Runge-Kutte method [40].

2.4. Surface Resistances

The resistances to the transport of water vapor from within the crop and upper soil layer to the adjacent exterior air are defined as the bulk crop stomatal resistance, $\overline{r_c}$, and soil surface resistance, r_{surf}, respectively. Combining dependence of $\overline{r_c}$ on solar radiation, air temperature, atmospheric water vapor pressure deficit and water stress [35] is parameterized as

$$\overline{r_c} = \frac{r_{s\,\min}}{LAI} \frac{1 + \dfrac{1.1\langle F_f \rangle}{R_{ol} LAI}}{\dfrac{1.1\langle F_f \rangle}{R_{ol} LAI} + \dfrac{r_{s\,\min}}{r_{s\,\max}}} \left[1.0 - 0.0016(298 - T_r)^2\right]^{-1} \left\{1 - \eta\left[e_*(T_f) - e_r\right]\right\}^{-1} \Phi_2^{-1} \tag{36}$$

where $r_{s\ min}$, $r_{s\ max}$ are the minimum and maximum of stomatal resistance (s m^{-1}); R_{0l} is limit value of 100 W m^{-2} for crops; and η the crop-dependent empirical parameter that is equal to 0.025 hPa^{-1}. In this model the value of 5000 s m^{-1} for $r_{s\ max}$ is used. The factor Φ_2 takes into account the effect of water stress on the stomatal resistance and is parameterized in the following way

$$\Phi_2 = \begin{cases} 1 & \vartheta_a > \vartheta_{fc} \\ 1 - \left(\dfrac{\vartheta_{wil}}{\vartheta_a}\right)^{1.5} & \vartheta_{wil} \le \vartheta_a \le \vartheta_{fc} \\ 0 & \vartheta_a < \vartheta_{wil} \end{cases} \tag{37}$$

where ϑ_a is the mean volumetric soil water content in the first and second soil layers (m^3 m^{-3}); ϑ_{wil} is volumetric soil water content at wilting point (m^3 m^{-3}); and ϑ_{fc} volumetric soil water content at field capacity (m^3 m^{-3}).

The soil surface resistance, r_{surf}, is parameterized using the empirical expression given by Sun (1982) [41], i.e.,

$$r_{surf} = d_1 + d_2 \langle \vartheta_1 \rangle^{-d} \tag{38}$$

where d_1, d_2 (s m^{-1}) and d_3 are empirical constants [42], while ϑ_1 is the top volumetric soil water content (m^3 m^{-3}).

2.5. Hydrological Module

The parameterization of the volumetric soil moisture content is based on the following three governing equations

$$\frac{\partial \vartheta_1}{\partial t} = \frac{1}{D_1}\left[P_1 - F_{1,2} - \frac{E_g + E_{tf,1}}{\rho_w} - R_0 - R_1\right] \tag{39}$$

$$\frac{\partial \vartheta_2}{\partial t} = \frac{1}{D_2}\left[F_{1,2} - F_{2,3} - \frac{E_{tf,2}}{\rho_w} - R_2\right] \tag{40}$$

$$\frac{\partial \vartheta_3}{\partial t} = \frac{1}{D_3}\left[F_{2,3} - F_3 - R_3\right] \tag{41}$$

where ϑ_i is the volumetric soil water content ($m^3\ m^{-3}$) in the *ith* layer; P_1 is infiltration rate of precipitation into the upper soil moisture store ($m\ s^{-1}$); D_i (m) is thickness of the *ith* soil layer; $F_{i,i+1}$ is water flux between i and $i+1$ soil layer ($m\ s^{-1}$); F_3 is gravitational drainage flux from recharge soil water store ($m\ s^{-1}$); $E_{tf,1}$ and $E_{tf,2}$ are canopy extraction of soil moisture by transpiration from the rooted first and second soil layers ($kg\ m^{-2}\ s^{-1}$), respectively; R_0 the surface runoff ($m\ s^{-1}$); and R_i is subsurface runoff from the *ith* soil layer ($m\ s^{-1}$) (Figure 5). In the model Eqs. (39)-(41) are solved using an explicit time scheme.

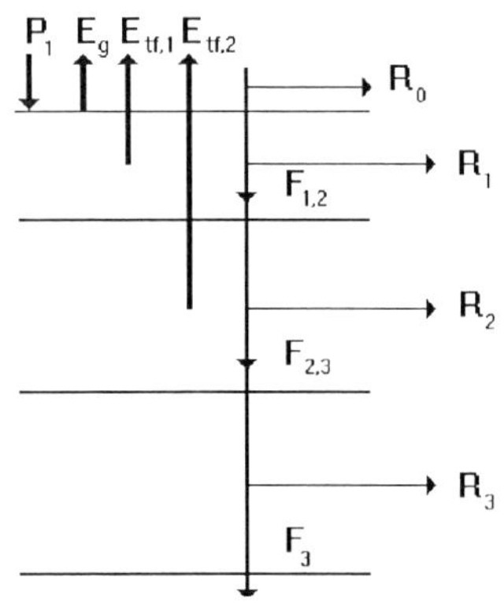

Figure 5. Schematic diagram of the LAPS scheme hydrology.

The precipitation P_1 that infiltrates into the top soil layer is given by

$$P_1 = \begin{cases} \min(P_0, K_s) & \vartheta_1 < \vartheta_2 \\ 0 & \vartheta_1 = \vartheta_2 \end{cases} \tag{42}$$

where K_s is the saturated hydraulic conductivity ($m\ s^{-1}$), ϑ_s is saturated volumetric soil water content and P_0 the effective precipitation rate on the soil surface ($m\ s^{-1}$) given by

$$P_0 = P - (P_f - D_f) \tag{43}$$

The rate of interception (inflow) for the canopy, P_f is given by

$$P_f = P\left(1 - e^{-v}\right)\sigma_c \tag{44}$$

where P is the precipitation rate ($m\ s^{-1}$) above the canopy and v is a constant depending on the leaf area index assuming that the interception of the rainfall can be considered via the expression describing the exponential attenuation [19] (Sellers et al., 1986). The rate of drainage of water stored on the vegetation (outflow) for the canopy ($m\ s^{-1}$), D_f, is given by

$$D_f = \begin{cases} 0 & W_f = w_{\max} \\ P_f & W_f = w_{\max} \end{cases} \tag{45}$$

The transfer of water between adjacent layers (m s^{-1}), $F_{I,i+1}$, is given by

$$F_{i,i+1} = K_{ef}\left[2\Psi_s \frac{\left(\frac{\vartheta_i}{\vartheta_s}\right)^{-B} - \left(\frac{\vartheta_{i+1}}{\vartheta_s}\right)^{-B}}{D_i + D_{i+1}} + 1\right] \tag{46}$$

where B is the Clapp Hornberger [43] constant and F_{ef} is effective hydraulic conductivity (m s^{-1}) between soil layers given by

$$K_{ef} = \frac{D_i K_i + D_{i+1} K_{i+1}}{D_i + D_{i+1}} \tag{47}$$

where F_i is the hydraulic conductivity (m s^{-1}) of the i*th* soil layer determined by the empirical formula

$$K_i = K_{sat,i}\left(\frac{\vartheta_i}{\vartheta_s}\right)^{2B+3} \tag{48}$$

where $K_{sat,i}$ is the hydraulic conductivity at saturation of the i*th* soil layer. The gravitational drainage from the bottom soil layer is defined as

$$F_3 = K_{si}\left(\frac{\vartheta_3}{\vartheta_s}\right)^{2B+3} \sin(x) \tag{49}$$

where x is the mean slope angle [44,45].

The surface runoff R_0 is computed as

$$R_o = P_I - \min(P_I, K_{sat}) \tag{50}$$

while the subsurface runoff, R_i (Figure 4), for each soil layer is calculated using the expressions

$$R_1 = F_{1,2} - \min(F_{1,2}, K_{sat,1}) \tag{51}$$

$$R_2 = F_{2,3} - \min(F_{2,3}, K_{sat,2}) \tag{52}$$

$$R_3 = F_3 - \min(F_3, K_{sat,3}). \tag{53}$$

At the end of every time step, Δ_i, the variable, Γ_i is calculated as

$$\Gamma_i = \frac{D_i}{\Delta t}\left[\vartheta_i^k + A_i + \Delta t - \vartheta_c\right] \tag{54}$$

where ϑ_i^k is the volumetric soil moisture content at the beginning of time step, A_i, representing the terms on the right-hand side of Eqs. (51)–(53), and ϑ_c is a wet reference parameter like field capacity or volumetric soil moisture content at saturation, depending on soil texture. If the condition Γ_i >0 is satisfied, then the Γ_i becomes runoff, which is added to the corresponding subsurface runoff, R_i. Consequently, at the end of the time step, the calculated value of the volumetric soil moisture content ϑ_i^{k+1} takes the value ϑ_c.

3. Simulation of Soybean Crop Water Balance

3.1. Description of Data Sets Used in Simulation

For simulations by the LAPS scheme we selected the following locations where the soybean is cultivated: Marchfeld plain, Austria (48° 12' N, 16° 34' E, mean altitude 153 m) during May-September 1995; Caumont, France (SAMER No. 3, 43^0 41' N, 0^0 06' W, mean altitude 113 m) during May-September 1986; and Paragominas, Brasil (2°58' S, 47°28' W, mean altitude 100m) during February–June 2007. The main features of the data sets, and corresponding land, morphological, physiological parameters, forcing data and initial conditions are described in Mihailovic and Eitzinger (2007), Cajic (2003), Mihailovic et al. (2006) for Marchfeld plain [46,47,39]. For Paragominas soil parameters are described in El-Husny et al. (2003) [48], while soybean characteristics during growing season provided by Jose Paulo (personal communication).

We have considered in more details HAPEX-MOBILHY experiment at Caumont because it includes a full year of atmospheric forcing and weekly soil moisture measurements up to 1.6 m depth, with 0.1 m interval [49]. Therefore we were able to perform the sensitivity test for the LAPS hydrological module estimating its capabilities for other runs. The HAPEX data set in its present form was prepared by Shao et al. (1995) [50]. The data were obtained from the HAPEX-MOBILHY. Detailed information on the SAMER network and the site can be found in Goutorbe and Tarrieu (1991) [51]. Most of the forcing data were taken from Caumont, particularly during the intensive observation period (May–July, 1986). If data at Caumont were missing, measurements from neighboring meteorological stations were used. The atmospheric forcing data (downward short wave radiation, downward infrared radiation, precipitation, air temperature at 2 m, wind speed at 2 m, atmospheric pressure at 2 m and specific humidity at 2 m) were available on 30 min intervals for one continuous. The chosen location was a soybean crop field, where soybean plants start to grow in May and are harvested at the end of September. Although the HAPEX data set was collected in a heterogeneous area, the immediate surroundings of Caumont can be considered as uniform on the scale of several hundred meters. For the HAPEX area at large, surface fluxes reveal the signature of the two main ecotypes: coniferous forest and crops (SAMER-3 represents one of the crops). Analysis of soil moisture also splits soil texture into two broad categories: sand and loam. The soil type at Caumont is loam. The parameters used for characterizing the land surface are summarized in Table 1 [52]. Monthly leaf area index, LAI, fraction vegetation cover, bare soil roughness length, canopy roughness length, zero plane displacement, canopy height and root distribution used in the one year integration, are presented in Table 2 [52]. In the same table, the monthly distribution of the soybean root system during the considered year is listed. Crop height and zero plane displacement have been estimated from the measurements. Albedo is based on measurements of radiation, which revealed a nearly constant value of 0.20 for the albedo during the year. Other albedos and emissivities used in the LAPS scheme are listed in Table 3 [52]. Because of the focus of this chapter, we used the observations concerning the water balance components which were available from the HAPEX-MOBILHY programme. We had available weekly volumetric soil moisture measurements through the year, based on neutron sounding probes for the top 1.6 m soil layer, with 0.1 m intervals. Another reference data set, that we have used for estimating the partitioning of the land surface water simulated by the LAPS, was an approximate water budget for the first four months (days 0–120), which was generated using the observed weekly root zone (1.6 m), soil moisture content, accumulated precipitation, and evaporation estimated by the Penman–Monteith formula [50]. Estimated evaporation was 149.6 mm for the period indicated. The total precipitation was 368.5 mm. The available observations from the first four months show very few changes in total soil moisture. The total root zone water content change was estimated at 22.2 mm, while the generated runoff plus drainage was 241.1 mm during the four months.

3.2. Validation of Hydrological Module of the LAPS

In order to test the performance of the hydrological module we run the LAPS for one year period with a time step of 1800 s. Atmospheric forcing data, validation data and parameters representing the land surface properties are obtained from the afore mentioned HAPEX-MOBILHY experiment. The run was initialized by setting the prognostic variables as follows: all water stores as saturated, canopy water as zero, snow mass as zero and all temperatures at 279.0 K. After initialization the scheme was running to equilibrium by looping through the one year forcing data. The equilibrium was reached when the conditions:

$$|\lambda E_m(n+1)-\lambda E_m(n)|<\upsilon,\ |\lambda E_{st}(n+1)-\lambda E_{st}(n)|<\upsilon,\ |H_m(n+1)-H_m(n)|<\upsilon \text{ and } |H_{st}(n+1)-H_{st}(n)|<\upsilon$$

were satisfied; here $\nu=0.10$ W m^{-2}. The used symbols have the meaning: $\lambda E_m(n)$, $H_m(n)$, $\lambda E_m(n+1)$, $H_m(n+1)$ are the annual means of latent and sensible heat fluxes for year n and $n+1$, while $\lambda E_{st}(n)$, $H_{st}(n)$, $\lambda E_{st}(n+1)$, $H_{st}(n+1)$ are the standard deviations for year n and $n+1$, respectively. The LAPS was converging after the third iteration. Additionally, the LAPS was tested using "Milly criteria" which requires that the condition $|P_a-D_a-R_a-E_a|<1$ mm has to be satisfied. Here, P_a, D_a, R_a and E_a denote annual cumulated values of the: precipitation, drainage, runoff and evapotranspiration. The residual, in this criteria, obtained by the LAPS was 0.6 mm. A comparison of the predicted total soil water during soybean growing season with the HAPEX measurements is shown in Fig 6. It shows that there is a general agreement between the simulation and the observations. Moreover, the LAPS correctly describes the trend of the soil moisture in a qualitative sense. The curve representing simulated values of the total soil water content over 1.6 m depth is very close to the observations. At the beginning of the growing season (early May), the soil was loosing the water intensively since the evapotranspiration was the dominant process. The good agreement between the simulated and observed values is extended up the end of the growing season.

It has been increasingly apparent that the water fluxes from a natural surface (covered by vegetation or a bare soil) simulated by the land surface scheme and their correct partitioning into horizontal and vertical water fluxes, and evaporation, are sensitive to the procedure for calculating its temperature [53]. Specification of the deep soil temperature cnnot be easy done when a land surface scheme provides the lower boundary condition in atmospheric models covering the domain of various soil textures. A wrong specification of deep soil temperature introduces an error in calculating the ground temperature and evaporation [24] and finally the incorrect partitioning of the surface water into water balance components. These errors are even more pronounced during long term integration. Mihailovic et al. (1998) [52] have discussed how the changes of the deep soil temperature in the restore term of the "force-restore" equation affect the partitioning of: (i) the surface energy into the sensible and latent heat fluxes; and (ii) the land surface water into water balance components. Basically, there are several possibilities for determining the deep soil temperature. For instance, it can be (i) specified as a constant, (ii) calculated from a prognostic equation or (iii) calculated as a running mean of ground temperature from the previous day as it is done in the LAPS,. Integrated water balance components (cumulative values) and soil water change for the first 120 days of integration at Caumont (France) are given in Figure 7. Calculated, cumulative values are: 151.8 mm for the evaporation, 239.5 mm for the horizontal and vertical flows through the corresponding borders and -22.3 mm for the soil water. Comparison these values with the observed ones (in the brackets of this figure) shows that the LAPS correctly simulate water balance components.

3.3. Simulation of Water Balance Components

Model outputs of latent heat flux (evaporation from bare soil fraction and water intercepted by leaves and transpiration) for Day of Year (DOY) 150-155 were compared with measurements in a soybean field at Coumont (France) using the aforementioned HAPEX data set. The parameters used in this simulation are given in Mihailovic et al. (1998) [52]. Monthly mean leaf area index, *LAI*, the fractional vegetation cover, σ_f, overall surface roughness length, z_0, the canopy roughness length z_{0c}, the zero-displacement height, d, and the canopy height, H, are as specified in the Mihailovic et al. (1998) [52]. These quantities are interpolated to smaller time intervals when required. The roots of soybean plants were assumed to be shallow and are distributed mainly in the top 0.5 m. The year was divided into the bare soil period (January-April, October-December), the transition period (May) and the growing season (June-September). For the bare soil period, there were no roots; for the transition period, it was assumed that the top 0.1 m soil layer contains 70% of the roots and the soil layer between 0.1-0.5 m contains the rest i.e. 30%; for the growing season, 60%, 30% and 10% roots were assumed to be in the soil layers 0-0.1, 0.1-0.5 and 0.5-1.6 m, respectively. Figure 8 depicts comparison between the calculated diurnal variations of latent heat flux for DOY 150-155. The simulated values are close to the observations correctly following the diurnal course of the latent heat flux.

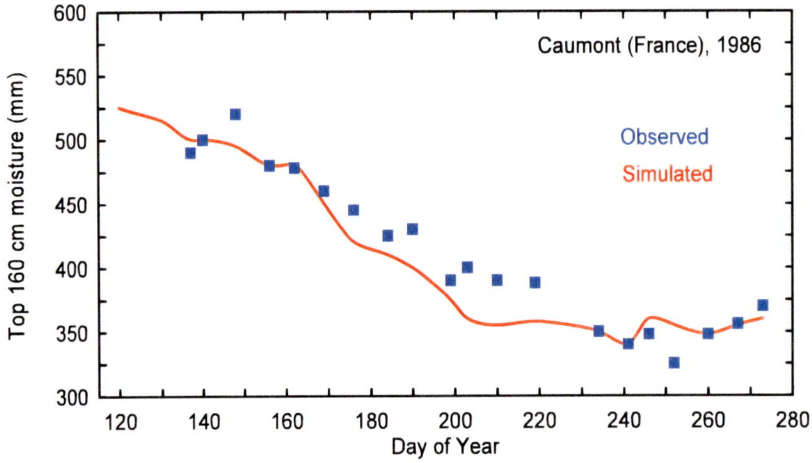

Figure 6. Daily averages of the total soil water content (mm) over a depth of 1.6 m simulated by the LAPS scheme compared with weekly measurements under a soybean field at Caumont (France) during its growing season in 1986.

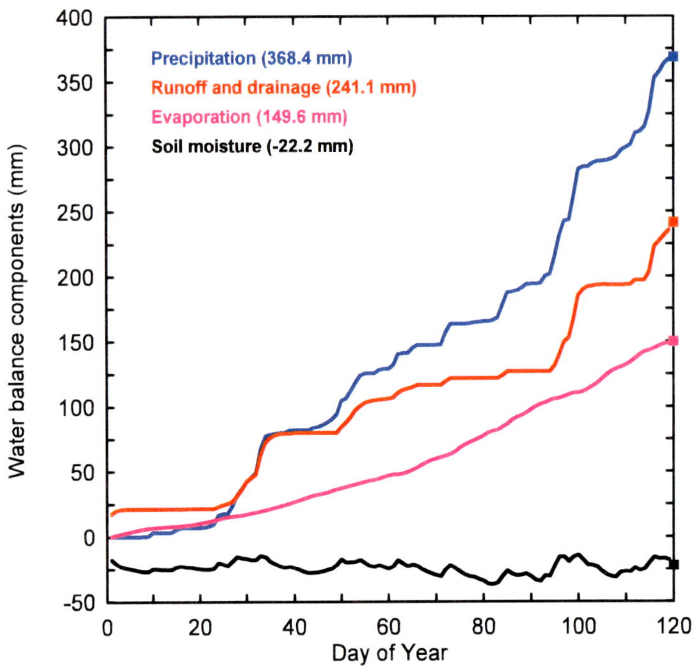

Figure 7. Simulated integrated water balance components (cumulative values) and soil water change for the first 120 days of integration at Caumont (France). The squares indicate the observations. The simulations were performed using the LAPS scheme.

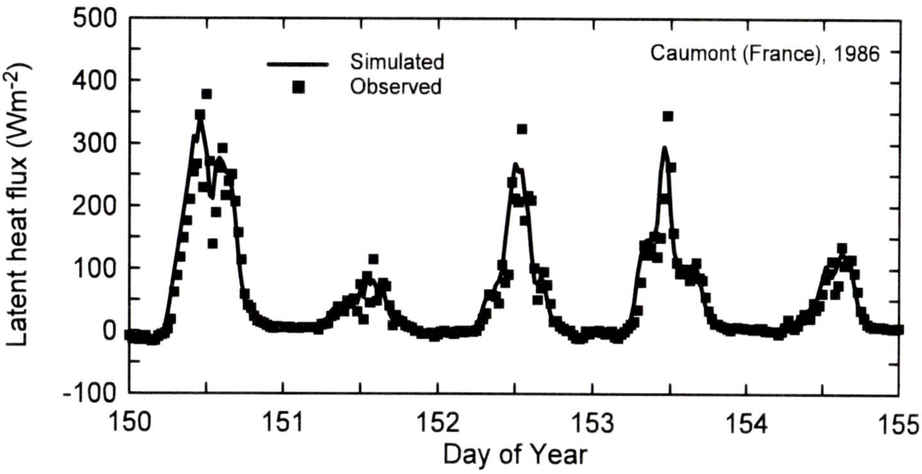

Figure 8. Temporal variation of latent heat flux for DOY 150-155 obtained by the LAPS scheme compared with the observations over a soybean field at Caumont (France) during its growing season in 1986.

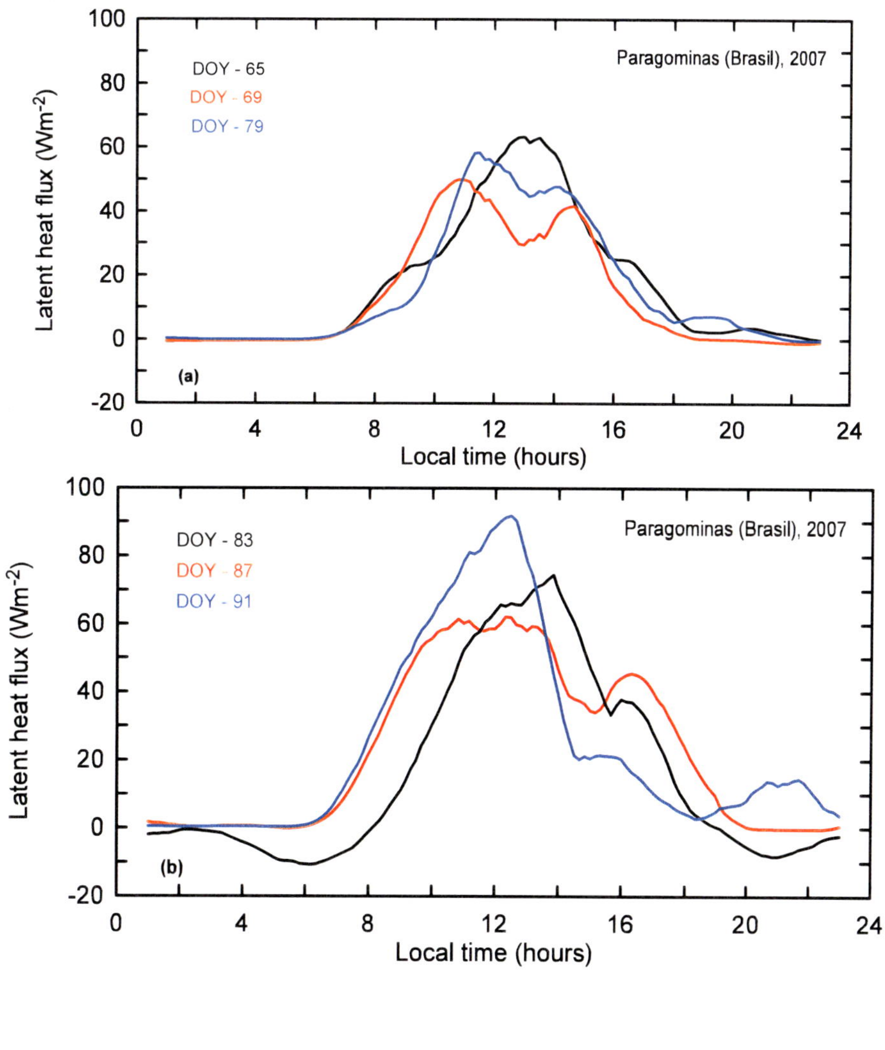

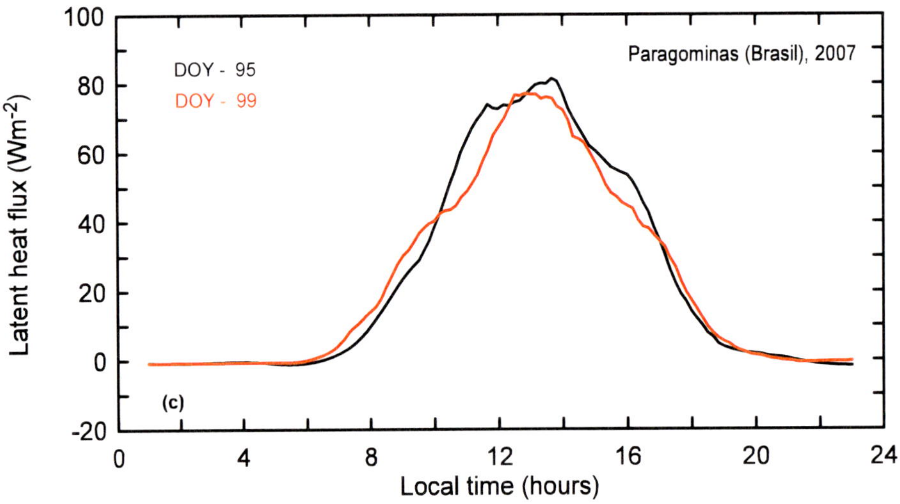

Figure 9. Daily variation of latent heat flux for DOY 65, 69, 79, 83, 87, 91, 95 and 99 obtained by the LAPS scheme over a soybean field in Paragominas (Brasil) during its growing season in 2007.

Figure 9 depicts diurnal variation of the latent heat flux for DOY 65, 69, 79, 83, 87, 91, 95 and 99 in Paragominas (Brasil). The evaporation during these days usually comes from the bare soil fraction since the fractional cover was still not so large. Also the humidity was very high. Therefore, the maximum amount of total evaporation is small not exceeding value of 100 Wm^{-2}.

The forcing data for simulation of latent heat flux, over a soybean during its growing season, were taken from the micrometeorological measurement program over a soybean field, that was located in Marchfeld plain, an area of intensive arable agricultural production north-east of Vienna. The soil is loamy sand and sandy silt loam, which is typical for the Marchfeld region. The soybean was grown at the experimental site on a 100 x 50 m plot with rows oriented north to south. The plant population was sown with 0.5 m row spacing and 0.07 spacing in the row (50 plants per m^2). More details of the experimental setup can be found in [47]. We run the LAPS for the DOY 150-240. For that period the canopy height H, and leaf area index were derived from the measurements using the cubic spline interpolation. The fractional vegetation cover σ_f was derived from LAI using the relationship, $\sigma_f = 1\text{-}e^{-cLAI}$ where the coefficient c is set to 0.6 which is appropriate for soybean crops. All other aerodynamic parameters are derived accordingly. Temporal variation of latent heat flux for DOY 150-240 obtained by the LAPS scheme compared with the observations over a soybean field at Marchfeld (Austria) during its growing season in 1995 is given in Figure 10.

Figure 11 depicts the soil moisture simulation of the top 0.1 m, 0.5 m and 1.6 m, respectively over the year obtained by the LAPS scheme It is seen that the during growing season the variations of the soil moisture in the top 0.1 m are less smoother than in two other top layers. This is because the first, the thin layer (0.1 m thin) has more instantaneous response on the input/output water comparing with two other layers.

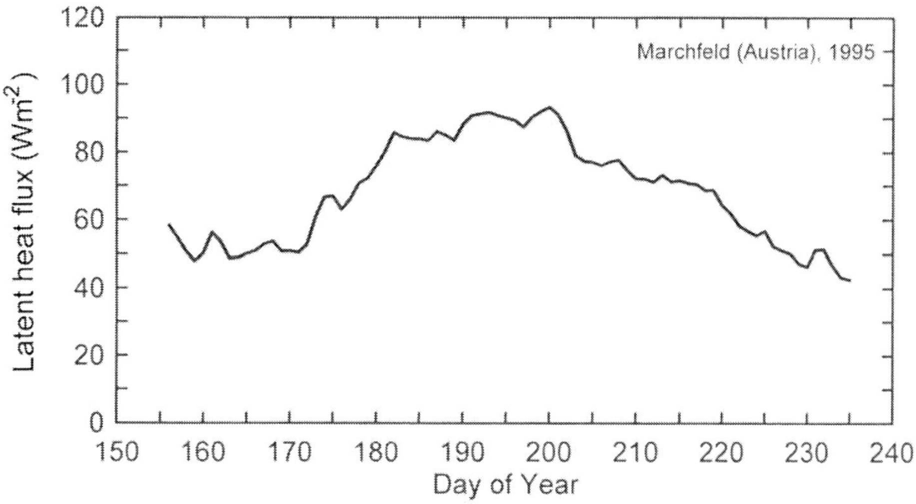

Figure 10. Temporal variation of latent heat flux for DOY 150-240 obtained by the LAPS scheme compared with the observations over a soybean field at Marchfeld (Austria) during its growing season in 1995.

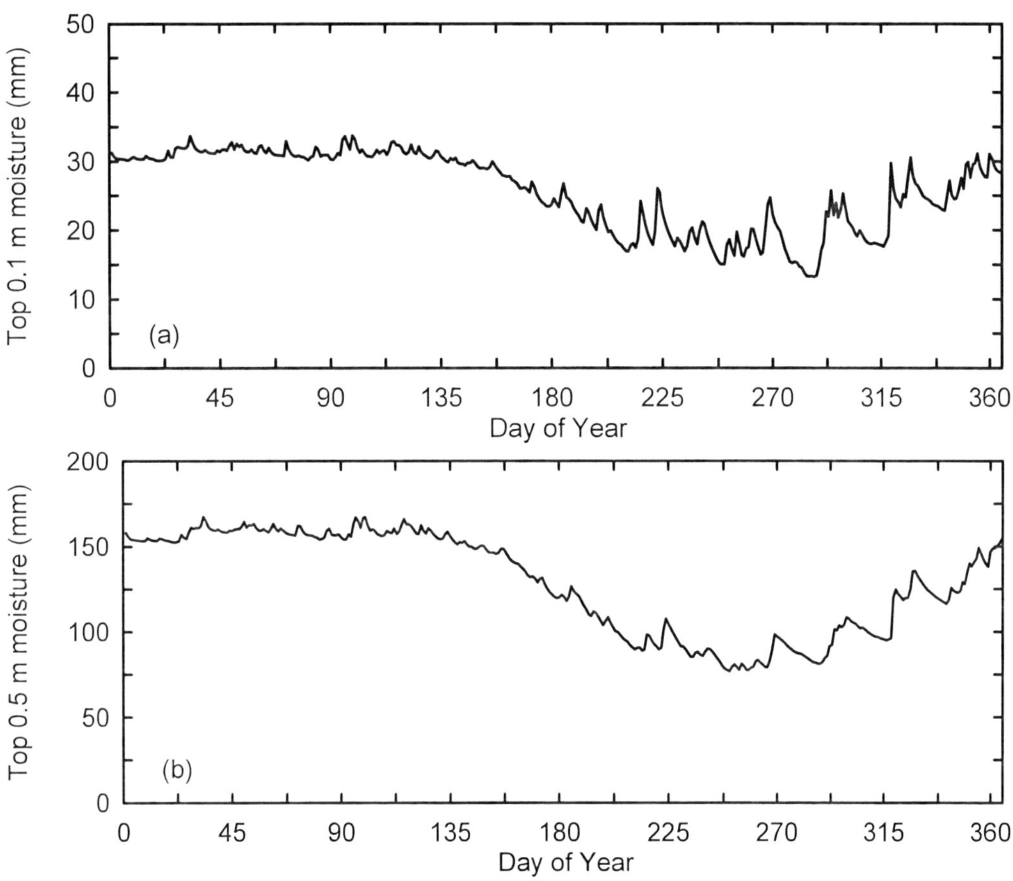

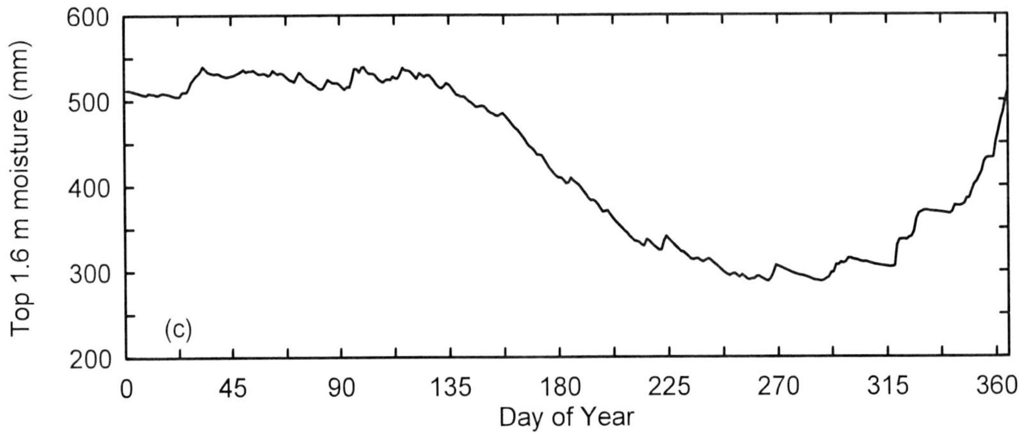

Figure 11. Simulation of soil moisture in the top: (a) 0.1 m (b) 0.5 m and (c) 1.6 m soil layer by the LAPS scheme under a soybean field at Caumont (France) during 1986.

References

[1] R.W. Elomore, J.E. Eisenhauer, Specht, J.H. Williams, *J.Prod.Agric.*, 1, 196-201 (1988).

[2] N.L. Klocke, D.E. Eisenhauer, J.E. Specht, R.W. Elmore, G.W. Hergert *Irrigation of soybean by growth stages in nebraska,* Transactions, American Society of Agricultural Engineers, 5(3), 361-366 (1989).

[3] S.W. Ritchie, J.J. Honway, H.E. Thompson, G.O. Benson, How soybean plant develops, Cooperative. Extension, Iowa State University, Ames, Iowa. Special report No. 55, 20 pp. (1994).

[4] W.L. Kranz, R.W. Elmore, J. Specht, Irrigating soybean, NebGuide, University of Nebraska-Lincoln Extensions Publications, Institute of Agriculture and Natural resources, 6 pp. (2005).

[5] D.T. Mihailovic, B. Kovacevic, F. Acs, Proceedings of 19th International Conference on Alpine Meteorology, 1-6 September, Rauris, Austria, 386-389 (1986).

[6] W.J .Jones, K.J. Boote, Simulation models for soybeans and other crops, Concepts of crop systems, Technical Bulletin No. 106. Food and Fertilizer Technology Center, Tapei City, Taiwan, pp. 1-7. (1987).

[7] K.J. Boote, J.W. Jones, G. Hoogenboom, Simulation of crop growth: CROPGRO, In: Peart, R.M., Curry, R.B. (Eds.), Agricultural Systems Modeling and Simulation. Marcel Dekker, New York, pp. 651–692 (1998).

[8] Piara Singh, D. Vijaya, K. Srinivas, S.P. Wani, Potential productivity, yield gap, and water balance of soybean-chickpea sequential system at selected benchmark sites in India. Global Theme 3: Water, Soil, and Agrobiodiversity Management for Ecosystem Health. Report no. 1. Patancheru 502 324, Andhra Pradesh, India: International Crops Research Institute for the Semi-Arid Tropics. 52 pp. (2002).

[9] J. Cavazzoni, T. Volk, G. Stutte, A modified CROPGRO model for simulating soybean in controlled environment, *Life Support Biosphere Science*, 4, 43-48 (1997)

[10] J. Cavazzoni, T. Volk, G. Stutte, B. Bugbee, T. Dougher, Phasic temperature and photo period control for soybean using modified CROPGRO model, *Life Support Biosphere Science*, 6, 273-278 (1999).
[11] D.T. Mihailovic, F. Acs, B. Kovacevic, 12th Conference on Carpathian Meteorology, 1-5 October 1985, Belgrade, Federal Hidromereorological Institute, Papers Compendium, 220-222. (1985).
[12] D.T. Mihailovic, B. Kovacevic, F. Acs, Proceedings of 13th International Conference on Carpathian Meteorology, 14-19 September 1987, Busteni, Romania, 635-642 (1987).
[13] K.J. Boote, J.W. Jones, W.D. Batchelor, Evaluating and improving CROPGRO-soybean and CERES-maize models for predicting growth and yield response to climate change, 2001–2002 NIGEC Annual Report. Great Plains Regional Center, pp. 16–22 (2001).
[14] R.M Ogoshi, G.Y. Tsuji, G. Uehara, N.P. Kefford, Simulation of best management practices for soybean production in Hawaii, College of Tropical Agriculture and Human Resources in cooperation with the U.S. Department of Agriculture. CTAHR, University of Hawaii at Manoa (1998).
[15] S.S Deepak, M. Agrawal, *Environ. Exp. Bot.*, 46, 81–91 (2001).
[16] F. Wang, C. Fraisse, N.R. Kitchen, K.A. Sudduth, Site-specific evaluation of the CROPGRO-SOYBEAN Model on Missouri Claypan Soil, University of Missouri, Columbia, MO 65211. TEKTRAN, USDA-ARS. Updated: 2001-08-17 (2001).
[17] R.J. Mera, D. Niyogi, G.S. Buol, G.G. Wilkerson, F.H.M. Semazzi, *Global Planet. Change*, 54, 163-182 (2006).
[18] A. Olioso, Y. Inoue, S. Ortega-FARIAS, J. Demarty, J.-P. Wigneron, I. Braud, F. Jacob, P. Lecharpentier, C. OttlÉ, J.-C. Calvet, N. Brisson, *Irrig. Drain.Systems,* 19, 377-412 (2005).
[19] W.G.M. Bastiaanssen, M. Menenti, R.A. Feddes, A.A. Holtslag, *J.Hydrol.,* 212–213: 198–212 (1998).
[20] F. Jacob, A. Olioso, X.F. Gu, Z. Su, B. Seguin, *Agronomie*, 22, 669–680 (2002).
[21] P.J. Sellers, Y. Mintz, Y. Sud, A. Dalcher, *Journal of Atmospheric. Sciences.*, 43, 506-531 (1986).
[22] D.T. Mihailovic G. Kallos, *Bound-Layer. Meteorol.*, 82, 283-315 (1997).
[23] D.T. Mihailovic, B. Rajkovic, B. Lalic, LJ. Dekic, *J. Appl. Meteorology*, 34, 2462-2475 (1995).
[24] D.T. Mihailovic, *Global Planet. Change*, 13, 207-215. (1996).
[25] D.T. Mihailovic, B. Rajkovic, *Meteorologishe Zeitschrift.*, 2, 239-243 (1993).
[26] W.J. Massman, J. C. Weil, *Bound-Layer. Meteorol.,* 91, 81-107 (1999).
[27] G.G. Katul, W.-H. Chang, *J. Appl. Meteorology.,* 38, 1631-1643 (1999).
[28] J. D. Jean-Paul Pinard, J. D. Wilson, *J. Appl. Meteorology,* 40, 1762-1768 (2001).
[29] B.J. Legg, I. F. Long, Turbulent diffusion within a wheat canopy II, *Q. J. Roy. Meteor. Soc.,* 101, 611-628 (1975).
[30] Denmead, Temperate cereals, Vegetation and the Atmosphere, 2d ed. J. L. Monteith, Ed., Academic Press, New York, 1-31 (1976).
[31] Y. Brunet, J. J. Finnigan, M. R. Raupach, *Bound-Layer. Meteorol.*, 70, 95-132 (1994).
[32] A.S. Dubov, L. P. Bikova, S. V. Marunich, Turbulence Inside a Canopy. Gidrometeoizdat, Leniningrad, 184 pp. (In Russian) (1978).

[33] Goudriaan, Crop micrometeorology, A simulation study, Wageningen Center for Agricultural Publishing and Documentation, Wageningen, 211 pp. (1977).

[34] Y.K. Xue, P. J. Sellers, J. K. Kinter, J. Shukla, *J. Climate*, 4, 345-364 (1991).

[35] D.T. Mihailovic, T. J. Lee, R. A. Pielke, B. Lalic, I. Arsenic, B. Rajkovic, P. L Vidale, *Theor. Appl. Climatol.*, 67, 135-151 (2000).

[36] J.D. Wilson, D. P. Ward, G. W. Thurtell, G. E. Kidd, *Bound-Layer. Meteorol.,* 24, 495-519 (1982).

[37] W.A.J. van Pul, The flux of ozone to a maize crop and underlying soil during a growing season, Ph.D. dissertation, Wageningen Agricultural University, 147 pp. (1992).

[38] Z. Uchijima, Radiation characteristics of maize and rice field, Vegetation and the Atmosphere. 2d. ed. J. L. Monteith, Ed., Academic Press, New York, 32-45 (1976)

[39] D.T. Mihailovic, B. Lalic, J. Eitzinger, S. Malinovic, I. Arsenic, *J. Appl. Meteorol. Clim.*, 45, 348–356 (2006).

[40] Jr.F. Ayers, Theory and Problems of Differential Equations, Shaum Publishing, New York, 296 pp. (1952)

[41] S.F. Sun, Moisture and heat transport in a soil layer forced by atmospheric conditions, M.Sc Thesis. Dep. Civ. Eng. Univ. Conn.., 72 pp (1982).

[42] D.T. Mihailovic, Implementation of Land-Air Parameterization Scheme (LAPS) in a limited area model, Final Report, The New York State Energy Conservation and Development Authority, Albany, NY, 110 pp. (2003).

[43] R.B. Clapp, G.M. Hornberger, *Water Resour. Res.*, 14, 601-604 (1978).

[44] F. Abramopoulos, C Rosenzweig., B. Choudhrry, *J. Climate*, 1, 921–941 (1988).

[45] D.T. Mihailovic, B. Rajkovic, B. Lalic, D. Jovic, I. Arsenic, *Phys. Chem. Earth*, 21, 201–204 (1997).

[46] D.T. Mihailovic, J. Eitzinger, *Ecol.Model.*, 202, 465-475 (2007).

[47] V. Cajic, Response of soybean yield to climate change using CROPGRO crop simulation model. PhD thesis, Univ. of Natural Resources and Applied Life Sciences (BOKU), Vienna (2003).

[48] J.C. El-Husny, E. B. De Andrade, A.S. Filho, L.A. de Almeida, D. Klepker, M.C. Meyer, *Recommendations for soybean cultivation in the micro region Paragominas (Para, Brasil)*, Technical Communication, Ministry of Agriculture, ISSN 1517-2244, Belem, PA, pp 6 (In Brasilian) (2003).

[49] J.F. Mahfouf, *Bound-Layer. Meteorol.*, 53, 210–222 (1990).

[50] Y. Shao, R.D. Anne, A. Henderson-Sellers, P. Irannejad, P. Thornton, X. Liang, T.H. Chen, C. Cirat, C. Desborough, O. Balachova, A. Haxeltine, A. Ducharne, *Soil moisture simulation*. A report of the RICE and PILPS workshop: GEWEX/GAIM Report. IGPO Publication Series, Washington DC, p. 179. (1995).

[51] J.P. Goutorbe, C. Tarrieu, *HAPEX-MOBILHY data base in land surface evaporation*. In: T. Schmugge, J.C. Andre, (Eds.), Land Surface Evaporation, Springer, Berlin, pp. 403–410 (1991).

[52] D.T. Mihailovic, B. Rajkovic, B. Lalic, D. Jovic, Lj. Dekic, *J. Hydrol.*, 211, 17–33 (1998).

[53] R. Avissar, R.A. Pielke, *Mon. Weather Rev.*, 117, 2113–2136 (1989).

In: Soybean and Wheat Crops
Editors: S. Davies, G. Evans
ISBN: 978-1-60741-173-4

Chapter 4

CHARACTERIZATION OF SOYBEAN CULTIVARS: RAPID HPLC PROFILING BASED ON PROTEIN MARKERS

Maria Luisa Marina and Maria Concepción García
Department of Analytical Chemistry, Faculty of Chemistry, University of Alcalá, Ctra. Madrid-Barcelona Km 33.600, 28871 Alcalá de Henares, Madrid, Spain

ABSTRACT

During the past 30 years, breeding programs have developed many soybean varieties for their adaptation to different geographical areas and for improving seed characteristics: increasing protein and oil concentrations, improving protein quality, reducing antinutritional compounds, etc. The differentiation among the increasing number of soybean cultivars is not an easy task since many of them are genetically very close. Traditional methodologies for the identification of soybean cultivars were based on phenotypic characters from the leaf, stem, and seed. Since many different soybean cultivars are indistinguishable based on these features, other methodologies have raised as alternatives for cultivar characterization.

The characterization of soybean cultivars through the analysis of proteins has been reviewed in this work. Special emphasis was made on the use of electrophoretic and chromatographic techniques. The discussion of the results obtained by our research team in relation with the differentiation of 91 soybean varieties through their protein profiles obtained by a rapid chromatographic methodology will also be included.

INTRODUCTION

Soybean [*Glycine max* (L.) Merr.] is a legume highly appreciated for its proteins, both for their quantity and quality. Soybean contains approximately 38-40% proteins divided into four groups: enzymes involved in the metabolism, structural proteins, membrane proteins, and storage proteins. Main soybean proteins are the storage proteins 11S globulin (glycinin) and 7S globulin (β-conglycinin).

Glycinin consists of a hexamer with a molecular weight of 320-380 kDa. Glycinin has five non-allelic genes, Gy1, Gy2, Gy3, Gy4, and Gy5, which code for five glycinin protein precursor molecules, G1, G2, G3, G4, and G5, respectively. Each protein precursor subunit consists of two or three chains which are cleaved post-translationally. Based on the homology in amino acid sequences, these five genetic variants are classified into two major groups: group I consisting of G1 ($A_{1a}Bx$), G2 (A_2B_{1a}), and G3 ($A_{1a}B_{1b}$) variants and group II that includes G4 ($A_5A_4B_3$) and G5 (A_3B_4) variants. Each glycinin variant consists of an acidic (Mr ~ 32 kDa) and a basic (Mr ~ 20 kDa) polypeptide linked together through disulfide bonds (A-S-S-B, where A represent the acidic polypeptide and B de basic polypeptide). For each subunit there is more than 84% homology within the group and 45-49% between groups [1, 2].

β-conglycinin is a trimeric glycoprotein with a molecular weight close to 180 kDa which is composed of three different subunits, α-, α'-, and β- of 76, 72, and 53 kDa, respectively. In addition, it contains a 5% of carbohydrates, mainly mannose type.

Glycinin and β-conglycinin differ in quantity observing a protein ratio of glycinin to β-conglycinin ranging from 1.6 to 2.5 among soybean varieties [3]. Moreover, despite soybean proteins are, in general, limited in sulfur containing amino acids, glycinin presents a higher content in this kind of amino acids than β-conglycinin which is practically devoid of methionine content [4]. Nevertheless, the composition and content in these proteins not only varies with the cultivar but also with the environmental and growing conditions [3, 5].

The differences regarding the structure, composition, and quantity between these two main protein constituents of soybean made their contribution to nutritional and physical properties in different soybean cultivars was not identical. Thus, while glycinin is more advantageous from a nutritional point of view due to their higher methionine content, from a functional perspective, glycinin is a better gel former and β-conglycinin possesses greater emulsifying properties [6]. Different works have contributed to the discovery of the differences in content and subunit composition of glycinin and β-conglycinin in different soybean cultivars [7-15].

Soybean Breeding

Breeding consists of combining different cultivars with desirable characteristics to obtain a new one with improved features. During the past 30 years, soybean breeding programs have developed varieties for their adaptation to different geographical areas [16, 17] and for improving seed characteristics: seeds with a higher protein or oil contents, with improved protein quality, with improved oil quality, with reduced allergenicity, with genetic resistance to plagues, with higher yield, etc. [18-22].

Important efforts have been performed for the production of high protein seeds of more commercial value. Cultivars with a 10 to 20 % higher protein content have been developed at expense of cultivar yield and oil content [23, 24]. On the other hand, despite soybean proteins are superior in quality to other vegetable proteins, they are of lower quality than animal proteins mainly due to their limitation in sulfur containing amino acids being the improvement of soybean protein quality another important breeding goal. An increase in methionine level could be achieved by a reduction in β-conglycinin (devoid of methionine) content or by an increase in the proportion of other soybean proteins with higher content in methionine such as glycinin, trypsin inhibitor, and urease [4, 18, 25-28].

Regarding soybean oil, main breeding goals have been to improve oil flavor and shelf life by the reduction of lipoxygenase activity and the decrease in linoleic acid concentration. In addition, the reduction of the content in saturated fatty acids such as palmitic acid is another pursued goal in soybean breeding programs [19].

Soybean is classified as one of the eight main foods causing allergenic reactions and the development of new soybean varieties with lower allergenicity has been of great consideration [29, 30]. Despite there are more than twenty different allergens in soybean, three proteins can be considered as main responsible for allergenicity in soybean: the Gly m Bd 60 K (α subunit of β-conglycinin), the Gly m Bd 28 K (7S globulin), and the Gly m Bd 30 K (P34) which constitutes the major soybean allergen. Although the reduction and removal of some of these allergens has been possible, the methodology employed has not been successful with the main allergen, the P34. Recently, recombinant techniques have been applied to create a genetically modified soybean without this allergen [20].

On the other hand, the increasing demand for specific soybean foodstuffs such as soybean milk, miso or tofu has forced the development and identification of soybean cultivars that were more suitable for the production of every product [31]. As example, Japanese processors, the world's largest importers of food quality soybean, are demanding special soybean varieties grown under contract with their identity preserved from point of production to the manufacturing of soybean product [32]. This is the case of tofu, a gelatinous soybean food whose quality significantly depends on its textural properties greatly influenced by the properties of the storage protein subunits. Despite the contribution of 11S and 7S globulins in tofu's yield and quality is not really clear and published results differed significantly, most studies seem to point towards a positive relationship between 11S globulin and tofu textural properties [33, 34]. In this sense, more productive tofu type cultivars with superior quality for tofu (e.g., Harovinton variety) have been developed [3]. Poysa et al. [35] developed 20 new null genotypes with different glycinin and β-conglycinin subunit composition by the crossing of Harovinton and other Japanese cultivars lacking either glycinin subunits or the α' subunit of β-conglycinin. They observed that soybean genotypes lacking the α' subunit of β-conglycinin and the G4 variant of glycinin and containing increasing levels of the G5 subunit of glycinin produced firmer tofus.

Another breeding goals have been the development of soybean varieties with reduced level of undesirable compounds (e. g., lipoxygenase isozymes and trypsin inhibitors) [36] or with higher level of compounds presenting health benefits (e. g., isoflavone, compounds with antioxidative effects (phenolic compounds, vitamins such as lutein, vitamin E, etc.), and saponins) [32, 37-40].

The existence of all these different genotypes joint to different plant growing conditions seem to have a great impact on the protein composition and profile of soybean cultivars [12, 13, 41, 42]. Consequently, these soybean varieties will differ in their nutritional and physico-chemical characteristics which influence their performance and the quality of products prepared with them [33, 34, 43]. Obviously, the development of characterizing analytical methodologies enabling to guarantee genetic purity (seed certification for purity), estimate genetic relationship (important for organizing germplasm collections, parent selection for crossing purposes, etc.), and identify germplasms (to avoid false labeling of crops with the name of a superior variety or a particular brand name) are required.

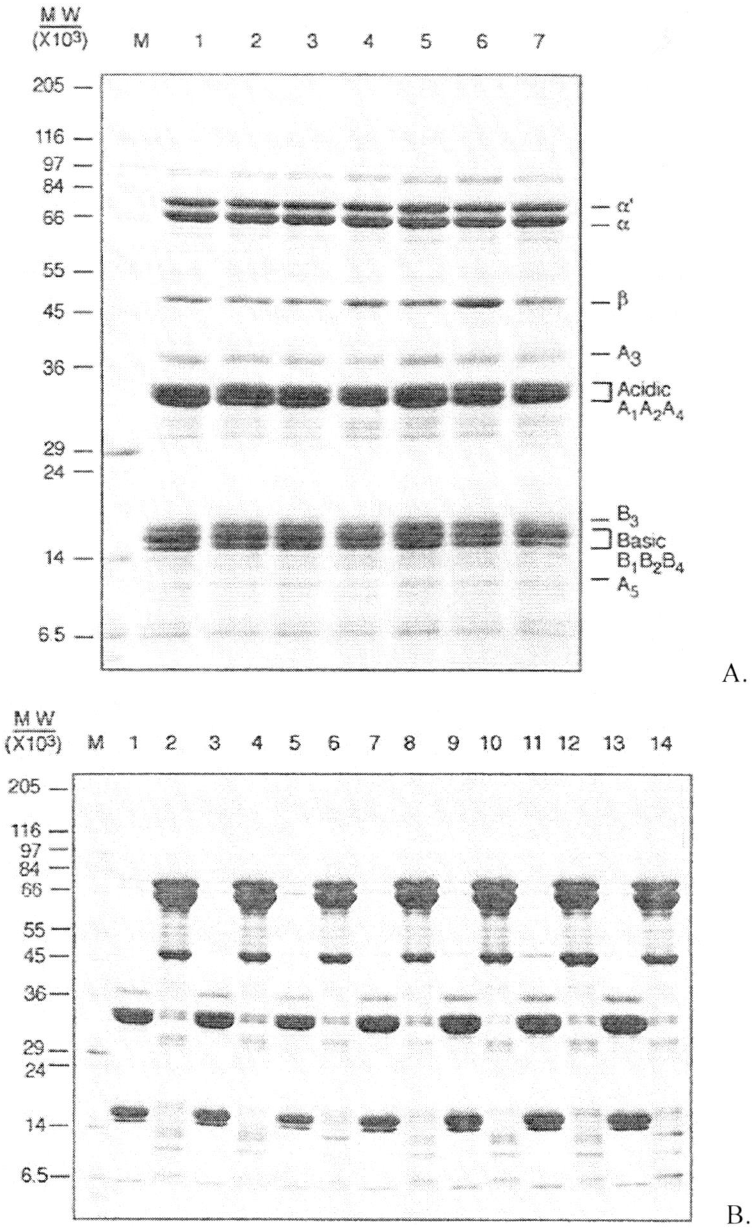

Figure 1. SDS-PAGE profiles of total proteins and 11S and 7S protein fractions from seven soybean varieties. Figure 1A: lane 1, Vinton-81; 2: S-20F8; 3: HP-204; 4: IA-2034; 5: Steyer; 6: IA-2020; 7: S-2020. Figure 1B: lanes 1 and 2: 11S and 7S protein fractions from Vinton-81; 3 and 4: S-20F8; 5 and 6: HP-204; 7 and 8: IA-2034; 9 and 10: Steyer; 11 and 12: IA-2020; 13 and 14: S-2020 (from Mujoo et al., with permission).

Characterization of Soybean Cultivars

The differentiation among the increasing number of soybean cultivars is not an easy task since many of them are genetically very close. Traditional methodologies for the identification of soybean cultivars are based on phenotypic characters [14, 43, 44]. Since many different soybean cultivars are indistinguishable based on these characteristics, other methodologies have raised as alternatives for cultivar characterization [45].

In addition to morphological markers, different types of DNA-based markers have also been applied for studying genetic diversity in soybean: RFLP (Restriction Fragment Length Polymorphism), AFLP (Amplified Fragment Length Polymorphism), RAPD (Random Amplified Polymorphic DNAs), DAF (DNA Amplification Fingerprinting), and SSR (Simple Sequence Repeats) [13, 46-54]. In general, the results of these studies revealed the huge genetic diversity of soybean cultivars and the difficulty in their differentiation and only the application of microsatellite markers such as SSR resulted to be more suitable for genotypic identification [52-54]. Moreover, the application of these techniques requires complicated procedures and specialized techniques and laboratory equipment being both time-consuming and costly and, thus, not suitable for routine analysis [55].

The identification of soybean cultivars through the analysis of soybean proteins using different electrophoretic techniques has also been widely employed. Most electrophoretic systems employed either starch or polyacrylamide gels in which proteins were separated based on their molecular weight and charge density [6, 7, 12, 14, 22, 33, 41, 45, 56]. In addition to the low reproducibility, tediousness, and time cost of these methodologies, one-dimensional electrophoresis, in general, did not enable the differentiation among soybean cultivars. As example, Figure 1 shows the sodium dodecyl sulphate polyacrilamide gel electrophoresis (SDS-PAGE) profile of total proteins and 11S and 7S fractions for seven different soybean varieties not observing differences in profiles among cultivars [33]. The use of isoelectrically focused gels and blotting were also tried in gel electrophoresis but in no case successful results regarding cultivar differentiation were reported [57, 58].

The application of two-dimensional SDS-PAGE has resulted advantageous in the selection of high quality soybean varieties [1, 3, 13, 25, 59]. The application of this technique requires a prefractionation step to reduce the complexity of the sample and minimize protein degradation due to the presence of proteases, and to remove interfering compounds such as lipids, salts, nucleic acids, polyphenols, alkaloids, pigments, terpenes, organic acids, and other compounds. The extraction with trichloroacetic acid/acetone is very useful for minimizing protein degradation and removing interfering compounds while the introduction of an immobilized pH gradient has improved the separation of proteins. Natarajan et al. [1, 13, 60] observed variability in glycinin polypeptides, β-conglycinin, and soybean seed allergens (Gly m Bd 60K, Gly m Bd 30K, and Gly m Bd 28K) between wild and cultivated soybean but they could not find significant differences within the same group of soybean cultivars. Zarcadas et al. [3] observed differences in the proteome and subunit expression of glycinin and β-conglycinin among two tofu and eleven null soybean genotypes (lacking a subunit or part of a subunit of β-conglycinin). Nevertheless, they also reported [59] strong similarities in the overall distribution pattern of glycinin, β-conglycinin, and total proteins among other soybean cultivars. The application of image analysis to the spots using a scanning densitometer enabled to obtain the composition of glycinin and β-conglycinin in the different cultivars and to observe quantitative differences among them [59].

-Conglycinin Fractions

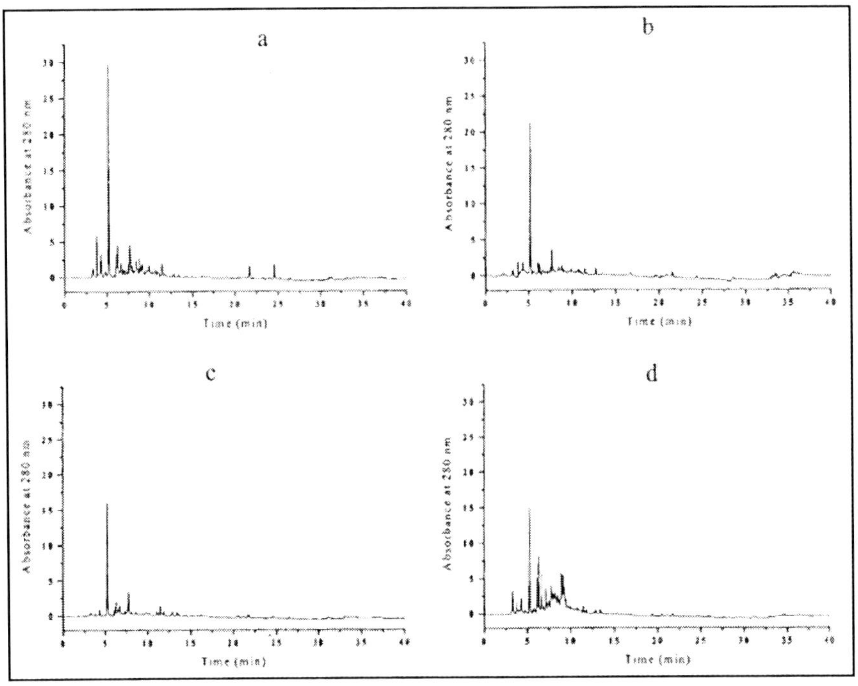

Glycinin Fractions

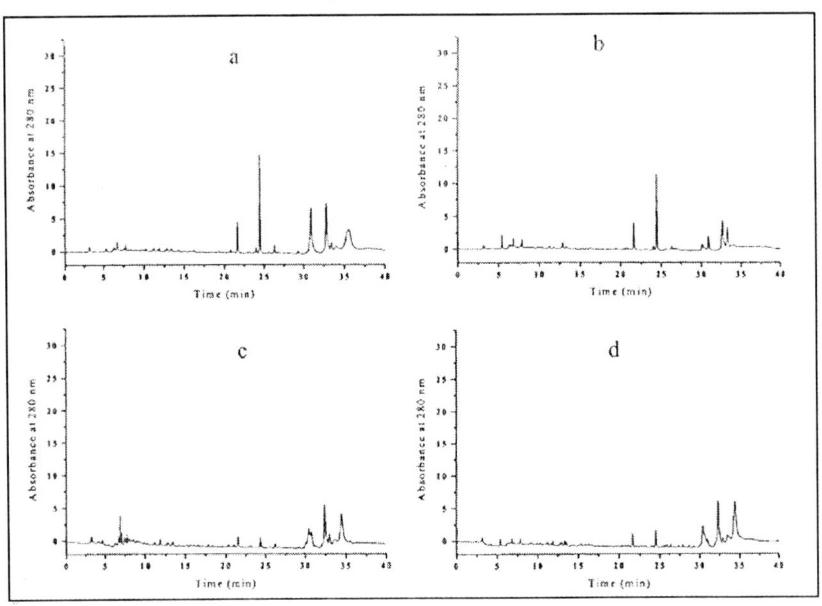

Figure 2. Reversed-phase high performance liquid chromatography profiles of the β-conglycinin and glycinin fractions in four soybean varieties: a, Macon; b, Enrei; c, Ohio FG1; d, IL2 (From Riblett et al., with permission).

Chromatographic techniques have also been used for soybean cultivar identification. Buelher et al. [61] developed a chromatographic method in the reversed-phase mode enabling the separation of soybean proteins in 55 min and Oomah et al. [62] used size-exclusion (with a separation time of 25 min) and reversed-phase (with a separation time of 40 min) chromatography for the same purpose. Both groups found promising the use of HPLC for cultivar identification despite differentiation among closely related soybean cultivars was difficult. Cole and Cousin [63] demonstrated the suitability of size-exclusion chromatography for profiling soybean proteins using separation times of 40 min. They also applied the proposed methodology to the analysis of two different soybean varieties observing profiles with the same number of peaks but with different peak size. More recently, Riblett et al. [6] and Mujoo et al. [33] have applied reversed-phase high performance liquid chromatography (RP-HPLC) for the separation of total proteins and the 11S and 7S fractions in different soybean varieties. Despite it was possible to observe differences among soybean cultivars studied, the times used for the completion of the separation were so high (40 and 90 min, respectively) that both methodologies can be considered of limited application for routine or field studies. Figure 2 shows the profiles obtained for glycinin and β-conglycinin in four different soybean cultivars revealing variations in the composition of each fraction.

Our research group has a great expertise in the use of chromatographic techniques in the reversed-phase and ion-exchange modes using conventional, perfusion, and monolithic stationary phases and capillary electrophoresis for the analysis of soybean proteins [64-68]. Some of these methodologies have been applied to the differentiation between soybean and other related legumes (mungbean, azuki bean, etc.) commonly commercialized as soybeans. Moreover, the differentiation between transgenic and non-transgenic soybeans has also been dealt with success [68]. We have focused the present work on the characterization of soybean cultivars. For that purpose, protein profiles obtained by HPLC with rapid perfusion stationary phases and multivariate classification techniques are proposed. Perfusion stationary phases enable the drastic reduction of analysis times due to their characteristic structure that alleviate the low diffusivity of proteins [69]. Moreover, due to the short time needed for the completion of every analysis a very comprehensive study consisting of the analysis of 91 different soybean cultivars could be performed.

Perfusion RP-HPLC Applied to the Rapid Characterization of Soybean Cultivars Based on Protein Markers

Materials and Methods

1. Chemicals and Samples

HPLC grade acetonitrile (ACN) (Merck, Darmstadt, Germany), HPLC grade water (Milli-Q system, Millipore, Bradford, MA, USA), and trifluoroacetic acid (TFA) (Sigma, St Louis, MO, USA) were used for the preparation of mobile phases and soybean solutions.

A total of 91 non-transgenic soybean varieties with different origins and corresponding to the germplasm collection of the CRF (Centro de Recursos Fitogenéticos del Instituto Nacional de Investigaciones Agrarias, Madrid, Spain) were analyzed. Most soybean varieties were from USA: 41 were commercial varieties (Ap-18, Gnome, Elf, Cumberland, Amcor,

Wilkin, Woodworth, Williams, Wayne, Traverse, Swift, Steele, Sm-ie, Rampage, Provar, Protana, Norman, Mokapu Summer, Magna, Mack, Kahala, Kailua, Hawkeye 63, Harosoy 63, Harosoy, Forrest, Evans, Essex, Disoy, Davis, Corsoy, Coles, Clay, Calland, Bragg, Bonus, Beeson, Anoka, Amsoy 71, Ada, and Hodgson) and 19 were crossing varieties (Ap-225 C, Clark Z, Clark E, Harosoy, Harosoy Dense, J-114, A 73-19084, Ix1-17, Clark Dt1e2, Harosoy F, Harosoy Fe, Harosoy Lf1, Harosoy Dt1, Harosoy Dt2, Clark Dt1e1te2, Clark Fe, Clark Lo, Clark Rj, Clark S, and Clark Dt1). Moreover, one commercial variety was from Canada (Bellatti-4p), eleven were from different countries (France, Poland, Rumania, Germany, Austria, Russia, and Old Yugoslavia) of Europe (10 commercial varieties (Ns-gr, Fred, Srecka, Zolta Brzebedowska, Wilenska Brutnana, Mazowiecka II, Giesso, Flora, Caloria, and A-100) and one crossing variety (Ns-mm)), 17 commercial varieties were from Japan and Taiwan (Wasedaizu No. 1, Wase shiroge, Tokachi, Toyosuzu, Tokachi nagaha, Tainung No. 4, Shiroge-9, Shinanomejiro, Shih-shih, Oshimashirome, Osage, Nagaba jiro, Koganejiro, Kita musume, Kitamishiro, Kakushin 1, and Kaohsiung No. 3), and one commercial variety was from Uganda (Merit). All these soybean cultivars presented an almost identical aspect being not possible their differentiation based on seed color (yellow), shape (round) or size.

Soybean solutions were prepared by pulverizing 50 mg of every variety with a domestic miller, dissolving in 10 mL of extracting solution (80:20 (v/v), water:ACN), sonicating for 5 min in a bath sonicator (150W, 50Hz, Ultrasons-H, Selecta, Barcelona, Spain) and centrifuging at 3662g (Heraeus Instruments, Megafuge 2.0R, Osterode, Germany). The supernatant was removed and injected into the chromatographic system [69].

2. High-Performance Liquid Chromatography

A Hewlett-Packard 1100 series liquid chromatograph (Hewlett-Packard, Pittsburgh, PA, USA) with a degassing system, a binary pump, a thermostated compartment for the column, an injection system, and a diode-array detector was employed to carry out the separation. HP-Chemstation software was used for data acquisition and processing. The chromatographic separations were accomplished with the reversed-phase POROS R2/10 perfusion column (4.6 mm I.D. x 50 mm, 10 μm particle size) (Applied Biosystems, Foster City, CA, USA). The separation conditions were: injection volume, 20μL; flow-rate, 3 mL/min; temperature, 60 °C; mobile phase A, 0.1% (v/v) TFA in Milli-Q water; mobile phase B, 0.1% (v/v) TFA in ACN; binary gradient: 5-25% B in 1.7 min, 25-45% B in 0.8 min, and 45-5% B in 1 min; UV detection, 280 nm [69].

3. Data Treatment

The area percentage for every peak was calculated as the average of two replicates (each one injected in duplicate). The integration was performed by setting the baseline from valley to valley. Multivariate classification techniques (cluster analysis and linear discriminant analysis), Box-and-whisker diagram, and multiple linear regression (MLR) were performed with the computer program Statgraphics Plus for Windows 4.0 (Statistical Corp., Rockville, MD, USA). Principal components regressions (PCR) and partial least squares regressions (PLS1) were performed with the computer program Unscrambler 9.7 (Camo Software, Oslo, Norway).

Table 1. Retention times and peak area percentages obtained in the analysis of 91 different soybean varieties from America, Europe, Asia, and Africa

Origin	Peak 1	Peak 2	Peak 3	Peak 4	Peak 5	Peak 6	Peak 7	Peak 8	Peak 9	Peak 10	Peak 11	Peak 12	Peak 13	Peak 14
Retention time (min)														
	0.77	0.90	1.06	1.20	1.36	1.44	1.53	1.62	1.74	1.81	2.15	2.58	2.71	2.82
Peak area (%)														
America	5.0 ± 1.6	4.6 ± 1.0	10.2 ± 2.2	9.9 ± 3.2	1.1 ± 0.3	10.7 ± 3.7	1.8 ± 0.7	1.5 ± 0.6	0.4 ± 0.2	0.8 ± 0.3	0.8 ± 0.2	0.9 ± 0.3	4.8 ± 1.5	47.5 ± 7.4
Asia	5.0 ± 2.1	4.1 ± 1.2	10.0 ± 2.9	7.6 ± 3.5	0.7 ± 0.2	11.7 ± 4.4	1.2 ± 0.6	1.1 ± 0.3	0.6 ± 0.3	0.4 ± 0.2	0.8 ± 0.3	1.2 ± 0.3	2.2 ± 1.4	53.4 ± 9.4
Europe	5.4 ± 2.6	5.0 ± 1.1	11.0 ± 2.4	11.0 ± 4.8	0.9 ± 0.2	12.7 ± 4.5	1.6 ± 0.7	1.3 ± 0.3	0.3 ± 0.2	0.5 ± 0.2	0.7 ± 0.3	1.1 ± 0.4	3.1 ± 0.9	45.4 ± 9.9
Africa	3.5	7.2	10.4	9.7	1.4	11.5	1.9	0.9	0.5	1.3	1.2	1.1	6.0	43.5

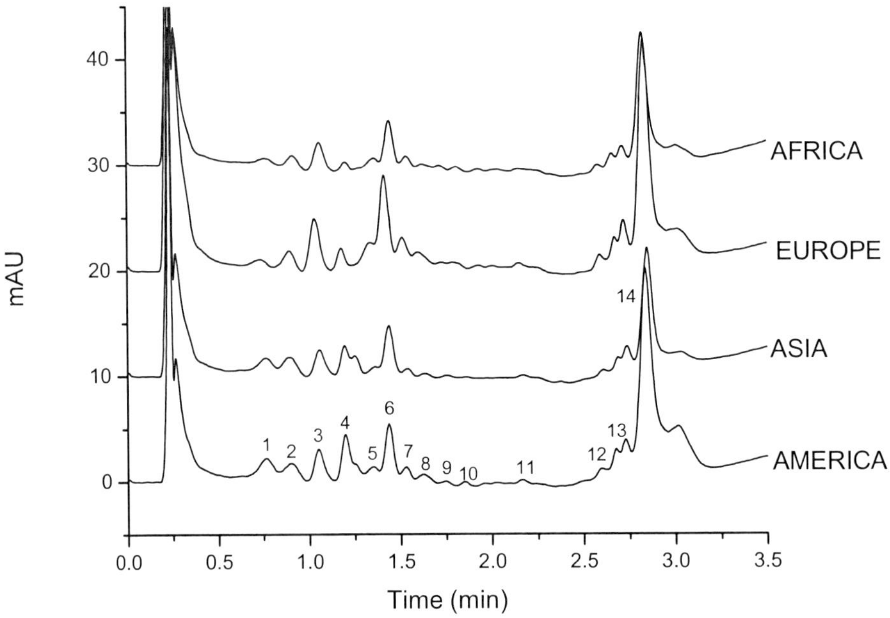

Figure 3. Protein profiles obtained from one American (Harosoy), one Asian (Osage), one European (Fred), and one African (Merit) soybean cultivars. Experimental conditions detailed in the text. Peaks observed have been numbered from 1 to 14.

Results and Discussion

1. Separation of Soybean Proteins from Different Soybean Cultivars

A pool of 91 soybean cultivars with different origins have been analyzed with a reversed-phase HPLC method already described [69]. As example, the chromatographic profiles of one soybean cultivar from North America, one from Europe, one from Africa (Uganda), and one from Asia are grouped in Figure 3. All soybean varieties showed 14 peaks consisting peak 13 of two unresolved peaks and being peaks 6 and 14 the majority ones in all cases. Despite these similarities among the protein profiles observed for these soybean varieties, there were some differences that could be useful for cultivar characterization. In addition to a different peak size, some peaks were resolved in a different way such as peak 4 that was partially resolved in two peaks in some cases. Table 1 shows the retention time and the average and standard deviation of the area percentage of every peak in all American, European, Asian, and the African cultivars studied. The only African cultivar employed (Merit original from Uganda) presented peak area percentages very different to the observed for the other varieties, especially for peaks 1, 2, 5, 10, 11, and 13. Peaks having the most differentiating capability among American, Asian, and European soybeans were peaks 4, 10, and 13 while peaks presenting the highest capability for the within group differentiation were peaks 1, 4, 7, and 9 (peaks presenting the highest RSD values).

In order to facilitate the interpretation of the data, multivariate classification methods were applied to the area percentages obtained for every peak in each of the 91 soybean cultivars by perfusion RP-HPLC.

2. Application of Multivariate Techniques for the Classification of Soybean Varieties

Cluster analysis was used for searching natural grouping among the studied soybean lines. This analysis will allow an association of samples based on their similarities in protein profiles. Hierarchical agglomerative clustering was performed by means of the Ward method on raw data using squared Euclidean distances as measure of similarity. Three main groups were observed: two of the clusters consisted of a miscellaneous of soybean varieties from different continents while the third group (in the middle) was the most homogeneous being mainly constituted by American varieties.

For the application of discriminant analysis, 64 of the 91 soybean varieties were employed since they were the ones whose origin data could be checked. 54 of these 64 soybean samples were employed as training samples while the other 10 were used for the validation of the model. The application of discriminant analysis techniques enabled to obtain two mathematical functions (DF1 and DF2) that were statistically significant at the 95% of confidence. These discriminant functions classified the 54 soybean varieties in four established categories: America, Europe, Asia, and Africa (Figure 4). Soybeans from the same continent showed certain tendency to present similar chromatographic profiles observing a good separation among the samples according to their origin. The African cultivar was located in the corner corresponding to the most negative value of DF1 and the most positive value of DF2. American and Asian soybean cultivars were differentiated by DF1 while European cultivars presented intermediate DF2 values. A total of 49 soybean cultivars from the 54 were correctly classified (percentage of correct classification, 90.7%) observing the best classification for European and African soybeans while Asian ones were the worst classified (see Table 2). The most influencing variables for the classification of soybean cultivars (determined by the stepwise method (forward selection)) were peaks 2 and 5 being possible to obtain a percentage of correct classification of 72.2% using only these two variables.

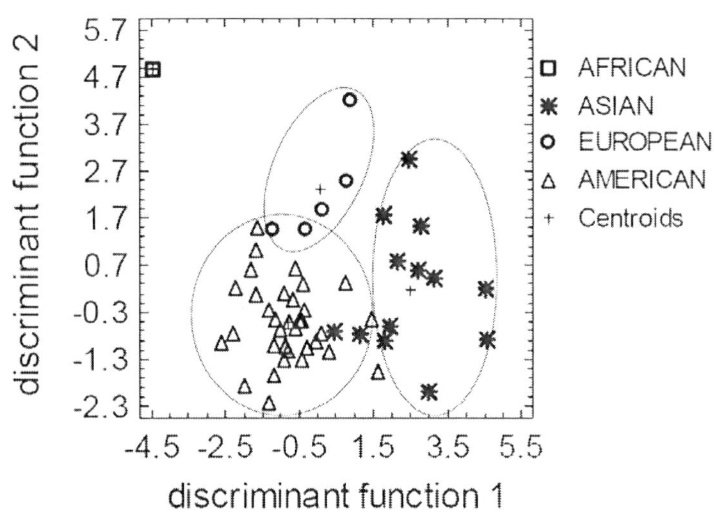

Figure 4. Application of discriminant analysis to the area percentages corresponding to 64 different soybean cultivars from America, Asia, Europe, and Africa.

Table 2. Classification of soybean varieties in the training, evaluation, and cross-validation of the discriminant model proposed.

	Size	American	European	African	Asian
Training					
American	35	32 (91.4%)	0	0	3 (8.6 %)
European	5	0	5 (100%)	0	0
African	1	0	0	1 (100%)	0
Asian	13	1 (7.7%)	1 (7.7%)	0	11 (84.6 %)
Evaluation					
American	5	4 (80%)	0	0	1 (20 %)
European	2	1 (50%)	1 (50%)	0	0
Asian	3	0	0	0	3 (100 %)
Cross validation					
American	35	28 (80%)	2 (5.7%)	1 (2.8%)	4 (11.4%)
European	5	3 (60%)	1 (20%)	0	1 (20%)
African	1	0	1 (100%)	0	0
Asian	13	3 (23.1%)	3 (23.1%)	0	7 (53.8%)

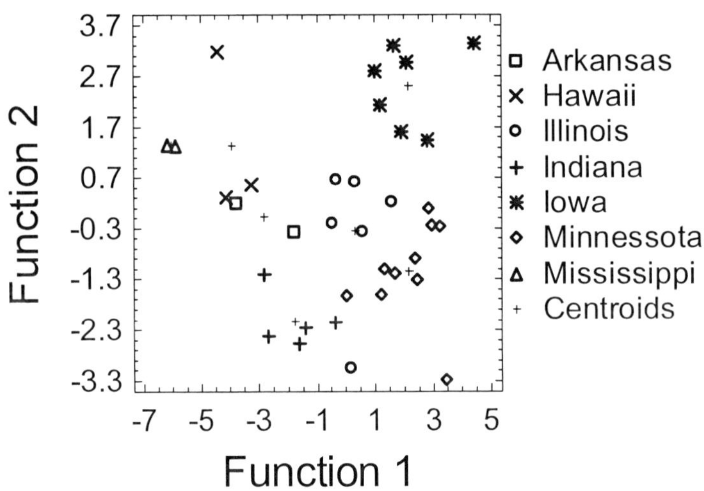

Figure 5. Application of discriminant analysis to the area percentages corresponding to 35 different soybean cultivars from different USA states: Arkansas, Hawaii, Illinois, Indiana, Iowa, Minnesota, and Mississippi.

The validation of the method was performed by the prediction of the origin of ten soybean samples observing that eight of them were correctly classified (80% of classification capability). Moreover, a cross-validation procedure by the treatment of *n-1* out of *n* observations as training dataset to determine the discrimination rule and the use of that rule for the classification of the observation left out was conducted with all 14 variables. 36 of 54 soybean varieties were correctly classified observing the best classification for the American seeds (see Table 2).

In order to get more precise information of soybean cultivars, only USA varieties were subjected to a discriminant analysis for differentiating among cultivars from different American states. Seven different categories, corresponding to seven different American states from where the soybeans were original, were established: Arkansas (2 varieties), Hawaii (3 varieties), Illinois (6 varieties), Indiana (5 varieties), Iowa (7 varieties), Minnesota (10 varieties), and Mississippi (2 varieties). Two discriminant functions with P-values less than 0.05 (at the 95% confidence level) enabled the correct classification of all varieties (97.1%) with the exception of one Minnesota cultivar that was incorrectly classified as Illinois variety. Figure 5 shows how those soybean cultivars original from the warmest regions such as Mississippi, Hawaii or Arkansas were displayed in the left and up side of the diagram while varieties grown in colder regions appeared in the right and down side.

3. Estimation of Different Chemical and Physical Characteristics

The chromatographic data collected from the protein profiles obtained from the 91 different soybean cultivars were compared with different chemical and physical characteristics of cultivars in order to explore possible correlations. Thus, the area percentages obtained were used for the estimation of the 7S and 11S globulin contents and the ratio 11S/7S. These parameters vary among soybean cultivars and are very important for the characterization of the seed since they affect significantly to the protein quality and food processing properties. The estimation of 7S and 11S globulin fractions in every cultivar was performed from the peak area percentages taking into account a previous research work [64] in which peaks 1 to 7 were mainly identified as the 7S globulin fraction while peaks 12-14 were attributed to the 11S globulin fraction. Figure 6 shows the box-and-whisker diagram for the 11S/7S ratio in the American, European, and Asian soybeans. Results confirmed the variability of this ratio with the soybean cultivar. In fact, 11S/7S ratios ranged from 0.46 to 3.45 observing the highest 11S/7S ratios for Asian soybean cultivars while European varieties presented the lowest ratios. Taking into account the characteristics of 7S and 11S globulin fractions, European cultivars, in general, could be most suitable when certain functional properties of soybean are required while Asian soybeans, in general, presented higher 11S globulin contents and, therefore, could be more useful for their nutritional properties.

Moreover, trying to explore the possibilities of using protein profiles as "fingerprints" of soybean cultivars, their capability for predicting different characteristics of soybean cultivars was studied. Eleven different parameters including compositional parameters (contents in oil, protein, linoleic acid, linolenic acid, oleic acid, palmitic acid, and stearic acid) and physical ones (height of the plant from ground to stem tip, weight of 100 seeds, time needed by 50% of the plants for flowering, and time needed by 95% of the pods for reaching its final color) were studied. All these data corresponding to 64 of the 91 soybean cultivars employed in this work were taken from the Agricultural Research Service from the United States Agricultural Department (http://www.ars-grin.gov/npgs/urbana.html). Highest variability among soybean cultivars were observed for the plant height (85.1 ± 16.8 cm), the 100-grain weight (2.45 ± 0.48 g), the linolenic acid content (1.51 ± 0.28 g/100g), the oleic acid content (4.12 ± 0.99 g/100g), the palmitic acid content (2.21 ± 0.48 g/100g), and the stearic acid content (0.62 ± 0.16 g/100g) while those parameters showing the lowest variability among the soybean cultivars studied were the protein content (43.5 ± 2.5 g/100g), the oil content (19.4 ± 1.5 g/100g), the linoleic acid content (10.5 ± 0.9 g/100g), the flowering time (221 ± 15 days), and the maturation time (290 ± 26 days).

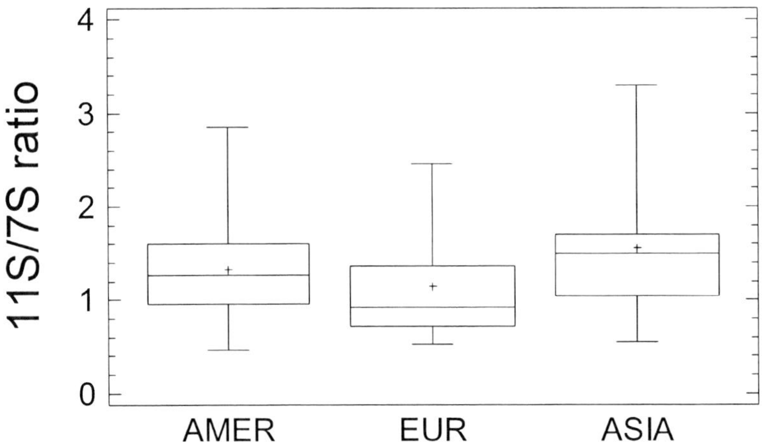

Figure 6. Box-and-whisker diagram for the 11S/7S ratios in the American, European, and Asian soybean cultivars studied.

Different mathematical models (simple linear regression, multiple linear regression (MLR), principal components regression (PCR), and partial least squares regression (PLS)) were applied. Simple linear regression was applied for every peak area with every parameter. The highest correlations were observed between the plant height and the area percentages corresponding to peaks 5 and 13, between the linoleic acid content and the area percentages corresponding to peaks 7 and 8, and between the oil content and the area percentage corresponding to peak 7. The application of multiple regression models demonstrated that best correlations were always obtained by MLR analysis. Despite this, no correlation by MLR was observed for the protein content, the stearic acid content, the maturation time, and the oleic acid content while a statistical significant relationship (P-value < 0.10) was detected for the other parameters studied, especially for the palmitic acid and the linoleic acid contents which presented correlation coefficients of 0.720 and 0.785, respectively, by MLR.

CONCLUSION

Soybean cultivar characterization is essential for maintaining genetic purity, estimating genetic relationship, and identifying soybean germsplams. Nevertheless, differentiation among soybean cultivars is not an easy task due to their diversity. Different methodologies based on different molecular markers and phenotypic characters have been tried but no satisfactory results have always been obtained. Moreover, rapid methodologies are needed to fulfill requirements of routine and field characterization. This is the first time a study comprising the analysis of the protein profiles of 91 different soybean cultivars from America, Europe, Asia, and Africa has been performed. Protein profiles were obtained by perfusion HPLC in only 3 min. Despite all soybean cultivars showed the same 14 peaks, the peak area percentages were different. Indeed, significant differences among soybean cultivars could be observed by the comparison of peak area percentages being possible to build a mathematical model enabling the differentiation of soybean varieties in relation to their protein profiles. This discriminant model allowed the correct classification of most soybean varieties (90.7%) being peaks 2 and 5 those showing the highest discriminant capabilities. Moreover, the discriminant analysis methodology was also useful for the differentiation among soybean varieties original from different USA states observing the correct classification of 97.1% of the American soybeans studied. Peak area percentages also allowed the estimation of the 11S/7S ratios in all 91 soybean cultivars observing that they ranged from 0.46 to 3.45 and detecting the highest ratios in Asian varieties and the lowest in European ones. Certain correlations among the 14 chromatographic variables (peak area percentages) and the palmitic and linoleic acids contents were also observed. All these results seem to demonstrate the chromatographic protein profiles are characteristic of a cultivar and could be useful for characterization purposes.

ACKNOWLEDGMENTS

The authors gratefully acknowledge financial support from the Comunidad Autónoma de Madrid, Spain (project S-0505/AGR/0312) and the Ministerio de Ciencia y Tecnología, Spain (project AGL2005-05320-C02-01/ALI). Dr. M. C. García also thanks financial support received from the Comunidad Autónoma de Madrid (project CCG07-UAH/AMB-1831). J. N. Ovalles is acknowledged for technical assistance. Finally, the authors thank the kind donation of the 91 soybean varieties by Dr Lucía de la Rosa (CRF-INIA, Madrid, Spain).

REFERENCES

[1] Natarajan, S., Xu, C., Bae, H., Caperna, T. J., Garrett, W. M. Characterization of storage proteins in wild (Glycine soja) and cultivated (Glycine max) soybean seed using proteomic analysis. *J. Agric. Food Chem.*, 2006, 54, 3114-3120.

[2] Wang, C.; Wu, X.; Jia, F.; Zhang, J.; Chen, S. Genetic variations of glycinin subunit genes among cultivated and wild type soybean species. *Progress Nat. Sci.*, 2008, 18, 33-41.

[3] Zarcadas, C. G.; Gagnon, C.; Poysa, V.; Khanizadeh, S.; Cober, E. R.; Chang, V.; Gleddie, S. Protein quality and identification of the storage protein subunits of tofu and null soybean genotypes, using amino acid analysis, one- and two-dimensional gel electrophoresis, and tandem mass spectrometry. *Food Res. Int.*, 2007, 40, 111-128.

[4] Clarke, E. J.; Wiseman, J. Developments in plant breeding for improved nutritional quality of soya beans I. Protein and amino acid content. *J. Agric. Sci.*, 2000, 134, 111-124.

[5] Krishnan, H. B.; Bennett, J. O.; Kim, W. S.; Krishnan, A. H.; Mawhinney, T. P. Nitrogen lowers the sulfur amino acid content of soybean (*Glycine max [L] Merr.*) by regulating the accumulation of Bowman-Birk protease inhibitor. *J. Agric. Food Chem.*, 2005, 53, 6347-6354.

[6] Riblett, A. L.; Herald, T. J.; Schmidt, K. A.; Tilley, K. A. Characterization of β-conglycinin and glycinin soy protein fractions from four selected soybean genotypes. *J. Agric. Food Chem.*, 2001, 49, 4983-4989.

[7] Mori, T.; Utsumi, S.; Inaba, H.; Kitamura, K.; Harada, K. Differences in subunit composition of glycinin among soybean cultivars. *J. Agric. Food Chem.*, 1981, 29, 20-23.

[8] Mori, T.; Utsumi, S. Purification and properties of storage proteins of broad bean. *Agric. Biol. Chem.*, 1979, 43, 577-583.

[9] Mori, T.; Utsumi, S.; Inaba, H. Interaction involving disulfide bridges between subunits of soybean seed globulin and between subunits of soybean and sesame seed globulins. *Agric. Biol. Chem.*, 1979, 43, 2317-2322.

[10] Nakamura, T.; Utsumi, S.; Kitamura, K.; Harada, K.; Mori, T. Cultivar differences in gelling characteristics of soybean glycinin. *J. Agric. Food Chem.*, 1984, 32, 647-651.

[11] Nakamura, T.; Utsumi, S.; Mori, T. Interactions during heat-induced gelation in a mixed system of soybean-7S and 11S-globulins. *Agric. Biol. Chem.*, 1980, 50, 2429-2435.

[12] Hughes, S. A.; Murphy, P. A. Varietal influence on the quantity of glycinin in soybeans. *J. Agric. Food Chem.*, 1983, 31, 376-379.

[13] Natarajan, S.; Xu, C.; Bae, H.; Bailey, B.; Cregan, P.; Caperna, T. J.; Garret, W. M.; Luthria, D. Proteomic and genetic analysis of glycinin subunits of sixteen soybean genotypes. *Plant Physiol. Biochem.*, 2007, 45, 436-444.

[14] Camps, G.; Vernetti, F. J.; Augustin, E.; Irigon, D. Morphological and electrophoretical characterization of twenty soybean cultivars. *Pesq. Agropec. Bras.*, 1994, 29, 1779-1787.

[15] Khatib, K. A.; Herald, T. J.; Aramouni, F. M.; MacRitchie, E.; Schapaugh, W. T. Characterization and functional properties of soy beta-conglycinin and glycinin of selected genotypes. *J. Food Sci.*, 2002, 67, 2923-2929.

[16] Tiwari, S. P.; Bhatnagar, P. S.; Prabhak, A. R. Identification of suitable soybean (glycine-max) varieties for nontraditional regions of India. *Indian J. Agric. Sci.*, 1994, 64, 872-874.

[17] Fairbanks, D. J. Development of genetic resistance to iron-deficiency chlorosis in soybean. *J. Plant Nutr.*, 2000, 23, 1903-1913.

[18] Burton, J. W. Breeding soybeans for improved protein quantity and quality. *World Soybean Research Conference*, 1984, 3, 361-367.

[19] Burton, J. W. Recent developments in breeding soybeans for improved oil quality. *Fett Wissenschaft Technologie-Fat Science Technology*, 1991, 93, 121-128.

[20] Hermann, E. M.; Helm, R. M.; Jung, R.; Kinney, A. J. Genetic modification removes an immunodominant allergen from soybean. *Plant Physiol.*, 2003, 132, 36-43.

[21] Hartwig, E. E. Breeding soybeans with a higher percentage of proteins and quality, New approaches to breeding for improved plant protein (Proc. Panel Röstanga, 1968), IAEA, Vienna, 1969, 67-70.

[22] Staswick, P. E.; Nielsen, N. C. Characterization of soybean cultivar lacking certain glycinin subunits. *Arch. Biochem. Biophys.*, 1983, 223, 1-8.

[23] Hartwig, E. E. Breeding productive soybeans with higher percentage of protein, International symposium on seed protein improvement in cereals and grain legumes, Neuherberg, Germany, 1978, 59-66.

[24] Leffel, R. C.; Cregan, P. B.; Bolgiano, A. P. Nitrogen-metabolism of soybean genotypes selected for seed composition, fascinated stem, or harvest index. *Crop Sci.*, 1992, 32, 1428-1432.

[25] Krishnan, H. B.; Natarajan, S. S.; Mahmoud, A. A.; Nelson, R. L. Identification of glycinin and beta-conglycinin subunits that contribute to the increased protein content of high-protein soybean lines. *J. Agric. Food Chem.*, 2007, 55, 1839-1845.
[26] Zarcadas, C. G.; Voldeng, H. D.; Yu, Z. R.; Choi, V. K. Determination of the protein quality of three new northern adapted cultivars of common and miso type soybeans by amino acid analysis. *J. Agric. Food Chem.*, 1997, 45, 1161-1168.
[27] Tsubokura, Y.; Hajika, M.; Harada, K. Molecular characterization of a beta-conglycinin deficient soybean, *Euphytica*, 2006, 150, 249-255.
[28] Hayashi, M.; Harada, K.; Fujiwara, T.; Kitamura, K. Characterization of a 7S globulin-deficient mutant of soybean (Glycine max (L.) Merrill). *Mol. General Genet.*, 1998, 258, 208-214.
[29] L'Hocine, L; Boye, J. I. Allergenicity of soybean: new developments in identification of allergenic proteins, cross-reactivities and hypoallergenization technologies. *Crit. Rev. Food Sci. Nutr.*, 2007, 47, 127-143.
[30] Wilson, S.; Blaschek, K.; Gonzalez de Mejia, E. Allergenic proteins in soybean: processing and reduction of P34 allergenicity. *Nutr. Rev.*, 2005, 63, 47-58.
[31] Achouri, A.; Boye, J. I.; Zamani, Y. Soybean variety and storage effects on soymilk flavour and quality, *Int. J. Food Sci. Technol.*, 2008, 43, 82-90.
[32] Chen, P. Developing high-quality identity-preserved soybeans for the specialty soyfood market in Technical aspects of identity-preserved systems for soybeans in the United States, chapter 3, U.S. Soybean Export Council. http://www.ussoyexports.org/buying_u.s._soy/ipbeans.htm
[33] Mujoo, R.; Trinh, D. T.; Ng, P. K. W. Characterization of storage proteins in different soybean varieties and their relationship to tofu yield and texture. *Food Chem.*, 2003, 82, 265-273.
[34] Tezuka, M.; Taira, H.; Igarashi, Y.; Yagasaki, K.; Ono, T. Properties of tofus and soy milks prepared from soybeans having different subunits of glycinin, *J. Agric. Food Chem.*, 2000, 48, 1111-1117.
[35] Poysa, V.; Woodrow, L.; Yu, K. Effect of soy protein subunit composition on tofu quality, Effect of soy protein subunit composition on tofu quality, *Food Res. Int.*, 2006, 39, 309-317.
[36] Krishnan, H. B. Characterization of a soybean [Glycine max (L.) Merr.] mutant with reduced levels of Kunitz trypsin inhibitor. *Plant Sci.*, 2001, 160, 979-986.
[37] Lin, P. Y.; Lai, H. M. Bioactive compounds in legumes and their germinated products. *J. Agric. Food Chem.*, 2006, 54, 3807-3814.
[38] Hubert, J.; Berger, M.; Dayde, J. Use of a simplified HPLC-UV analysis for soyasaponin B determination; study of saponin and isoflavone variability in soybean cultivars and soy-based health food products. *J. Agric. Food Chem.*, 2005, 53, 3923-3930.
[39] Kanamaru, K.; Wang, S. D.; Abe, J.; Yamada, T.; Kitamura, K. Identification and characterization of wild soybean (Glycine soja Sied. et Zecc.) strains with high lutein content. *Breeding Sci.*, 2006, 56, 231-234.
[40] Ujiie, A.; Yamada, T.; Fujimoto, K.; Endo, Y.; Kitamura, K. Identification of soybean varieties with high α-tocopherol content. *Breeding Sci.*, 2005, 55, 123-125.
[41] Murphy, P. A.; Resurrección, A. P. Varietal and environmental differences in soybean glycinin and β-conglycinin content. *J. Agric. Food Chem.*, 1984, 32, 911-915.

[42] Zarkadas, C. G.; Gagnon, C.; Gleddie, S.; Khanizadeh, S.; Cober, E. R.; Guillemette, R. J. D. Assessment of the protein quality of fourteen soybean [Glycine max (L.) Merr.] cultivars using amino acid analysis and two-dimensional electrophoresis. *Food Res. Int.*, 2007, 40, 129-146.

[43] Cui, Z.; Carter, T. E., Jr.; Burton, J. W.; Wells, R. Phenotypic diversity of modern Chinese and North American soybean cultivars. *Crop Sci.*, 2001, 41, 1954-1967.

[44] Gizlice, Z.; Carter, T. E., Jr.; Burton, J. W. Genetic diversity in North American soybean: I. Multivariate analysis of founding stock and relation to coefficient of parentage. *Crop Sci.*, 1993, 33, 614-620.

[45] Wagner, C. K.; McDonald, M. B. Rapid laboratory tests useful for differentiation of soybean (Glycine max) cultivars. *Seed Sci. & Technol.*, 1982, 10, 431-449.

[46] Jianhua, Z.; McDonald, M. B.; Sweeney, P. M. Soybean cultivar identification using RAPD, *Seed Sci. & Technol.*, 1996, 24, 589-592.

[47] Grabau, E. A.; Davis, W. H.; Phelps, N. D.; Gengenbach, B. G. Classification of soybean cultivars based on mitochondrial DNA restriction fragment length polymorphisms. *Crop Sci.*, 1992, 32, 271-274.

[48] Vilela, R.; Goncalves, E.; Alves, M. Determination of genetic diversity within Brazilian soybean germplasm using random amplified polymorphic DNA techniques and comparative analysis with pedigree data. *Rev. Brasil. Genet.*, 1995, 18, 265-273.

[49] Ude, G. N.; Kenworthy, W. J.; Costa, J. M.; Cregan, P. B.; Alvernaz, J. Genetic diversity of soybean cultivars from China, Japan, North America, and North American ancestral lines determined by amplified fragment length polymorphism. *Crop Sci.*, 2003, 43, 1858-1867.

[50] Prabhu, R. R.; Webb, D.; Jessen, H.; Luk, S.; Smith, S.; Gresshoff, P. M. Genetic relatedness among soybean genotypes using DNA amplification fingerprinting (DAF), RFLP, and pedigree. *Crop Sci.*, 1997, 37, 1590-1595.

[51] Satyavathi, C. T.; Bhat, K. V.; Bharadwaj, C.; Tiwari, S. P.; Chaudhury, V. K. AFLP analysis of genetic diversity in Indian soybean [*Glycine max* (L.) Merr.] varieties. *Genetic Resources and Crop Evolution*, 2006, 53, 1069-1079.

[52] Doldi, M. L.; Vollmann, J.; Lelley, T. Genetic diversity in soybean as determined by RAPD and microsatellite analysis. *Plant Breeding*, 1997, 116, 331-335.

[53] Giancola, S.; Marcucci Poltri, S.; Lacaze, P.; Hopp, H. E. Feasibility of integration of molecular markers and morphological descriptors in a real case study of a plant variety protection system for soybean. *Euphytica*, 2002, 127, 95-113.

[54] Rongwen, J.; Akkaya, M. S.; Bhagwat, A. A.; Lavi, U.; Cregan, P. B. The use of microsatellite DNA markers for soybean genotype identification, *Theor. Appl. Genet.*, 1995, 90, 43-48.

[55] Emura, K.; Yamanak, S.; Isoda, H.; Watanabe, K. N. Estimation for different genotypes of plants based on DNA analysis using near-infrared (NIR) and Fourier-transform infrared (FT-IR) spectroscopy. *Breeding Sci.*, 2006, 56, 399-403.

[56] Cardy, B. J.; Kannenberg, L. W.; Beversdorf, W. D. Genetic identification of maize and soybean cultivars using electrophoretic techniques. *Canadian J. Plant Sci.*, 1983, 63, 342.

[57] McDonald, M. B. Blotting of seed proteins from isoelectrically focused gels for cultivar identification. *Seed Sci. & Technol.*, 1991, 19, 33-40.

[58] Blogg, D.; Imrie, B. C. Starch gel electrophoresis for soybean cultivar identification. *Seed Sci. & Technol.*, 1982, 10, 19-24.
[59] Zarcadas, C. G.; Gagnon, C.; Gleddie, S.; Khanizadeh, S.; Cober, E. R.; Guillemette, R. J. D. Assessment of the protein quality of fourteen soybean [*Glycine max* (L.) Merr.] cultivars using amino acid analysis and two-dimensional electrophoresis. *Food Res. Int.*, 2007, 40, 129-146.
[60] Xu, C. P.; Caperna, T. J.; Garrett, W. M.; Cregan, P.; Bae, H. H.; Luthria, D. L.; Natarajan, S. Proteomic analysis of the distribution of the major seed allergens in wild, landrace, ancestral, and modern soybean genotypes. *J. Sci. Food Agric.*, 2007, 87, 2511-2518.
[61] Buehler, R. E.; McDonald, M. B., Jr.; Van Toai, T. T.; St. Martin, S. K. Soybean cultivar identification using high-performance liquid-chromatography of seed proteins. *Crop Sci.*, 1989, 29, 32-37.
[62] Oomah, B. D.; Voldeng, H.; Fregeau-Reid, J. A. Characterization of soybean proteins by HPLC. *Plant Foods Human Nutr.*, 1994, 45, 251-263.
[63] Cole, K. D.; Cousin, S. L. Size exclusion chromatography of soybean proteins and isoflavones. *J. Agric. Food Chem.*, 1994, 42, 2713-2720.
[64] García, M. C.; Heras, J. M.; Marina, M. L. Simple and rapid characterization of soybean cultivars by perfusion reversed-phase HPLC: application to the estimation of the 11S and 7S globulin contents. *J. Sep. Sci.*, 2007, 30, 475-482.
[65] Heras, J. M.; Marina, M. L.; García, M. C. Development of a perfusion ion-exchange chromatography method for the separation of soybean proteins and its application to cultivar characterization. *J. Chromatogr. A*, 2007, 1153, 97-103.
[66] Castro-Rubio, F.; Marina, M. L.; García, M. C. Perfusion reversed-phase high-performance liquid chromatography/mass spectrometry analysis of intact soybean proteins for the characterization of soybean cultivars. *J. Chromatogr. A*, 2007, 1170, 34-43.
[67] García-Ruiz, C.; García, M. C.; Cifuentes, A.; Marina, M. L. Characterization and differentiation of diverse transgenic and nontransgenic soybean varieties from CE protein profiles. *Electrophoresis*, 2007, 28, 2314-2323.
[68] García, M. C.; García, B.; García-Ruiz, C.; Gómez, A; Cifuentes, A.; Marina, M. L. Rapid characterization of (glyphosate tolerant) transgenic and non-transgenic soybeans using chromatographic protein profiles. *Food Chem.*, 2009, 113, 1212-1217.
[69] García, M. C.; Marina, M. L.; Torre, M. Perfusion chromatography: an emergent technique for the analysis of food proteins. *J. Chromatogr. A*, 2000, 880, 169-187.

In: Soybean and Wheat Crops
Editors: S. Davies, G. Evans
ISBN: 978-1-60741-173-4

Chapter 5

EFFECT OF LIMING, N AND P FERTILISATION OF A LIXISOL ON THE GROWTH OF SELECTED SOYBEAN CULTIVARS UNDER SUB-HUMID TROPICAL CONDITIONS IN ZIMBABWE

J. Nyamangara*, C. Musharo, M. Matokwe
Department of Soil Science and Agricultural Engineering, University of Zimbabwe, P.O. Box MP 167, Mount Pleasant, Harare, Zimbabwe

ABSTRACT

Soybean (*Glycine max* (L) Merr) production in the smallholder farming areas of southern Africa is constrained by soil acidity and nutrient deficiency among other factors. A study was conducted to determine the performance of four soybean cultivars commonly grown in Southern Africa, in acid soil, their response to liming, and N and P fertilisation. Soybean was grown over two cropping seasons at a research station and in a sub-humid smallholder farming area in north-eastern Zimbabwe. Liming increased the number of nodules and nodule dry matter yield (NDMY) in both cropping seasons in all the four soybean cultivars tested but the differences were only significant in the second cropping season (nodule number, $p=0.004$; NDMY, $p=0.025$). In both seasons liming increased grain yield (season 1, $p=0.046$; season 2, $p=0.023$) but cultivar differences were not different. Addition of P fertiliser increased P uptake, grain and stover yield and liming further enhanced both P uptake and grain and stover yield. Addition of 30 kg N ha^{-1} as ammonium nitrate significantly ($p<0.05$) reduced nodules numbers and grain yield compared to treatments where N had not been applied. It was concluded that soybean productivity in acid soil prevalent in humid and sub-humid areas of SSA can be effectively increased through liming and P fertilisation. Application of mineral N not supported by soil testing reduces the effectiveness of biological N fixation thereby depriving farmers of a cheaper source of N where soil N is relatively high.

---

* Corresponding author: *Email address:* jnyamangara@agric.uz.ac.zw

Keywords: soil acidity, soybean, nodulation, liming, N and P uptake

1. INTRODUCTION

Soybean (*Glycine max* (L) Merr) accounts for 40% of the national edible vegetable oil and seed cake output in Zimbabwe (Whingwiri, 1996). On average a 100 g of soybean contains 1700 KJ energy, 40 g protein, 20 g fat and 10 g water. The inclusion of soybean in cropping systems significantly improves soil N status as the crop can fix up to half of its N requirements when in symbiosis with Rhizobia thus reducing the amount of N farmers have to apply as fertilizer (Wynch and Rains, 1978). Successful production of the crop in the smallholder sector can be a source of protein source for poor households most of which have been affected by the HIV/AIDS pandemic and therefore require a good diet to improve their health. Soybean can significantly contribute to sustainable soil fertility management if the stover in incorporated into the soil after harvesting grain.

Soil acidity is a major constraint to crop production in tropical and subtropical areas as well as cultivated temperate areas. About 27% of soils in tropical Africa are classified as acidic (Pandey *et al.,* 1994) and low soil pH is the number one growth-limiting factor (Jones, 1975; Copeland, 1976). Reduced plant growth in acid soils is mainly due to Al and Mn toxicities as well as deficiencies of P, Ca and Mg (Mugwira and Haque, 1993). In Zimbabwe the majority of smallholder farming areas are located on sandy soils which are low in organic matter (<1%) and weakly buffered, and in high rainfall zones the soils are also acidic and often deficient in P, Ca and Mg (Mugwira and Nyamangara, 1998). Like any other legume, soybean is sensitive to soil acidity (Al and Mn toxicity) and poor growth has been reported on acid sandy soils in Zimbabwe (Mpepereki, 1994).

Despite the documented positive effects of lime, its adoption by smallholder farmers has been poor mainly because of inaccessibility and high transportation costs although the lime itself is relatively cheap. The financially constrained farmers opt to buy seed and mineral fertilizers before they consider lime (Dhliwayo et al., 1998).

The breeding or selection of crop cultivars tolerant to soil acidity offers an opportunity to sustain crop production in high rainfall areas of Zimbabwe, and related farming systems in sub Saharan Africa.The growth of legumes under acid conditions depends on the tolerance by both the host plant and the bacteria species it forms a symbiosis relationship with. Reduced performance by a legume and/or rhizobia under acidic conditions can be a result of the negative environmental conditions induced by acidity such as toxicities of Al, Fe, Mn and deficiencies in Ca, Mg, P and Mo both of which hinder growth, survival and development of the rhizobia and the plant (Giller, 2001). Deficiencies of the major cations Ca and Mg are common due to the small amounts of these cations in acid soil and to the secondary effect of high concentrations of H^+ inhibiting uptake of these cations by plants (Andrew, 1978). Al decreases the initiation of nodules at concentrations that have no discernible effect on growth of the legume plant or functioning of nodules (Robson, 1983). In acid soil rhizobia fails because it cannot grow and colonise the host rhizosphere (Keyser et al., 1979). Few rhizobia strains can grow under conditions more acidic than pH 4.5 (Graham, 1992).

A corresponding problem, with acidity, particularly when exchangeable Al^{3+} is high is high P fixation (Graham, 1992) and soybean has relatively high P requirements. Phosphorus is a structural element required by Rhizobia for synthesis of nucleic acids and phospholipids (O'Hara and Glenn, 1994). Acute deficiency of P can prevent nodulation and nitrogen fixation of legumes (Giller, 2001). Mullen et al. (1988) observed delayed infection of the primary root and reduced yield benefit from biological N fixation when soybean was grown under low P conditions. However, high soil nitrate has been reported to inhibit biological N fixation (Abaidoo and van Kessell, 1989).

The objectives of this study were (i) to assess the tolerance of soybean cultivars widely grown in southern Africa to soil acidity and the response of these cultivars to liming, and (ii) to determine the effect of N and P fertilisation on the growth of a selected soybean cultivar grown in acidic or limed soil. It was hypothesized that smallholder farmers, who do not have access to lime, can successfully grow soybean if a cultivar tolerant to soil acidity is available.

2. Materials and Methods

2.1. Site Description, Soil Sampling and Land Preparation

The field experiment was conducted over two cropping seasons (2002/3 and 2003/4) at Domboshawa Training Centre (DTC), Zimbabwe (17°35′S; 31°14′E, altitude 1574 m). DTC is located in agroecological natural region II characterised by unimodal rainfall (750-1000 mm $annum^{-1}$, October-April) and mean annual temperature of 18.8 °C. Soils at the site are relatively highly weathered, leached and well-drained sandy clay loams derived from granodiorite gneiss and classified as Haplic Lixisols (FAO) (Nyamapfene, 1991). The site had been under a grass fallow for at least five years.

The experimental site was disc-ploughed at the beginning of each cropping season. Soil sampling was done before ploughing and at the end of the first cropping season. Ten sub-samples were randomly collected from each plot using a soil auger (0-15 cm) and sub-samples mixed to make one composite sample. The soils were taken to the laboratory, air-dried and passed through a 2-mm sieve before basic characterisation.

2.2. Experimental Layout and Planting

Two experiments were set up, one to assess the tolerance of selected soybean cultivars to soil acidity, and another to determine the effect of liming, N and P fertilisation on the growth of a selected soybean cultivar. For the first experiment four soybean cultivars (Magoye, Safari, Solitaire and Storm) and two lime rates (0 and 1500 kg ha^{-1} applied in the first year) were used. A split-split-plot design was used, with lime as the main plot and cultivar as the sub-plot, with four replications. A basal application of compound L fertiliser (5%N, 8% P, 8%K, 8%K and 0.25% B) was banded in rows at 225 kg ha^{-1} in both cropping seasons.

In the second experiment only one cultivar, Solitaire, was used. The treatments were 2 lime rates (0 and 1500 kg ha^{-1} applied in the first year), two N rates (0 and 30 kg ha^{-1}) and four P rates (0, 7.5, 15 and 22.5 kg ha^{-1}) arranged in a split-split plot design with lime as the main plot, N as subplot and P and sub-sub plot, with four replications. All the N and P were banded into planting rows at planting as ammonium nitrate (AN) (34.5% N) and single super phosphate (8% P and 12% S), respectively.

In both experiments lime was broadcasted on the soil surface and incorporated into the 0-15cm soil horizon. Plots were 5m x 5m in size and planting rows were marked out using hand hoes. Soybean seed was inoculated with *Bradyrhizobium* strain MAR 1491 (USDA 110) (Kasasa et al., 1998) in both seasons. Seed was banded in rows 45 cm apart and plants were then thinned to 7 cm intra-row at two weeks after germination to give a target plant population of 320 000 plants ha^{-1}. Weed control was done by hand-hoeing throughout the cropping season.

2.3. Assessment of Nodule Number and Effectiveness

Nodulation was assessed at eight weeks after planting by carefully digging up root systems of ten randomly selected plants in each plot. A 3x3m net plot at the centre of each plot was reserved for yield determination at harvesting. The nodules were removed from the roots, dried and weighed. To determine nodule effectiveness, the internal colour of five fresh nodules per plant was recorded. A red colour depicted an effective nodule, pink colour a weakly effective nodule, and green or white colour depicted an ineffective nodule (Zengeni, 2004).

2.4. Grain and Stover Yield Determination

Grain and stover yield were determined at physiological maturity (16 weeks after planting) using plants in the 3mx3m net plot. Total aboveground dry matter yield was determined by weighing, and after threshing and winnowing grain yield was determined by weighing and adjusted to 11% moisture content. Stover yield was obtained by subtracting grain yield from total aboveground dry matter yield. Leaf litter was not included in the stover yield as the leaves had fallen to the ground by the time of harvesting. In the second experiment grain and stover sub-samples were taken for N and P analysis (Anderson and Ingram, 1993).

2.5. Data Analysis

Analysis of variance (ANOVA) using Genstat (2002) was used to identify treatment effects. Variables on which statistical analyses were performed are stover and grain yield, nodule numbers and effectiveness. Least significant differences ($P<0.05$) were used to separate treatment means. The effects of lime, cultivar, N and P were quantified as well as interaction effects.

Table 1. Soil properties of the experimental sites

	Textural class	Clay (%)	pH ($CaCl_2/H_2O$)	OC (%)	N (%)		Olsen P (mg kg^{-1})	Cation exchange ($cmol_c\ kg^{-1}$soil)					
					Mineral	Total		CEC	Ca	Mg	K	Na	Exchangeable acidity
Site 1	mSaCL	22	3.9/4.6	0.47	26	0.035	4.86	4.0	2.00	0.50	0.26	0.14	1.10
Site 2	mSaL	12	4.6/5.3	0.40	24	0.030	1.00	2.7	1.10	0.20	0.13	0.13	1.14

mSaCL – medium grained sandy clay loam; mSaL – medium grained sandy loam, OC – organic carbon.

3. Results

3.1. Site Description and Weather Conditions

The 2002/03 season recorded 911 mm of rainfall and was similar to the long-term average for DTC, whereas the 2003/04 season was drier (643 mm). The rainfall was well distributed in both seasons although brief periods of moisture stress occurred during the 2003/04 season. The soils were moderately leached (58-73% base saturation), strongly acidic (pH<5.0), and organic matter (<1% organic C) and mineral N were low (Table 1; Nyamangara et al., 2000). Available P was more deficient at site 2 compared to site 1, and at both sites exchangeable Mg was relatively low compared to Ca (Table 1).

3.2. Effect of Soil Acidity on Nodulation of Selected Cultivars

There were significant (P<0.001) differences in the number of nodules produced by the different soybean cultivars and over the two seasons (Table 2). In the first season, Safari produced the largest number of nodules (87 per 10 plants) in acid soil (unlimed) compared to other cultivars (≤52 per ten plants) whereas in the second season Solitaire had the highest (122 per ten plants) compared to others (≤101 per ten plants). Increase in the number of nodules due to liming was only significant (P=0.004; 72-454%) in the second season (Table 2). Magoye, a promiscuous variety widely grown in southern Africa, had the largest response in both seasons (91 and 454 %, respectively). Lime x cultivar interaction was only significant in the second season (P<0.001).

There were no significant nodule dry matter yield (NDMY) differences between cultivars in unlimed soil for both seasons. Responses were higher in the second season (>56 %) compared to the first season (5-48 %) and liming significantly increased NDMY in the second season for Magoye (368 %) and Safari (194 %) (Table 3).

Table 2. Effect of cultivar and liming acid soil on soybean nodule numbers

Cultivar		Nodule numbers (Nodules 10 plants^{-1})					
		2002/2003			2003/2004		
		- Lime	+Lime	% Response	-Lime	+Lime	% Response
Magoye		32	61	90.6	82	454	453.7
Safari		87	96	10.3	73	286	291.8
Solitaire		33	57	72.7	122	192	57.4
Storm		52	64	23.1	101	174	72.3
LSD	Lime		36.5			68.8	
(P<0.05)	Cultivar		21.0			44.4	
	Interaction		36.8			72.8	

Table 3. Effect of cultivar and liming acid soil on soybean nodule dry matter yield

Cultivar		Nodule dry matter yield (g 10 plants^{-1} nodules)					
		2002/2003			2003/2004		
		- Lime	+Lime	% Response	-Lime	+Lime	% Response
Magoye		0.42	0.48	13.7	0.21	0.96	368.3
Safari		0.55	0.58	5.5	0.19	0.55	193.6
Solitaire		0.31	0.42	26.2	0.29	0.45	56.1
Storm		0.38	0.56	47.9	0.17	0.32	92.3
LSD	Lime		0.362			0.274	
(P<0.05)	Cultivar		0.133			0.241	
	Interaction		0.334			0.346	

Nodules produced in the two seasons in both limed and unlimed plots were highly effective (>95%) (data not presented). However, the number of effective nodules was relatively low for Magoye (and Safari on unlimed soil) in both unlimed and limed soil. In the second season nodule effectiveness for Magoye was significantly ($LSD_{0.05}$ = 3.11) lower compared to the other three cultivars.

3.3. Effect of Liming, N and P Fertilisation on Nodulation

Liming improved nodule number in both the first and second seasons by an overall average of 21% and 56%, respectively ($p < 0.05$). The increase in nodule number due to liming was in treatments where AN fertilizer had not been applied (Table 4). The application of 30 kg N ha^{-1} significantly reduced nodule number in both seasons by an overall average of

34% in first season and 96% in the second season, compared to treatments where N was not applied.

In the first season addition of 22.5 kg P ha^{-1} in unlimed plots increased nodule number in comparison to the treatment where no P had been applied (Table 4). In the second season addition of 7.5 kg P ha^{-1} increased nodule number in limed treatments with no N addition (Table 4), but a further increase in P application had no significant effect. Increasing P rate had no significant effect on nodule number in treatments where AN had been applied (Table 4).

Liming improved NDMY by an overall average percentage increase 31% and 81% in the first and second seasons, respectively (Table 5). Application of 30 kg AN ha^{-1} reduced NDMY by an overall average of 60% and 98% in the first and second season, respectively, compared to treatments where AN had not been applied. Phosphorus rate had no significant ($p > 0.05$) effect on NDMY in the first season. In the second season in treatments where lime and AN had not been applied, a significant increase in NDMY occurred when 22.5 kg P ha^{-1} had been applied compared to the treatment where no P had been applied (Table 5). In the limed plots, a change in P application rate from 0 to 7.5 kg P ha^{-1} increased yield, but a further increase in P rate had no significant effect.

Table 4. Effect of liming, N and P fertilisation of nodule numbers of a selected soybean cultivar

		Nodule numbers (Nodules 10 $plants^{-1}$)							
		2002/2003				2003/2004			
P application rate (kg ha^{-1})		-Lime		+Lime		-Lime		+Lime	
		-N	+N	-N	+N	-N	+N	-N	+N
0		6.3	7.1	8.9	5.2	9.5	0	12.5	1.0
7.5		8.4	7.2	10.3	5.3	11.2	0.7	22.1	1.9
15		3.0	7.5	9.5	5.6	16.9	0.9	17.9	0.5
22.5		10.3	5.2	11.5	5.7	11.3	0.2	23.4	0
LSD	Lime	1.24				2.15			
(P<0.05)	N	1.33				1.90			
	P	NS				2.57			
	Lime x N	1.56				2.04			
	N x P	2.44				NS			
	Lime x P	NS				NS			
	N x P x Lime	NS				NS			

N rates: -N = 0 kg N ha^{-1}; +N = 30 kg N ha^{-1}.
NS – Not significant.

Table 5. Effect of liming, N and P fertilisation of nodule dry matter yield of a selected soybean cultivar

		Nodule dry matter yield (g 10 plants^{-1} nodules)							
		2002/2003				2003/2004			
P application rate (kg ha^{-1})		-Lime		+Lime		-Lime		+Lime	
		-N	+N	-N	+N	-N	+N	-N	+N
0		0.060	0.054	0.111	0.046	0.012	0	0.016	0.001
7.5		0.064	0.032	0.082	0.066	0.020	0.001	0.044	0.001
15		0.037	0.052	0.095	0.051	0.021	0.001	0.057	0
22.5		0.097	0.045	0.080	0.047	0.034	0.001	0.041	0
LSD	Lime	0.016				0.008			
(P<0.05)	N	0.014				0.005			
	P	NS				0.008			
	Lime x N	NS				NS			
	N x P	NS				0.011			
	Lime x N x P	NS				NS			

N rates: -N = 0 kg N ha^{-1}; +N = 30 kg N ha^{-1}.
NS – Not significant.

3.4. Effect of Soil Acidity of Grain and Stover Yield of Selected Cultivars

In the first year, Solitaire yielded significantly higher than other cultivars in unlimed soil but in the second season yields were similar (Figure 1). Liming significantly increased grain yield in first (P=0.046) and second (P=0.023) seasons. Lime x cultivar interaction was significant with liming responses ranging from 8.1% (Soliatire) to 19.1% (Safari) in the first season, and 5.9% (Storm) to 27.6% (Magoye) in the second season.

In both seasons, stover yield was significantly different across cultivars in acid soil (Figure 1). Safari produced the highest stover yield in acid (unlimed) soil in the first season, and in the second season stover yield for Magoye was significantly lower compared to the other three cultivars which were similar. Liming significantly increased yield in both years (P=0.008, 2002/03; P=0.036, 2003/04) and lime x cultivar interaction was not significant.

3.5. Effect of Liming, N and P Fertilization on Soybean Grain and Stover P Uptake

Liming, N and P rate had no effect on P uptake by grain in the first season (p<0.05). In the second season, liming increased by uptake of P by grain (P<0.001) by an overall average of 22%. Total P uptake increased with increasing P by an overall average of 19%, 23 % and 26% at 7.5, 15 and 22.5 kg P ha^{-1} respectively. Treatments in which lime and P fertilizer were applied together had higher grain P uptake compared to treatments where P fertilizer was applied without liming (Figure 2). Application of 30 kg N ha^{-1} reduced grain P uptake by an overall average of 14% in comparison to treatments where N was not applied (Figure 2).

In the first season, liming improved stover P uptake by an overall average 20%. Increasing P fertilizer rate had no effect on stover P uptake in treatments where lime had not been applied (Figure 3A and 3C). In the limed plots, the application of 7.5 kg P ha^{-1} did not improve uptake compared to the control, while application of 15 kg P ha^{-1} improved P uptake but a further increase had no effect (Figure 5B).

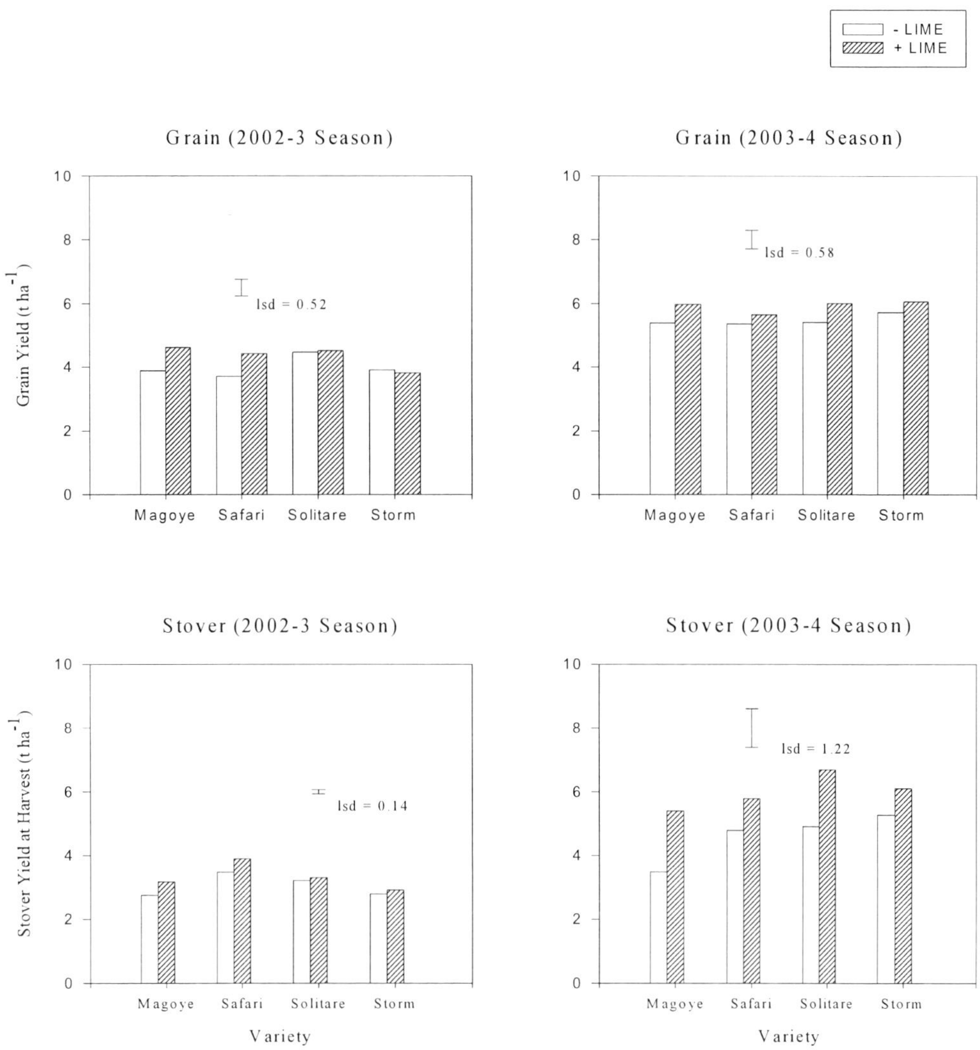

Figure 1. Effect of soil acidity and liming on grain and stover yield of selected soybean cultivars. (Error bars represent least significant difference of means at 95% confidence).

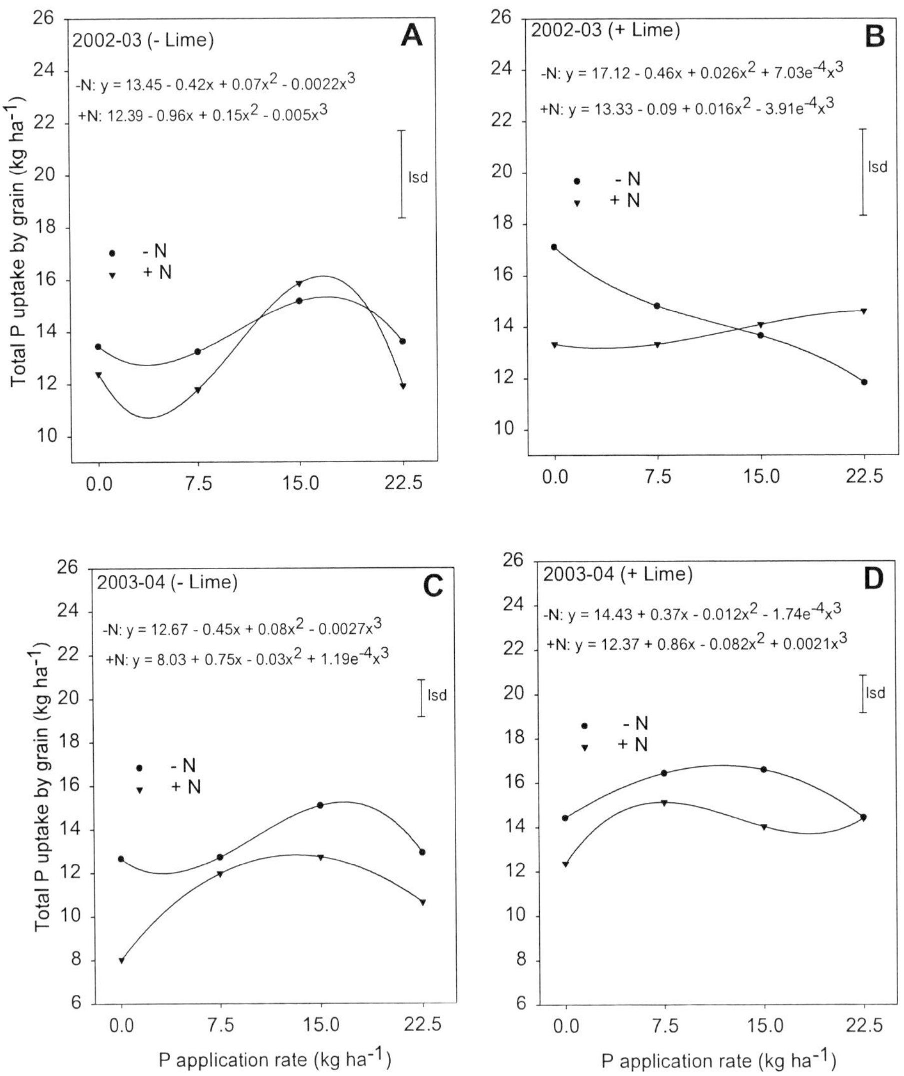

Figure 2. Effect of liming, N and P fertilization on P uptake by soybean grain. (Error bars represent least significant difference of means at 95%confidence).

The application of 30 kg N ha^{-1} generally reduced stover P uptake by an overall average of 15% compared to no N addition. In the second season liming improved (overall average 32%) stover P uptake in treatments where N fertilizer had not been applied, while there was no response to liming in treatments where N fertilizer had been applied (Figure 3 C and 3D).

3.6. Effect of Liming, N and P Fertilization on Grain and Stover Yield

Liming, N and P fertilizer rate had no significant effect on grain yield in the first season ($p>0.05$). In the second season liming also had no significant effect on grain yield ($p > 0.05$). However, increasing P rate from 0 to 7.5 kg P ha^{-1} increased grain yield for unlimed treatments which received N fertilizer but a further increase in P rate did not improve grain yield (Figure 4C). Without addition of N fertiliser, P addition had no effect on grain yield for the unlimed treatments in the same season (Figure 4C). In treatments where lime had been applied, increasing P fertilizer rate had no significant effect on grain yield (Figure 4D). Application of 30 kg N ha^{-1} reduced grain yield (average 13%) in the unlimed plots at 0 and 7.5 kg P ha^{-1} compared to treatments where N was not applied (Figure 4C). In the limed plots grain yield was higher in plots where N had not been applied but the differences were not significant ($p<0.05$).

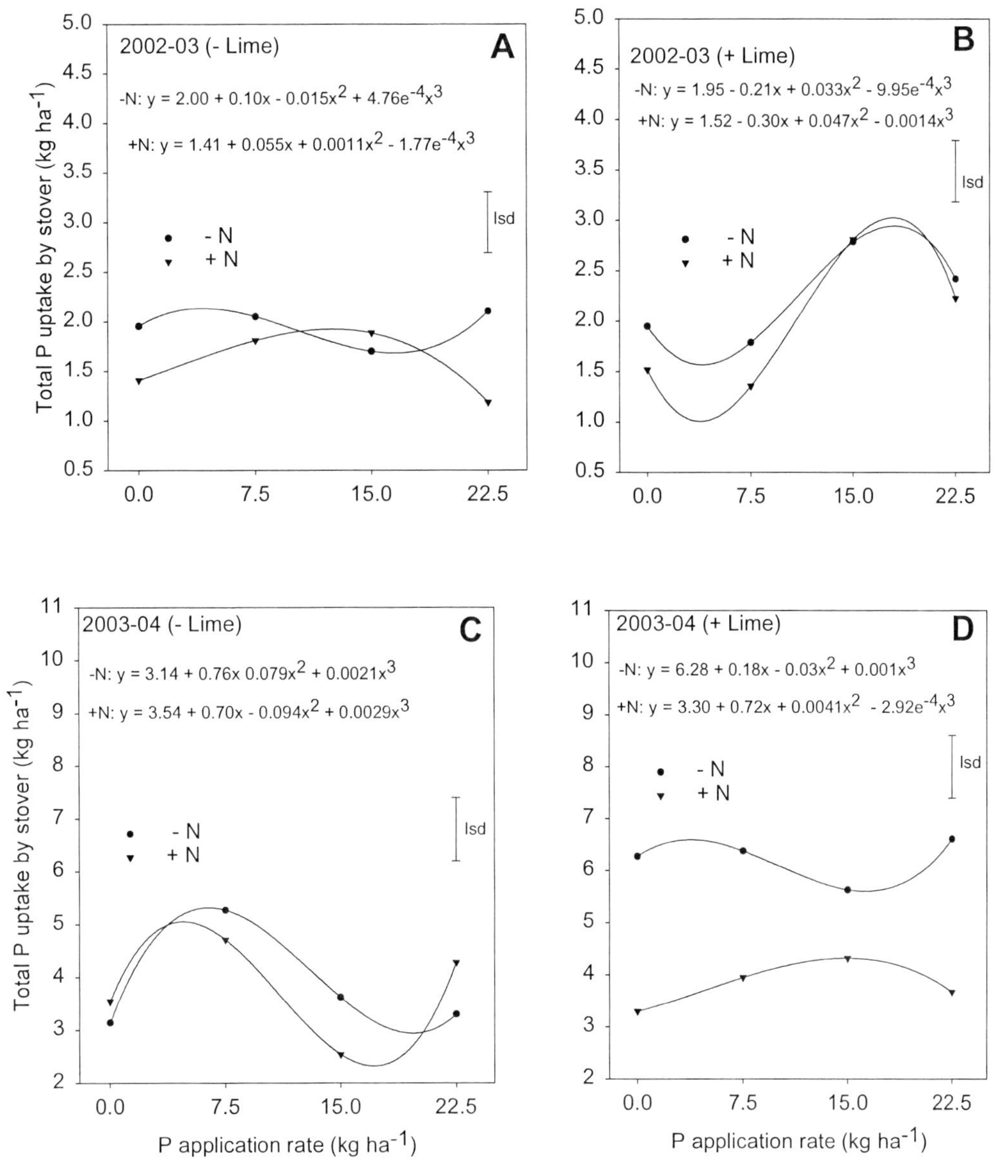

Figure 3. Effect of liming, N and P fertilization on P uptake by soybean stover at harvesting. (Error bars represent least significant difference of means at 95%confidence).

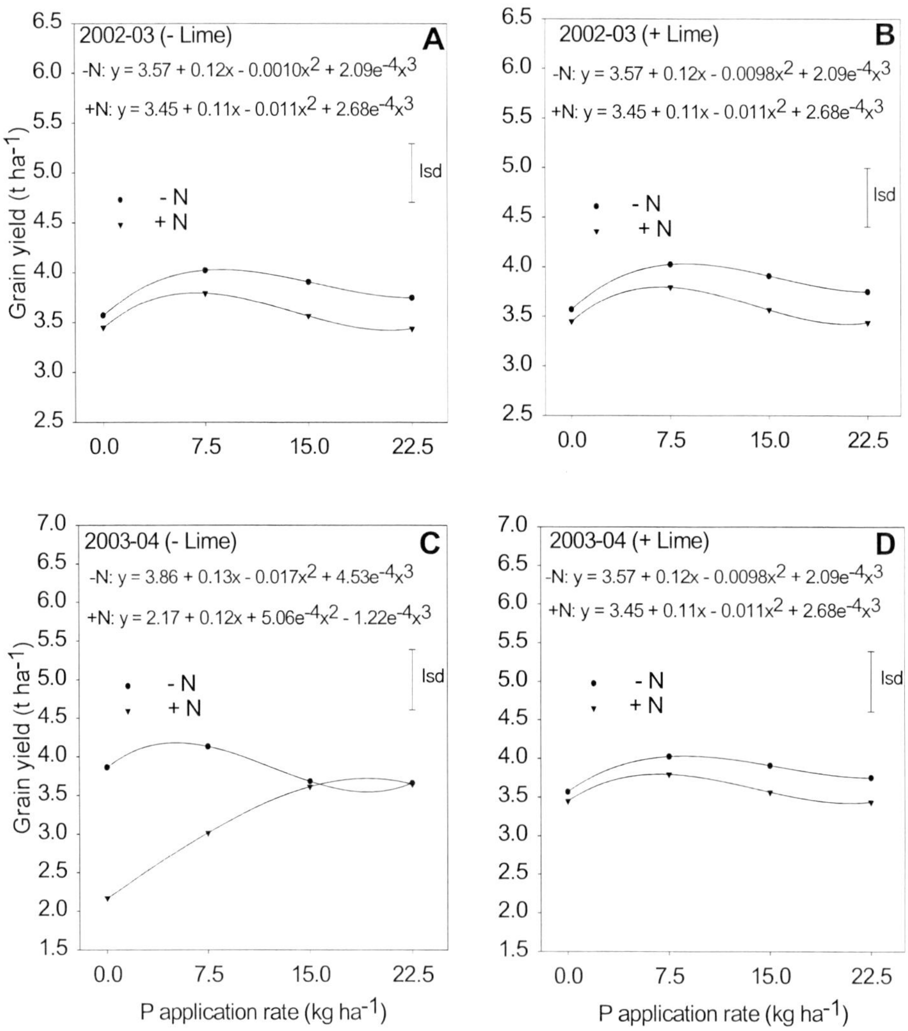

Figure 4. Effect of liming acid soil (B and D), and N and P fertilizer application on the grain yield of an acid-sensitive soybean cultivar. (Error bars represent least significant difference of means at 95% confidence).

Liming and increasing P rate had no significant effect on stover yield at harvesting in both cropping seasons ($p > 0.05$) (Figure 5). In the limed plots, application of 30 kg N ha^{-1} reduced stover yield (overall average 10% and 19% in first and second season, respectively) compared to treatments where no N was applied (Figure 5B and 5D). In the unlimed plots there was no significant difference between application of 0 and 30 kg N ha^{-1} (Figure 5A and 3C).

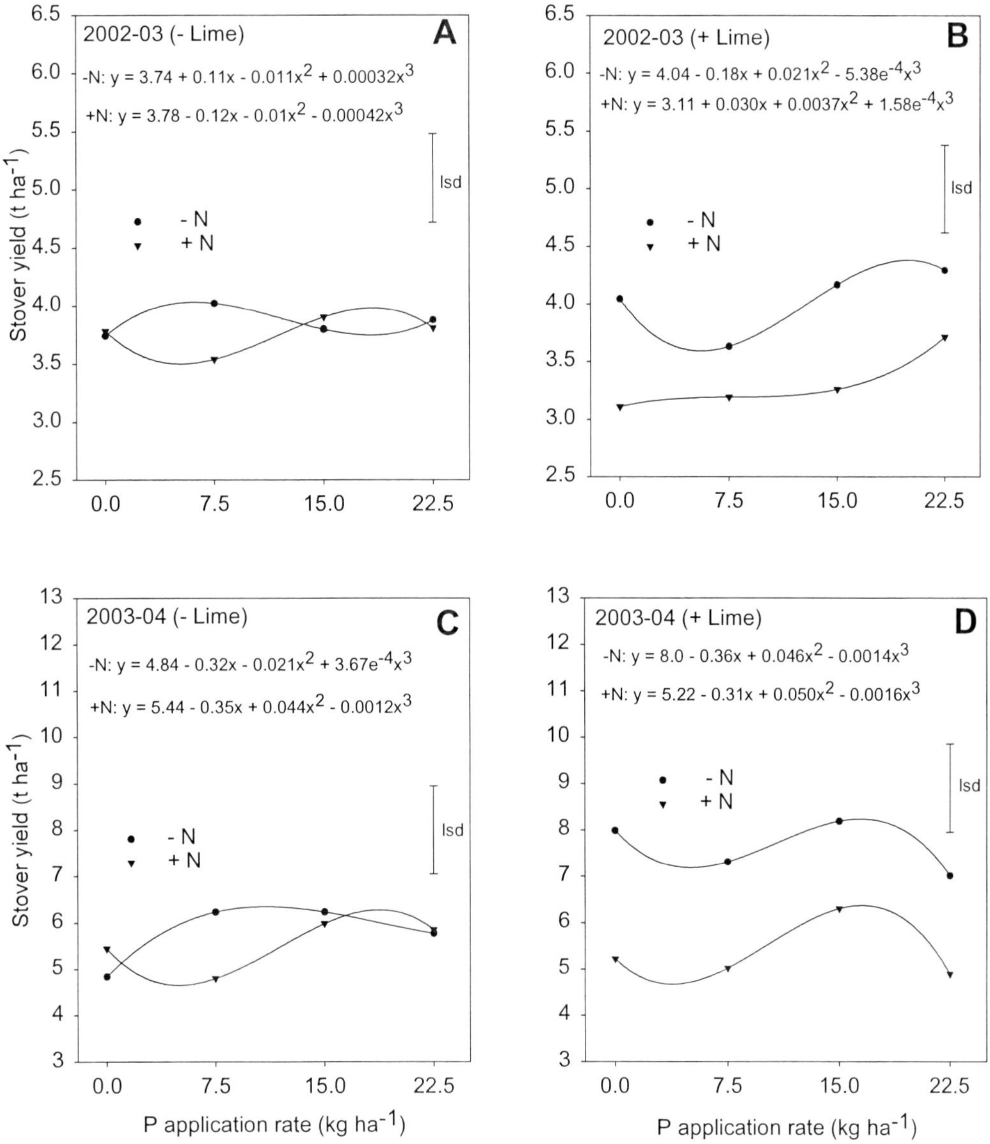

Figure 5. Effect of liming acid soil (B and D), and N and P fertilizer application on the yield of soybean stover. (Error bars represent least significant difference of means at 95%confidence).

4. Discussion

4.1. Effect of Liming and Mineral N Addition on Nodulation

Increase in nodule numbers and weights due to lime could be attributed to an increase in *Rhizobia* survival (Raychaunduri et al., 1997). Lime creates favorable conditions for survival and growth of root nodule bacteria by increasing pH (Watkin et al., 1997). The weight of nodules is directly related to the nitrogen fixing activity of the plant, while the number of nodules gives an indication of the conditions of the root infection by *Rhizobia*. Liming effects on nodulation are more pronounced in the second season after lime application (Raychaunduri et al., 1997) and in this study increases as high as 454% were observed in the second season (Table 2). The increase in nodule development could also be attributed to improved Ca supply through liming. Results from soil analysis showed that exchangeable Ca was higher (ave. 41%) in treatments where lime had been applied compared to treatments where lime had not been applied.

The pH of soil has a direct influence on the performance of the legume plant host, *Rhizobia*, and the legume-*Rhizobia* symbiosis (Richardson *et al*, 1988; Sprent and Sprent, 1990). Acidity affects root hair infection of the legume by *Rhizobia*, consequently reducing nodule formation, plant growth and yield (Keysar *et al*., 1979). In a study by Balatti *et al.* (1991), liming increased Ca levels in the soil which enhanced the root penetration of soybean into deeper soil layers and also induced the normal distribution of nodules on the tap root and lateral root by *Rhizobium*. Soil pH values below 5 are unfavourable for the legume-*Rhizobia* symbiosis leading to formation of ineffective nodules.

The lower nodulation response to liming by Safari in the first season compared to the other varieties in terms of nodulation implied that Safari was more tolerant to soil acidity than Magoye, Solitare and Storm (Tables 2 and 3). However, this tolerance was not shown in the second season implying that other factors could also be involved in the ability of Safari to tolerate soil acidity. Possibly the soil chemical environment (e.g. pH, Ca, micronutrients) had become more unfavourable in the unlimed soil after two years of soybean cropping (reduced nodulation) resulting in a wider gap when compared to limed plots. However, it is noteworthy that the cultivars that performed relatively well in acid soil in the first year (Safari and Magoye) had the highest response to liming, implying that farmers can chose these cultivars ahead of Solitare and Storm, whether they use lime or not. Although the effectiveness of nodules in Magoye was relatively low, it was still high (>95%) enough to ensure optimal N fixation.

Liming improved nodule number and NDMY in treatments where N had not been applied. This suggests that while the increase in soil pH due to liming enhanced nodule development, but N application restricted nodule development. Previous investigations have shown that high available N (inherent in soil or added) in the soil inhibits nodule development and nitrogenase activity, and accelerates nodule senescence in the short term (Jones, 1985). When a legume has two sources of nitrogen NO_3^- and N_2, it preferentially chooses NO_3^- and fixation is reduced. Thus adding N fertilizer reduces N_2 fixation. The plant utilizes the N already present in the soil instead of relying on the energy consuming symbiotic N_2 fixation process for its N requirements (Wolfgang and Martin, 1988). This implies that uncontrolled addition of mineral N (no soil testing) reduces the potential benefits that a farmer can derive from biological nitrogen fixation.

4.2. Effect of Liming on the Growth of Selected Soybean Cultivars

This study has clearly demonstrated the importance of applying lime to acid soil in order to optimized biological N fixation and yield of soybean. The response of Magoye to liming in terms of grain yield was highest and significant in both first (19%) and second (28%) seasons, thus out-performing the other three commercially available cultivars. Of the commercial cultivars, Safari significantly responded to liming in the first year, and in the second year all cultivars responded positively to liming (Figure 1).

Although nodulation in Solitare and Storm was affected by soil acidity in the first 8 WAP, the cultivars recovered and produced similar yields to Magoye and Safari. Since soil acidity only affected nodulation and thus N availability, these varieties could have performed well due to the inherent high N supplying power of the soil (Table 1) which had been under a grass fallow for at least five years. However, the cultivars may not do well in the majority of smallholder areas in sub Saharan Africa where the acid soils are also inherently low in N making N fixation essential. Mpepereki (1994) reported poor growth of commercial soybean cultivars in sandy acid soils in the smallholder areas of Zimbabwe and attributed it to inhibition of *Rhizobia* proliferation and host plant inhibition.

Magoye, a promiscuous nodulating cultivar bred to obviate the need of inoculation, has generally been classified as a higher stover and less grain yielding compared to more specific cultivars such as Solitaire, Storm and Safari (Tattersfield, 1996; Javaheri, 1981; Zengeni 2004). However, in this study stover yields at harvest for Magoye and Solitare were not significantly different in first season and in the second season stover yield for Magoye was actually less than that for Solitaire on both limed and unlimed plots (Figure 1).

Soybean grain P uptake results indicate that liming was essential in order to benefit from P fertilizer application when the soil is strongly acid. The results also showed that for the soil used in this study, the optimum P rate with liming would be 7.5-15 kg P ha^{-1}. Applying a higher P fertilizer rate will not have any benefit in terms of grain P uptake. In treatments where lime had not been applied increasing P rate had no effect on grain P uptake. This suggests that the applied P was not available for plant uptake. This could be attributed to P fixation under the acidic regime. In the limed plots increasing P application had no effect on dry matter yields and P uptake. This suggests that the increase in pH also improved the availability of native P in the soil.

Increasing P fertilizer rate improved stover yield but it had no effect on grain yield. However increasing P fertilizer application rate improved uptake by both stover and grain. This implies that the increase in stover yield and P uptake due to P fertilizer application did not enhance grain development. Soybeans are known to have relatively high P requirements (Mabika and Mariga, 1983). When growing under low P conditions, significant delays in the infection of the primary root have been observed, resulting in reduced yield benefit from biological nitrogen fixation (Mullen *et al.*, 1988).

5. Conclusions

Soybean productivity in acid soil prevalent in humid and sub-humid areas of SSA can be effectively increased through liming. Liming increases the availability of soil and applied P and also supplies Ca, and the latter is often low in these soils due to leaching. Liming increased stover relative to grain yield, an important aspect of soil fertility management if the stover is incorporated into the soil. Application of mineral N not supported by soil testing reduces the effectiveness of biological N fixation thereby depriving smallholder farmers of a cheaper source of N. This study did not confirm the notion that Magoye produces more stover and relatively less grain compared to specifically nodulating cultivars. Instead, the cultivar can be grown in both acid and limed soils achieving yields comparable to specifically nodulating cultivars. There is need for further studies to determine threshold N limits where mineral N has to be applied without causing the observed negative effects on N fixation and ultimately yield of soybean.

Acknowledgments

We thank the Rockefeller Foundation for funding this research through the Management of Soil Acidity Project (Grant no. 2001FS140).

References

Aibadoo, R.C. and van Kessel, C. N^{15} uptake, N_2-fixation and rhizobial interstain competition in soyabean and bean, intercropped with maize. *Soil Biol. Biochem.,* 1989, 21, 155-159.

Anderson, J.M. and Ingram, J.S.I. Tropical Soil Biology and Soil Fertility: *A Handbook of Methods.* C.A.B. International. Wallingford, UK, 1993.

Andrew, C.S. Legumes and acid soils. In: J. Dobereiner, R.H. Burris and A. Hollaender (Eds.) *Limitations and potentials of biological nitrogen fixation in the tropics.* Plenum Press, New York, 1978, 135-160.

Balatti, P. A., Krishnan, H. B. and Pueppke, S. G. Calcium regulates growth of *Rhizobium fredii* and its ability to nodulate soybean cv. Peking. *Can. J. Microbial,* 1991, 37, 542.

Copeland, K. *Soil acidity robbing Alabama farmers of $250 million each year.* Southeast Farm Press, 1976.

Dhliwayo D. K. C., Sithole T. and H. Nemasasi H. Soil acidity-is it a problem in maize-based production systems of the communal areas of Zimbabwe? In: Waddington, H. K. Murwira, J. D. T. Kumweda, D. Hikwa and F. Tagwira (Eds.) *Soil fertility research for maize-based farming systems in Zimbabwe.* Mutare, Zimbabwe, 1998, 217-221.

Giller, K.E. *Nitrogen fixation in Tropical Cropping Systems 2nd Edition.* CABI Publishing, Wallingford, 2001.

Graham, P.H. 1992. Stress tolerance in Rhizobium and Bradyrhizobium, and nodulation under adverse soil conditions. *Canadian Journal, Microbial.* 38:475-481.

Javaheri, F. *Release of four new soyabean varieties.* Mimeo, Government of Zambia. Lusaka, 1981.

Jones, J.B. *Acid soils reduce stand and yields.* Southeast Farm Press, 1975.

Jones, R.M. Effect of nitrogen and innoculum levels on establishment and nodulation of leucaena. Leucaena Research Reports, 1985, 6, 8-10.

Kasasa, P., Mpepereki, S., Giller, K.E. Nodulation and yield of promiscuous soybean (Glycine max (L.) Merr) varieties under field conditions. In: Waddington, S., Murwira, H.K., Kumwenda, J., Hikwa, D., Tagwira, F. (Eds.), *Soil fertility for maize-based farming systems in Malawi and Zimbabwe.* CIMMYT, Harare, Zimbabwe, 1998, 99-103.

Keyser H.H, Munns D.N and Hohanberg J.S. Acid tolerance of rhizobium in culture and in symbiosis with cowpea. *Soil Sci. Soc. Am. J.*, 1979, 43, 719-722.

Mugwira, L.M. and Haque, I. Screening forage and browse legume germplasm to nutrient stress. *J. Plant Nutr.*, 1993, 16, 17-35.

Mugwira, L.M. and Nyamangara, J. Organic carbon and plant nutrients in soils under maize in Chinamhora Communal Area. In: L Bergstrom and H. Kirchman (eds). Carbon and nutrient dynamics in natural and agricultural tropical systems. *C.A.B International, Wallingford,* UK, 1998, 15-21.

Mullen M.D, Isarael, D.Wand Wollum II. Effects of Bradyrhizobium japonicum and soyabean (*Glycine max (L) Merr*) phosphorus nutrition on nodulation and dinitrogen fixation. *Applied Environ. Microbiol.,* 1998, 42, 272 – 276.

Mpepereki S. Characterization Studies of Rhizobia in Some Zimbabwean Soils. *DPhil Thesis.* University of Zimbabwe, Harare, Zimbabwe, 1994.

Nyamangara J., Mugwira L.M. and Mpofu S.E. Soil fertility status in communal areas of Zimbabwe in relation to sustainable crop production. *J. Sust. Agric.,* 2000, 16, 15-29.

Nyamapfene, K. Soil of Zimbabwe. *Nehanda Publishers,* Harare, Zimbabwe, 1991.

O'Hara, G.W. and Glenn, A.R. The adaptive acid tolerance response in root nodule bacteria and Escherichia coli. *Archives Microbiol.* 1994, 16, 316-319.

Pandey, S., Ceballos, H., Magnavaca, R., Bahla Filha, A.F.C., Duque-Vargas, J. and Vinasco, L.E. Genetics of tolerance to soil acidity in tropical maize. *J.Crop Sci.*, 1994. 34: 1511-1514.

Richardson, A. E., Henderson, A. P., James, G. S. and Simpson, J. R. Consequences of soil acidity and the effect of lime on the nodulation of *Trifolium sunterraneum [L]* growing in acid soil. *Soil Biol. Biochem.,* 1988, 20, 439-445.

Roychaundauri, M, Kumar, K and Raychaundiri, S. Lime – rhizobium interaction on soyabean in Kanhaplohumolt of Manipur Hills. *Journal of the Indian Society of Soil Science,* 1997, 145, 739 – 742.

Sprent, J. I. And Sprent, P. *Nitrogen fixing Organisms – Pure and Applied Aspects.* Champman and Hall, London, 1990.

Tatterfield, J.R. Soyabean production and research in Zimbabwe. *Seed Co-op Company of Zimbabwe.* Harare, Zimbabwe, 1996.

Watkins, E., O'Hara, G. and Glenn, A. Calcium and acid stress interact to affect the growth of Rhizobium leguminousarum bv trifiolii. *Soil Biol. Biochem.,* 1997, 29: 1427 – 1432.

Whingwiri, E.E. Integrating soyabeans in smallholder cropping systems: lessons from Hurungwe District soyabean project 1986-1989. In: S. Mpepereki, K.E. Giller and Makonese, F. (Eds.) Soyabean in smallholder cropping systems of Zimbabwe: potential contribution of from BNF. *Soil Fertnet,* Harare, Zimbabwe, 1996.

Wolfgank, K. and Martin, P. Influence of nitrogen on the number of N2 fixing and total bacteria in the rhizosphere. *Soil Biol. Biochem.,* 1988, 20, 221-225.

Wynch, R.D. and Rains, D.W. Simultaneous measurements of nitrogen fixation estimated by the acetylene-ethylene assay and nitrate absorption by soyabeans. *Pl. Physiol.,* 1978, 62: 442 – 448.

Zengeni, R. Manure and soil properties effects on survival and persistence of soybean nodulating Rhizobia in smallholder field conditions of Zimbabwe. *Unpublished Master of Philosophy thesis,* University of Zimbabwe, Harare, Zimbabwe, 2004.

In: Soybean and Wheat Crops
Editors: S. Davies, G. Evans
ISBN: 978-1-60741-173-4

Chapter 6

NATURAL OCCURRENCE OF DEOXYNIVALENOL IN SOYBEAN GROWN IN SERBIA

***Biljana Abramović*[1,*] *and Igor Jajić*[2]**
[1]Faculty of Sciences, Department of Chemistry, Trg D. Obradovića 3, 21000 Novi Sad, Serbia
[2]Faculty of Agriculture, Trg D. Obradovića 8, 21000 Novi Sad, Serbia

ABSTRACT

Production of healthy food in sufficient amounts is, among the others, hindered by the activity of various plant pathogens, of which fungi have an especially negative impact. *Fusarium* species produce a broad spectrum of toxins including fumonisins, trichothecenes of the A- and B-type, and zearalenone. When contaminated plants are used as food and feed the toxins involved exhibit a lot of harmful effects on humans and animals. The occurrence of *Fusarium* toxins in cereals has been registered all over the world. However, in contrast to corn and wheat, no special attention has been paid to the study of *Fusarium* toxins in soybean. Because of that the aim of this study was to gain an insight into the presence of deoxynivalenol (DON) in soybean grown in Serbia, based on the analysis of 42 soybean and soybean meal samples collected during 2004–2007, as well as to compare the obtained results with data pertaining to a number of countries. Samples were analyzed by liquid chromatography on ODS Hypersil column with DAD detector. The DON content was above the limit of quantification (0.040 mg/kg) in 16.7%, with an average content in positive samples of 0.248 mg/kg (concentration range 0.10–0.45 mg DON/kg). However, none of the samples contained DON above the advisory level of 500 μg/kg of DON, passed by the European Union, which is not related strictly to soybean but to cereal products as consumed and other cereal products at retail stage.

Keywords: Deoxynivalenol, Soybean, LC, Serbia

* E-mail: abramovic@ih.ns.ac.yu

INTRODUCTION

Food safety continues to be an important issue in the whole world. Each year, a large number of crops are affected by fungal invasion, leading to considerable financial losses and impaired health in animals and humans. Toxicity is mainly caused by secondary metabolites of fungi, which are appropriately called mycotoxins [1]. Mycotoxin production can occur when fungi grow on crops in the field, at harvest, in storage or during the processing of feed, when conditions are favorable. No region of the word escapes these silent killers, and their negative impact on animal productivity and human health is enormous [2]. Mycotoxins are highly undesired substances that must not be present in food, a zero tolerance being an ideal. However, even the best agricultural, storage and processing practices cannot completely eliminate these contaminants, and hence it is impossible to achieve a truly mycotoxin-free food chain. Despite low consumer awareness of the problem, health risk related to mycotoxin ingestion has been quantified as exceeding risk from other food-related contaminants such as pesticides, additives, heavy metals, and microbial agents. For this reason mycotoxins have been called the 'hidden killers' [3]. According to the United Nation's Food and Agriculture Organization (FAO), approximately 25% of the world's grain supply is contaminated with mycotoxins. The economic loss due to mycotoxin contamination has been estimated to run into millions of dollars annually [4]. *Fusarium* fungi are among the most important agriculturally toxigenic fungi occurring in the moderate climatic zones of North America and Europe [5]. These species play an important role as plant pathogens, causing a wide range of diseases in a diversity of host plants such as vascular wilt, root and stem rots, cereal ear rot, head and pad blight, fruit rot, etc. Three of the most prevalent mycotoxins that occur in grain are deoxynivalenol (DON), fumonisins and zearalenone (ZEA) [6].

DON is a member of the trichothecene family of mycotoxins. Structurally, it is a polar organic compound which belongs to type B trichothecenes, and its chemical name is 12,13-epoxy-3α,7α,15-trihydroxytrichothec-9-ene-8-one (Figure 1). The occurrence of DON is associated primarily with *Fusarium graminearum* (teleomorph *Gibberella zeae*) and *Fusarium culmorum* (teleomorph unknown), both of which are important plant pathogens commonly found in cereals and other crops [7]. Although DON is among the least toxic of the trichothecenes, it is the most frequently detected one throughout the world, and its occurrence is considered to be an indicator of the possible presence of other, more toxic trichothecenes [8]. Consumption of contaminated feeds by livestock has been associated with a variety of adverse health effects including feed refusal (mainly by swine), reduced weight gain, diarrhea and emesis [9,10].

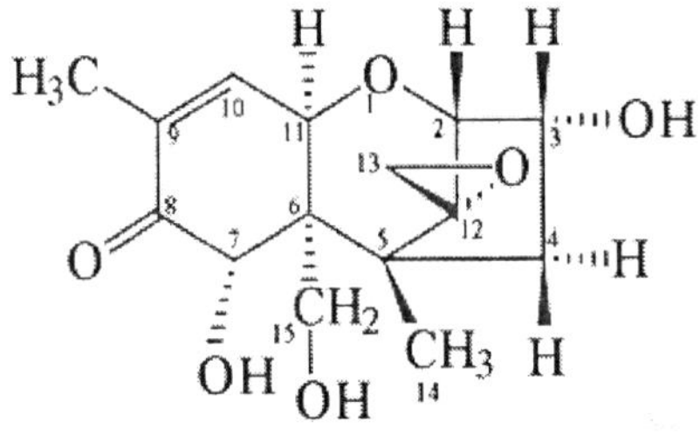

Figure 1. Chemical structure of deoxynivalenol (DON).

Soybean (*Glycine max*) represents an excellent source of high quality proteins; it has a low content of saturated fats, it contains a great amount of dietary fiber, and its isoflavone content makes it singular among other legumes [11]. Many researchers reported the benefits of legumes: chickpeas, beans, lentils and soy, among others. The characterization and positive health effects of soybeans have been studied recently and interest in this legume has increased because of its functional components. Most of the studies have been focused on soybean protein as a possible source of prevention against cardiovascular disease. This positive effect may be due to a decrease in serum cholesterol concentrations. In addition, there are many studies on isoflavones, non-nutritive substances, associated with prevention and treatment of different chronic diseases. Moreover, some studies have shown the health properties of soy dietary fiber. Therefore, it would be interesting to consider the replacement of animal-based foods for soybean foods in order to obtain some nutritional benefits. Soy products are also a good source of iron, and they contain vitamins B_1 and B_2 and an essential oil-linoleic acid, one of the omega-3 fatty acids [12].

In Serbia, a strong increase in the soybean areas that was registered in the eighties and beginning of nineties was followed by a significant drop to below 50,000 ha [13]. However, in the last several years, a rise in the soybean-sown areas has been observed, so that a record of 156,680 ha harvested area and the record yield of 429,639 metric tonas were registered in 2006 [14]. It is estimated that in Serbia in the year 2008 soybean was sown on about 145,057 ha [15], and the expected yield is 2.2-2.3 t/ha. Of the early-maturing soybean varieties prevail Afrodita and Proteinka, of middle-maturing Balkan and Ravnica, and of late-maturing varieties Vojvodjanka and Morava. Very early-maturing varieties Jelica, Krajina and Fortuna are sown as stubble crops [13].

Today soybean represents the main source of proteins and practically an unlimited component in the feedstuffs for all kinds of domestic animals. The proportion of soybean meal and other soybean products in pig feed in developed countries such as the USA, Japan and West Europe ranges from 14 to 25% and in poultry feed from 20 to 33%. In the West European countries, soybean meal satisfies 60% of the protein needs in livestock feeding [16].

In respect of the content of proteins and their biological value soybean represents the best high-quality protein feed, which can serve as the main and only source of proteins for feeding older pigs and poultry. The soybean amino acids composition is almost ideal and can satisfy the needs of domestic animals for essential amino acids. The only prerequisite is that the appropriate processing, involving not only oil extraction (sometimes oil is not extracted at all) but also the inactivation of harmful and inhibiting components that have an adverse effect on digestability and utilization of soybean proteins [16].

Mycotoxin contamination of soybeans has not been a significant problem as compared to commodities such as corn, cottonseed, peanuts, barley and other grains. In surveys conducted by the U.S. Department of Agriculture (USDA), 1046 soybean samples collected from different regions of the United States, including all grades and different crop years, were examined for aflatoxins. Aflatoxin was confirmed at low levels (7-14 μg/kg) in only two of the test samples analyzed. These findings, along with other observations, suggested that soybeans were not a good substrate for aflatoxin production. However, many of the test samples showed evidence of contamination with *Aspergillus flavus,* a main contributor to aflatoxin production under certain conditions. Therefore, the potential for aflatoxin formation during adverse storage conditions does exist [17]. Low contamination of soybean with aflatoxin registered Jacobsen *et al.* [18] and Beg *et al.* [19]. The former authors registered aflatoxin B_1 in hulls in only three of 24 samples with the largest content detected being 5.8 μg/kg, whereas the latter authors found only 0.20 μg/kg (range 0 – 1.27 μg/kg). However, the presence of aflatoxin in soybean in significantly higher concentrations was registered by Jajić *et al.* [20]. Also, it has been observed that 100% of the soy-based infant cereals (containing corn or rice as the probable sources of aflatoxin) had a detectable level of aflatoxin B_1 [21]. A survey of the pertinent literature shows that ochratoxin presence in soybean has not been identified frequently [18] or its content was relatively low 4.6 – 9.6 μg/kg [19] in contrast to soy-based product, probably because of the presence of other grains [22].

Relatively little information exist about the occurrence of *Fusarium* toxins in soybean and soybean meal. Thus the *in vitro* formation of HT-2 toxin, T-2 toxin, T-2 tetraol, neosolaniol and/or ZEA by *Fusarium* strains with soybeans as substrate was reported by Richardson *et al.* [23]. These authors suggested soybean products to present a mycotoxic hazard which warrants attention. In native beans as well as in some products used in agricultural practice trichothecenes and ZEA were detected by Jacobsen *et al.* [18], Jajić *et al.* [20], Clear *et al.* [24], Rafai *et al.* [25], Soković and Šimpraga [26] and Schollenberger *et al.* [27]. The occurrence of *Fusarium* toxins in soy-based foodstuffs was studied by Lombaert *et al.* [22], Scott [28] and Schollenberger *et al.* [29].

The legislations of some ten countries prescribe explicitly the maxumal allowed content of aflatoxins B_1, that is B_1, B_2, G_1 and G_2 for soybean and soybean products used as food and feed, whereas these quantities for ZEA and ergot are defined only in two countries each [30]. Advisory levels of 500 μg/kg of DON were passed by the European Union [31] for bread, cakes, biscuits, cereal-based snacks and breakfast cereals.

In view of all the above the aim of this study was to gain an insight into the presence of DON in soybean grown in Serbia during 2004–2007.

Experimental

Chemicals

All solvents used for the DON extraction from soybean and soybean meal samples, as well as for the mobile phase preparation, were of HPLC grade. All chemicals used in the investigation were of reagent grade. Solutions were prepared in doubly deionized water except when stated otherwise.

DON standard solutions. DON (Biopure, Tulln, Austria) was purchased as an analytical standard. Calibrant solution was prepared in ethyl acetate–methanol (19:1, v/v) at the concentration of 0.1–0.2 mg/ml from crystalline substance according to AOAC method 986.17. Stock solution was prepared by measuring 1.00 ml of calibrant solution of DON into a 5 or 10 ml volumetric flask and diluting to volume with ethyl acetate–methanol (19:1, v/v). Working calibrant solutions were prepared by evaporating the appropriate volume of the stock solution and diluting with 1.00 ml of methanol. Standard solutions were stored at 4 ^{0}C.

Samples

Samples of soybean from the 2004–2007 harvests were collected from different locations in the Republic of Serbia. Samples of the 2004 harvest were taken from barns, i.e., storages, where they had been kept for one year, while those of the 2005–2007 harvests were taken directly off fields immediately after the harvest. Soybean meal samples, originated from the same harvests, were taken from several oil factories after deoling by extraction. After sampling, 1000 g of each sample were ground in a laboratory mill in such a way that > 93% passed through a sieve with pore diameter of 0.8 mm. After that, the sample was homogenized by mixing. Samples thus prepared were packed in plastic bags and stored in a freezer at –20 °C until the analysis. Prior to each analysis, the samples were allowed to attain room temperature.

Extraction and Clean-up

Activated charcoal–alumina–Celite–cation exchange resin column was used for purification; 25.0 g of the sample were extracted with 100 ml of acetonitrile–water (84:16, v/v) and shaken on a magnetic stirrer for 60 minutes. After filtration through Advantec filter paper, 6.0 ml of the extract were applied to the prepared column. The column was then washed with 5 ml of the solvent mixture comprising of acetonitrile–water (84:16, v/v) at about 0.6 ml/min. The cleaned-up extract was evaporated to dryness, dissolved in 3.0 ml of ethyl acetate and quantitatively transferred to an evaporation vessel by triple washing with 1.5 ml ethyl acetate. The eluate was evaporated just to dryness.

Activated Charcoal–alumina–Celite–cation Exchange Resin (CACC) Column

The column was prepared in the following way (32): a plug of glass wool was inserted into the tapered end of a glass tube (9 cm x 1.5 cm i.d.), then 0.1 g of Celite, 1.5 g of activated charcoal, alumina, and Celite mixture (7:5:3) were added, loosely packed, and tapped to level. Two grams of prewashed cation exchange resin were added and lightly compacted above activated charcoal–alumina–Celite by pushing down a second glass wool plug.

Liquid Chromatographic Analysis

The equipment consisted of an HP 1090 Liquid Chromatograph (Hewlett Packard, Palo Alto, CA, USA) with a DAD detector (Hewlett Packard, Palo Alto, CA, USA) and a Hypersil ODS column (100 x 4.6 mm i.d., particle size 5 μm, Agilent Technologies, USA). DON analysis was performed after evaporation, the residue was redissolved in 300 μl of methanol, and a 15 μl aliquot of the solution was injected into the LC system. The mobile phase consisting of a mixture of acetonitrile–water (16:84, v/v) was used at a flow rate of 0.6 ml/min. UV detection was performed at 220 nm. The mobile phase was filtered through a 0.45 μm membrane (Aura industries, TFM, Hewlett Packard).

Analytical Quality Control

Calibration curves used for the quantitative determination were constructed on the basis of the area under the DON chromatographic peaks, using seven DON working standard solutions. The linearity of the method was assessed by standards ranging from 0.17–3.40 ng/μl. The correlation coefficient was 0.999. The limit of quantification (LOQ) for LC determination based on a signal-to-noise ratio of 10:1 was 0.20 ng/μl of DON, which is equivalent to 0.040 μg/g of DON in substrate. Recovery studies were performed on the blank samples of soybean meal spiked at the levels of 0.70 and 1.10 mg/kg of DON. The recovery in the former case was 89.7% (SD = 1.4) and 93.3% (SD = 1.6) in the latter. The results for the samples were not corrected for the recovery of the spike.

RESULTS

Many analytical procedures have been developed for the determination of DON in feed and food [33]. In this work the determination of DON in the collected soybean and soybean meal samples was carried out by liquid chromatography under previously determined optimal experimental conditions [32]. Figure 2 shows the chromatograms obtained for soybean meal sample. As can be seen, the procedure applied for the cleaning-up of the raw soybean meal extract and conditions used in the recording of chromatograms were satisfactory, i.e. the DON peak is well separated from those for the matrix.

In the present study a total of 42 soybean and soybean meal samples were analyzed for DON (Table 1).

Table 1. Occurrence of DON in soybean and soybean meal in Serbia from the 2004–2007 harvests

Year	No. of samples	No. of positive samples (%)	Concentration in samples (mg/kg)		
			Average	Range	Median
2004	13	1 (7.7)	0.109	0.109	0.109
2005	11	1 (9.1)	0.100	0.100	0.100
2006	12	4 (30.0)	0.346	0.162-0.450	0.385
2007	6	1 (16.7)	0.144	0.144	0.144
	42	7 (16.7)	0.248	0.100-0.450	0.162

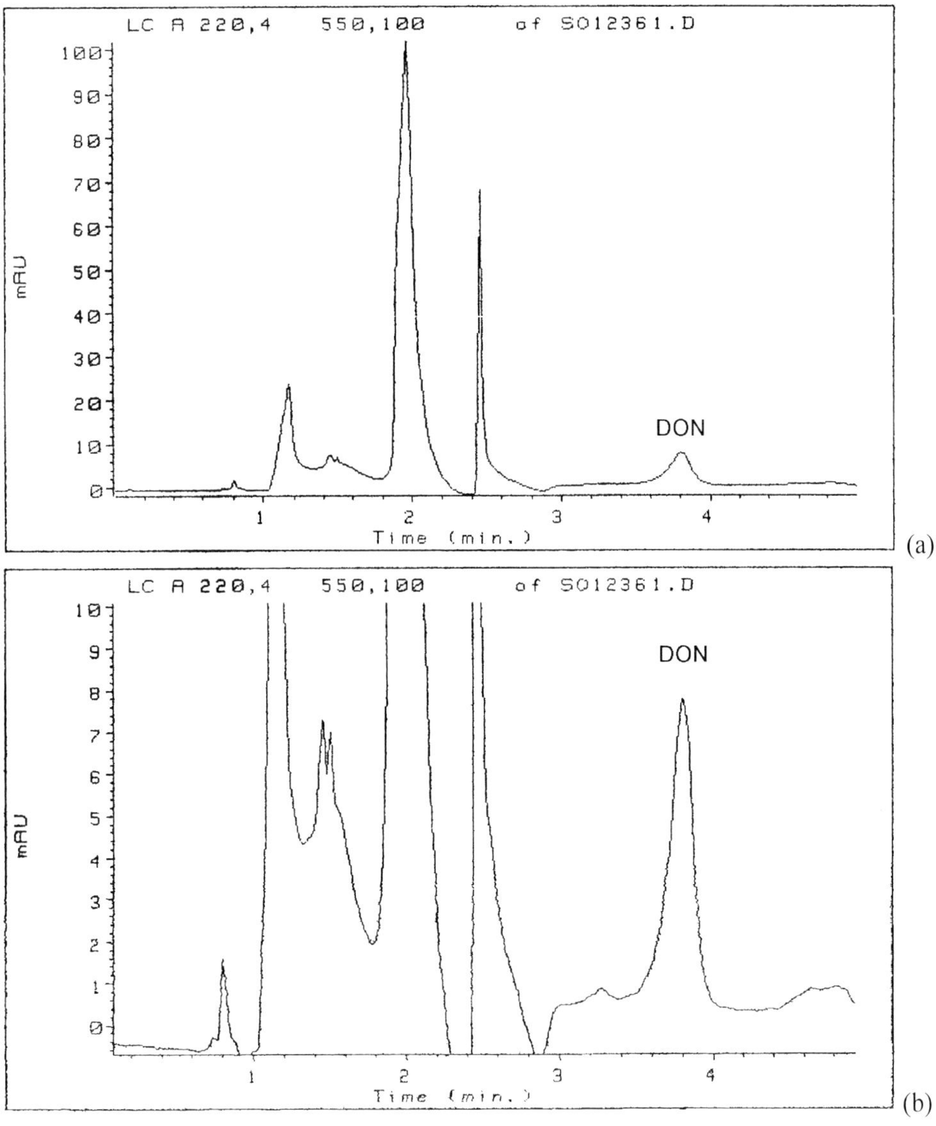

Figure 2. Chromatograms (a, b) of soybean meal sample containing 0.45 mg/kg of DON. Chromatogram b was recorded at ten times higher sensitivity.

Discussion

As can be seen from Table 1, the total number of contaminated samples, i.e. the samples in which DON content was above the limit of quantification (0.040 mg/kg) was 7 or 16.7%, the percentage of contaminated samples being highest for the 2006 harvest (4 out of 12). Average content in positive samples was 0.248 mg/kg (concentration range 0.10–0.45 mg DON/kg), the DON content being highest again in the 2006 samples. This can be explained by the climatic conditions of that year, which varied from the markedly wet spring months via drier beginning of the summer, to the extremely rainy August. The beginning of the autumn, that is the beginning of the harvest, was characterized by very hot weather and a very small amount of rain. Such variegated weather conditions did certainly influence the soybean resistance to *Fusarium* fungi and thus the occurrence of DON. However, none of the samples contained DON above the advisory level of 500 µg/kg of DON, passed by the European Union [31], which is not related strictly to soybean but to cereal products as consumed and other cereal products at retail stage.

The results obtained in this study were compared with data pertaining to a number of countries of the world, including EU, and especially with those of the neighboring countries. Unfortunately, as has already been said in Introduction, studies dealing with this issue are rather scarce. Of the neighboring countries there are only data for Croatia and Hungary.

In Hungary, Rafai *et al.* [25] analyzed DON content among the others in 119 soybean samples harvests 1991–1998 and found that 65.5% of the analyzed samples were contaminated with DON at an average level of 252.7 µg/kg. On the basis of these findings the authors concluded that soybean may be a good substrate for trichothecene producing fungi and the rate of contamination was considerable.

In Croatia, Sokolovic and Simpraga [26] analyzed crop samples of the 2001–2004 harvests, mentioning also soybean, but giving neither the number of its samples nor DON contamination rate. In their conclusion the authors pointed out that because of the rather small number of samples the data obtained could not be interpreted as the actual situation in the field conditions.

As far as we know, of the other European countries there are no data about DON occurrence in soybean except for that grown in Germany. Thus, out of 13 samples of soya meal Schollenberger *et al.* [27] found that 7 samples were contaminated with DON at a mean level of 64 µg/kg and maximal content of 237 µg/kg. The same group of researchers [29] analyzed also for *Fusarium* toxins a total of 45 samples of soy food including whole beans, roasted soy nuts, flour and flakes, textured soy protein, tofu, proteinisolate including infant formulas and fermented products (soy sauce), randomly collected in food and health food stores. Of the samples analyzed DON contamination was found in crisp, textured products, soy flour and soy sauce in a range of 11-260 µg/kg.

As has been noticed in Introduction, *Fusarium* fungi are among the most important agriculturally toxigenic fungi occurring in the moderate climatic zones of Europe and North America. Thus Lombaert *et al.* [22] investigated soy-based cereals from Canadian retail market for the presence of DON, and by analyzing 8 samples of soy-based infant cereals (which usually contain corn) they found the presence of DON in all samples, with a mean of 116 µg/kg and maximal content of 240 µg/kg. It may be pointed out that the DON content found in the analyzed samples was not very high as it was below the advisory level of 500 µg/kg of DON passed by the European Union, which is not related strictly to soybean but to cereal products as consumed and other cereal products at retail stage.

Further, Clear *et al.* [24], found that Canadian soybean seed discoloration was, among the others, a consequence of the presence of *Fusarium* toxins, identifying in the reddish discolored seeds contained DON and HT-2 toxin.

Scott [28] in its own multi-year monitoring of Canadian grains and grain-based foods for trichothecenes and zearalenone found that ZEA was rarely present in soybean.

Besides, Jacobsen *et al.* [18] investigated among the others the presence of DON in whole soybeans grown in the USA, as well as in the hulls, meal and oils. DON was detected in 16 of 24 whole soybean samples and in 15 of 17 hull samples, 14 of 17 meal samples, and one of three oil samples analyzed. Concentrations ranged from not detectable to 490 μg/kg in whole soybeans, from > 10 to 420 μg/kg in hulls, from < 5 to 600 μg/kg in meal, and from not detectable to 30 μg/kg in oil.

This survey shows that DON may be a significant contaminant of soybean and soybean products despite of the literature reports that contradict this conclusion.

Conclusion

On the basis of the findings presented it can be concluded that the overall DON contamination of samples of soybean and soybean meal (16.7%) was lower than the previously reported data. However, none of the samples contained DON above the advisory level of 500 μg/kg of DON, passed by the European Union [31], which is not related strictly to soybean but to cereal products as consumed and other cereal products at retail stage. The highest DON contamination was found for the 2006 harvest (30%), which can be explained by the specific climatic conditions of that year.

Acknowledgment

The work was financed by the Ministry of Science and Environmental Protection, Republic of Serbia (Project No ON142029).

References

[1] Grain Inspection, Packers and Stockyards Administration (GIPSA) Grain fungal diseases & mycotoxin reference. US Department of Agriculture (USDA), Kansas City, 2002.

[2] Devegowda, G.; Raju, M. V. L. N.; Swamy, H. V. L. N. *Feed Compounder* 1998, *18*, 22-27.

[3] Galvano, F.; Ritieni, A.; De Lorenzo, A.; Piva, G.; Pietri, A. (2006). Mycotoxins in the human food chain: what risks for the consumer? http://www.engormix.com/mycotoxins_in_the_human_e_articles_291_MYC.htm

[4] Devegowda, G. *Feeding Times* 2002, *7*, 2-3.

[5] Kos, G.; Lohninger, H.; Krska, R. *Anal. Chem.* 2003, *75*, 1211-1217.

[6] Schaafsma, A. W.; Nicol, R. W.; Savard, M. E.; Sinha, R. C.; Reid, L. M.; Rottinghaus, G. *Mycopathologia* 1998, *142*, 107-113.

[7] JECFA (Joint FAO/WHO Expert Committee on Food Additives), Fifty sixth meeting, February 2001. http://www.inchem.org/documents/jecfa/jecmono/v47je05.htm

[8] Lombaert, G. A. In: *Mycotoxins and Food Safety*; DeVries, J. W., Trucksess, M. W., Jackson, L. S. (Eds.), Kluwer academic/plenum publishers: New York, NY, 2002; pp. 141-153.

[9] Krska, R.; Baumgartner, S.; Josephs, R. *Fresenius J. Anal. Chem.* 2001, *371,* 285-299.
[10] Kuiper-Goodman, T. *Recent developments in the risk assessment of deoxynivalenol.* In session 2: Toxicology, quality and impact on industry. Second Canadian workshop on Fusarium head blight: Ottawa, CANADA, 2002. http://res2.agr.gc.ca/ecorc/fusarium01/session2.pdf.
[11] Mateos-Aparicio, I.; Redondo Cuenca, A.; Villanueva-Suárez, M. J.; Zapata-Revilla, M. A. *Nutricion Hospitalaria* 2008, *23,* 305-312.
[12] USDA, Agricultural research service, Nutritient data laboratory. http://www.nal.usda.gov/fnic/foodcomp/cgi-bin/list_nut_edit.pl
[13] Annon, *Soybean – Growing Handbook*; Institute of Field Crops and Vegetables: Novi Sad; Sojaprotein: Bečej; SERBIA, pp 1-2. http://www.victoriagroup.co.yu/pdf/p3.pdf (in Serbian)
[14] Statistical Office of the Republic of Serbia. Statistical Yearbook of Serbia, Belgrade. (2007). http://webrzs.statserb.sr.gov.yu/axd/en/god.htm
[15] Saopštenje PO13, Statistika poljoprivrede, Republički zavod za statistiku, Republika Srbija, SRB 164, 190608, http://webrzs.stat.gov.rs/axd/dokumenti/saopstenja/PO13/po13122008.pdf
[16] Kovčin, S.; Pejić, N.; Živković, S. *Soybean products in the nutrition of domestic animals;* Faculty of Agriculture: Novi Sad; SERBIA, 1988; pp 7-8. (in Serbian)
[17] Nesheim, S.; Wood, G. E. *J. Am. Oil Chem. Soc.* 1995, *72,* 1421-1423.
[18] Jacobsen, B. J.; Harlin, K. S.; Swanson, S. P.; Lambert, R. J.; Beasley, V. R.; Sinclair, J. B.; Wei, L. S. *Plant Dis.* 1995, *79,* 86-88.
[19] Beg, M. U.; Al-Mutairi, M.; Beg, K. R.; Al-Mazeedi, H. M.; Ali, L. N.; Saeed, T. *Arch. Environ. Contam. Toxicol.* 2006, *50,* 594–602.
[20] Jajić, I.; Terzić, V.; Jurić V.; Radanov-Pelagić, V. *IX Simpozium on Feed Technology with International Participation* Zlatibor, 2001, 194-200. (in Serbian)
[21] Tam, J.; Mankotia, M.; Mably, M.; Pantazopoulos, P.; Neil, R. J.; Calway, P.; Scott, P. M. *Food Addit. Contam.* 2006, *23,* 693–699
[22] Lombaert, G. A.; Pellaers, P.; Roscoe, V.; Mankotia, M.; Neil, R.; Scott, P. M. *Food Addit. Contam.* 2003, *20,* 494–504.
[23] Richardson, K. E.; Hagler, W. M.; Haney, C. A.; Hamilton, P. B. *J. Food Protec.* 1985, *48*, 240–243.
[24] Clear, R. M.; Nowocki, T. W.; Daun, J. K. *Can. J. Plant Pathol.* 1989, *11*, 308–312.
[25] Rafai, P.; Bata, Á.; Jakab, L.; Ványi, A. *Food Add. Contam.* 2000, *17,* 799–808.
[26] Sokolović, M.; Šimpraga, B. *Food Control,* 2006, *17,* 733–740.
[27] Schollenberger, M.; Müller, H.-M.; Rüfle, M.; Suchy, S.; Plank, S.; Drochner, W. *Mycopathologia* 2006, *161,* 43–52.
[28] Scott, P. M. *Food Addit. Contam.* 1997, *14*, 333–339.
[29] Schollenberger, M.; Müller, H.-M.; Rüfle, M.; Terry-Jara, H.; Suchy, S.; Plank, S.; Drochner, W. *Int. J. Food Microbiol.* 2007, *113*, 142–146.
[30] FAO Corporate document repository. Worldwide regulations for mycotoxins in food and feed in 2003 http://www.fao.org/docrep/007/y5499e/y5499e00.HTM
[31] Commission Regulation (EC) 1126/2007. (2007). Official Journal of the European Union L 255/14. http://eur-lex.europa.eu/LexUriServ/LexUriServ.do?uri=OJ:L:2007:255:0014:0017:EN:PDF

[32] Abramović, B.; Jajić, I.; Jurić, V.; Gaál, F. F. *J. Serb. Chem. Soc.* 2005, *70*, 1005-1013.
[33] Abramović, B.; Jajić, I.; Abramović, B.; Ćosić, J.; Jurić, V. *Acta Chim. Slov.* 2007, *54*, 859-867.

In: Soybean and Wheat Crops
Editors: S. Davies, G. Evans
ISBN: 978-1-60741-173-4

Chapter 7

STEM BORING OF SOYBEAN BY *DECTES TEXANUS* (COLEOPTERA: CERAMBYCIDAE) AND THE NATURE OF ITS IMPACT ON YIELD

J. P. Michaud, J. A. Qureshi*, A. K. Grant, and J. L. Jyoti
Kansas State University, Department of Entomology, Agricultural Research Center-Hays, Hays, Kansas, 67601, USA

ABSTRACT

The yield of Roundup Ready® soybean plants infested by larvae of *Dectes texanus* LeConte was compared to that of uninfested plants obtained from the same fields in two successive years in west-central Kansas. Plants were significantly larger in all respects in 2006 than in 2005, due largely to a lower plant population, but there was no reduction in either pod number or total seed weight as a consequence of larval boring in either year. Infested plants had marginally greater total seed weights than uninfested plants in 2005, an effect attributed to females preferring larger plants for oviposition when average plant size was small. Mean stem diameters were not notably different between infested and uninfested plants as observed in certain earlier studies, suggesting that *D. texanus* populations may now be exploiting smaller size classes of soybean plants to a greater extent. Although our results suggest no impact of larval boring on yield, the possibility of some yield impact under different growing conditions cannot be ruled out. However, any yield losses arising from *D. texanus* larval boring are likely small compared to those arising from pre-harvest lodging that results when mature larvae girdle plants prior to overwintering.

Keywords *Glycines max*, larval boring, yield loss

---

* Current address: University of Florida, Southwest Florida Research and Education Center, 2686 SR 29, Immokalee, FL, 34142.

INTRODUCTION

The longhorned beetle *Dectes texanus* LeConte, a.k.a. the soybean stem borer, has been a pest of soybeans throughout many eastern and midwestern states for more than forty years (Falter 1969; Patrick, 1973; Laster et al. 1981). Cultivated sunflowers are a preferred host plant (Michaud and Grant 2005) but *D. texanus* populations likely move between the two crops, depending on their availability. These beetles are monovoltine, overwintering as late instar larvae and pupating in spring. Adult females lay their eggs in leaf petioles in early summer and the eclosing larvae tunnel down into the main stalk of the plant where they feed on the pith core (Hatchett et al. 1975). As the plant matures and dries down, larvae cease feeding and prepare overwintering chambers in the base of plants at, or just above, soil level. Once a cavity is hollowed out, the larva girdles the stalk above the chamber and then seals itself off below by plugging the entrance with shredded fibers. It is this final act of cutting off the plant that causes economic losses as girdled plants break off at the slightest pressure and will often lodge under the force of wind alone. When harvesting girdled plants, farmers must reduce the speed of the combine substantially to avoid additional yield losses.

This pest has become increasingly problematic with the advent of no-till soybean production that likely favors the survival of overwintering larvae compared to regimes of tillage (Campbell and van Duyn 1977). Although yield losses due to the lodging of girdled plants are obvious, it is less clear whether plant productivity may be negatively impacted by larval boring during the course of the growing season. It is important to note that larval boring kills only the leaf attached to the petiole on which oviposition occurs; plants are not killed and stand establishment is never impacted, making it possible to measure yield impact at the level of individual plants. Various workers have suggested or inferred that larval boring may reduce yields by around 10% (Richardson 1975; Campbell 1980; Patrick et al. undated). However, trials that demonstrate yield increases in response to foliar or soil drench insecticide treatments directed at *D. texanus* (e.g. Buschmann and Sloderbeck 2005) are not necessarily reliable indicators of direct impact because collateral control of other potentially yield-limiting insects could be responsible for any portion of the observed effects. For example, Michaud et al. (2007a) found no impact of *D. texanus* larval boring in sunflower when they quantified yield at the level of individual plants in two successive growing seasons. Nevertheless, they observed a significant yield increase in one year in response to both foliar and soil drench insecticides that they attributed to control of other foliage- and root-feeding insects. Laster et al. (1981) managed to reduce *D. texanus* infestation of soybeans in Mississippi with eight treatments of methyl parathion, but treated plots did not yield more than untreated plots in either of two locations. Andrews and Williams (1988) conducted perhaps the most detailed study of *D. texanus* impact on soybean to date and found that infested plants yielded more than uninfested plants, ostensibly due to adult females avoiding plants in smaller size classes. In this study we examined the yield of soybean plants both with and without *D. texanus* larval infestation in two successive years in west-central Kansas.

Materials And Methods

On 29 September 2005 we sampled a rainfed commercial field of mature soybeans, cultivar Pioneer 9344 'Roundup Ready®', 10 miles south of Hays, Kansas that we had previously confirmed was infested with *D. texanus*. The crop was fully mature and the plant population was estimated at around 280,000 plants/ha. Individual plants were removed by snipping them off at ground level with a strong set of pruning shears. The stalk of each plant was then snapped open in several places to determine presence or absence of larval tunneling and the entire plant was then placed into a paper bag and labeled with a letter and number code. Plants were selected at random until a minimum of 100 plants of each type (infested / uninfested) were collected. Since uninfested plants were more scarce than infested ones, once 104 infested plants were obtained, all further infested plants were discarded until a similar number of uninfested plants had been obtained.

After harvest, the bags containing plants were placed in a series of cardboard boxes and dried at 40 °C for a 7-10 d in a large drying room. Boxes were then transported to the laboratory where each plant was carefully removed from its bag and placed on a large sorting tray. Since considerable abscission of leaf petioles had already occurred, the remaining petioles and leaves were stripped from plants to leave only stems and pods. The stalk diameter of each plant was estimated at the soil line by taking a maximum and minimum measurement with a digital caliper and averaging the two values. All pods containing filled seed were removed from the plant, counted, and then shelled. The stalks of infested plants were split lengthwise and the length of larval tunnels measured with a flexible tape measure. Stalks and seeds where then weighed separately on an analytical balance (Mettler model PL202-5).

On 23 September 2006 we sampled a 0.8 ha. field of Asgrow Roundup Ready® soybeans on the experimental farm at K-State Agricultural Research Center-Hays, in Hays, Kansas. The field was rain fed and plant population was estimated at around 165,000 plants/ha. Plants were fully mature and identical procedures were followed as in 2005.

The data were analyzed using PROC ANOVA (SAS Institute 2003) with 'year' and 'infestation' as fixed factors. Linear regression was then used to establish relationships among components of yield in both infested and uninfested plants.

Results

The data for all components of yield are reported for both years in Table 1. The effect of year was significant for all four dependent variables ($P < 0.001$ in all cases), with all values significantly larger in the second year. There were no significant interactions between year and infestation ($P > 0.05$ in all cases).

In 2005, infested plants tended to be slightly larger than uninfested plants with marginally larger stem diameters and greater seed weights, although differences in stem weight and pod number were not significant (Table 1). Given these apparent differences in plant size, we divided seed weight by stem weight to generate an index of plant productivity that would be independent of plant size and compared infested and uninfested plants again, but found no difference (F = 0.890; df = 1,205; P = 0.347). In 2006, there were no significant differences between infested and uninfested plants for any component of yield.

Among uninfested plants in 2005, stem diameter was linearly correlated with stem weight (F = 31.08; df = 101; P < 0.001; r^2 = 0.235), with pod number (F = 27.96; df = 101; P < 0.001; r^2 = 0.217) and with total seed weight (F = 49.63; df = 101; P < 0.001; r^2 = 0.329). Similarly, stem weight was linearly correlated with pod number (F = 48.20; df = 101; P < 0.001; r^2 = 0.323) and total seed weight (F = 98.87; df = 101; P < 0.001; r^2 = 0.495) and pod number was correlated with total seed weight (F = 83.70; df = 101; P < 0.001; r^2 = 0.453). Among infested plants, stem diameter was linearly correlated with stem weight (F = 42.92; df = 102; P < 0.001; r^2 = 0.296), with pod number (F = 39.63; df = 102; P < 0.001; r^2 = 0.280) and with total seed weight (F = 46.71; df = 102; P < 0.001; r^2 = 0.314) and stem weight with pod number (F = 59.18; df = 102; P < 0.001; r^2 = 0.367) and with total seed weight (F = 65.65; df = 102; P < 0.001; r^2 = 0.392) and pod number with total seed weight (F = 172.14; df = 102; P < 0.001; r^2 = 0.628).

Since there were no significant differences between infested and uninfested plants in 2006, data were pooled for regression analyses. Once again, stem diameter was linearly correlated with stem weight (F = 92.73; df = 121; P < 0.001; r^2 = 0.434), pod number (F = 81.90; df = 121; P < 0.001; r^2 = 0.404) and seed weight (F = 57.45; df = 121; P < 0.001; r^2 = 0.322) and stem weight with pod number (F = 133.68; df = 121; P < 0.001; r^2 = 0.525) and total seed weight (F = 106.77; df = 121; P < 0.001; r^2 = 0.469) and pod number with total seed weight (F = 418.07; df = 121; P < 0.001; r^2 = 0.776).

Among infested plants, tunnel length displayed no relationship with pod number (F= 0.81; df = 102; P = 0.371) in 2005 but was positively correlated with total seed weight (F = 4.06; df = 102; P = 0.047), although no significant variation was explained (r^2 = 0.038). In 2006, neither pod number (F = 0.98; df = 62; P = 0.326) nor total seed weight (F = 1.25; df = 62; P = 0.267) displayed any relationship with tunnel length.

Table 1. Mean (± SEM) values of various yield components (dry weights) for soybean plants with (Infested) and without (Control) larvae of *Dectes texanus* in two years

Treatment	Stem diameter (mm)	Stem weight (gm)	Pod number	Tunnel length (mm)	Seed weight (gm)
			2005		
Control	5.69 ± 0.10a	4.99 ± 0.13a	40.35 ± 1.07a	---	7.96 ± 0.25b
Infested	5.94 ± 0.08a	5.24 ± 0.13a	43.09 ± 1.14a	36.89 ± 1.13	8.72 ± 0.27a
$F_{1, 205}$	3. 74	1.83	3.07	N/A	4.30
P	0.053	0.178	0.081	N/A	0.039
			2006		
Control	8.14 ± 0.19a	7.47 ± 0.32a	54.64 ± 2.33a	---	21.31 ± 0.93a
Infested	7.82 ± 0.19a	7.67 ± 0.38a	52.54 ± 2.28a	21.62 ± 1.25	20.11 ± 0.93a
$F_{1,121}$	1.36	0.16	0.41	N/A	0.82
P	0.246	0.693	0.523	N/A	0.367

Values bearing the same letters were not significantly different (ANOVA, α > 0.05) between Infested and Control plants within years.

Discussion

There was no significant reduction in pod number or total seed weight as a result of *D. texanus* infestation in either year; rather, infested plants tended to have slightly larger stem diameters and higher seed weights than uninfested plants in 2005, the year in which plants were significantly smaller due to higher plant density. This need not imply a positive effect of larval boring on plant productivity, but could reflect a tendency for females to oviposit preferentially in larger plants, especially when average plant size is small, as inferred by others (Richardson 1975; Andrews and Williams 1988). Similarly, Michaud et al. (2007b) observed a difference of about 10% in *D. texanus* infestation rate between irrigated and unirrigated portions of the same soybean field that they attributed to a larger average plant size in the irrigated portion.

These data suggest that *D. texanus* larval boring may have no impact on soybean yield, at least under the growing conditions encountered in the present study. The same was concluded by Michaud et al. (2007a) with respect to the impact of larval boring by *D. texanus* on sunflower yields. Larval boring is confined to the central pith of the stalk and does not damage conductive tissues or reduce photosynthetic area, apart from the loss of a single leaf early in the season due to petiole boring by the first instar. However, yield is complex function of positive and negative factors influencing the plant and these factors may interact in complex ways. Thus it is possible that other plant stress factors or growing conditions not encountered in this study could interact with stalk boring to produce yield losses under other circumstances. Nevertheless, we are inclined to consider *D. texanus* a non-economic pest of soybeans unless harvest is delayed beyond the point where larvae girdle the plants.

Soybean represents a relatively recent and dramatic host shift for *D. texanus* whose ancestral host plants are all composite species native to North America (Hatchett et al. 1975). Soybean plants still represent a host of marginal suitability for *D. texanus*; adult body size averages 50% smaller and fitness is substantially reduced when larvae mature on soybean compared to cultivated sunflower (Michaud and Grant 2005; Niide et al. 2006). In contrast to sunflower that hosts a wide spectrum of boring insects, including competing cerambycid species (Michaud and Grant 2005), soybean is completely devoid of stem boring insects other than *D. texanus* in the New World and is therefore completely free of competitors for this niche. All indications suggest that intraspecific competition among *D. texanus* larvae can be intense within ancestral host plants – hence the girdling behavior thought to represent a defensive strategy for protecting the overwintering site. Thus, for genotypes of *D. texanus* that are able to survive in this plant, a field of soybeans likely represents a vast, untapped resource with reduced competition.

In the early 1970's, Richardson (1975) examined more than 10,000 soybean stalks of multiple cultivars and maturity groups infested by *D. texanus* in North Carolina and found significant effects of stem diameter – larger diameter stems were infested at higher rates and yielded superior larval survival than small diameter stems. In our study, we found larvae of *D. texanus* infesting all stem diameters and size classes of plants with approximately equal frequency (Figure 1). Although the average stem diameter of infested plants appeared slightly greater than that of uninfested plants in 2005 (Figure 1a), the difference was non-significant (Table 1). There was also no apparent shift toward larger stem diameters for infested plants in 2006 (Figure 1b) when plants were larger, but still not as large as many obtained by Andrews and Williams (1988). In contrast, the latter authors observed markedly disparate distributions of stem diameter for infested and uninfested plants in Mississippi in both 1984 and 1985, with plant populations of 236,822 and 146,175 plants per hectare, respectively.

Although large sunflower stalks may attain a girth that is well beyond the girdling radius of *D. texanus* larvae, the same is not true for soybean stalks that, regardless of size, are often visibly cut through to the outside leaving 'sawdust' visible on the soil surface. Thus, although irrigation and reduced planting density will increase plant size, neither are viable cultural approaches to reducing lodging losses to this pest in soybean as they are in sunflower. Given tendencies for larger soybean plants to have higher rates of infestation, increasing plant density to produce a larger number of smaller plants may actually reduce the proportion infested by reducing their average suitability to the pest, and by diluting the impact at field level.

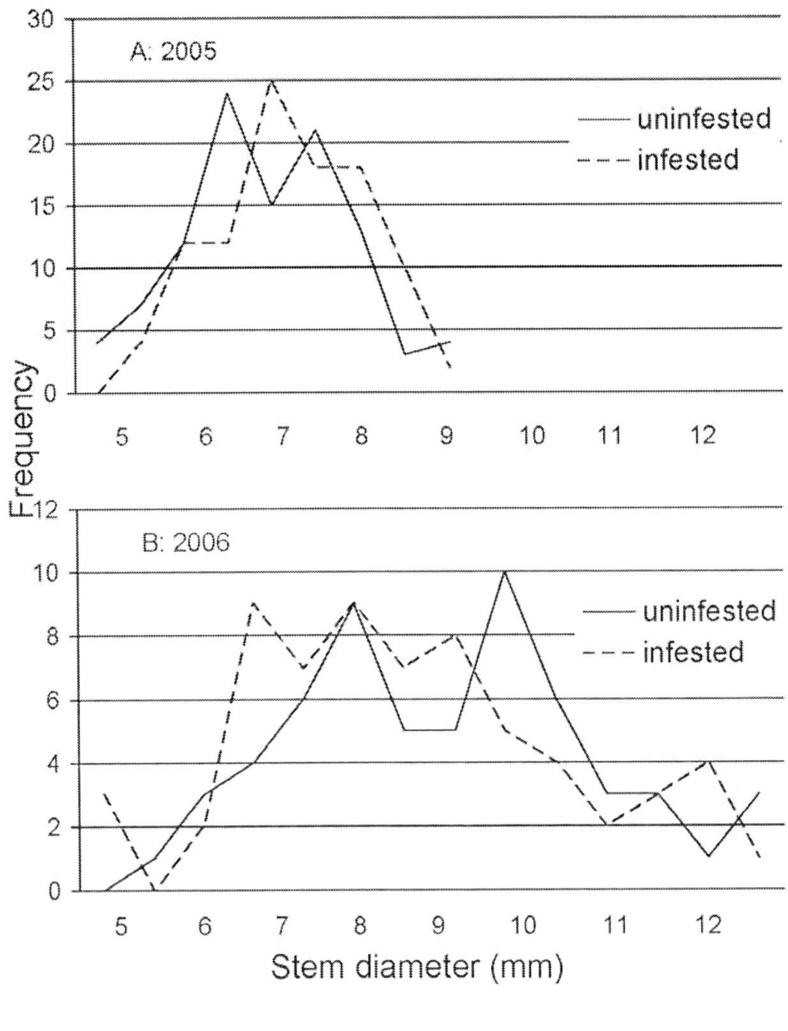

Figure 1. Frequency distributions of soybean plant stem diameters infested and uninfested by larvae of *Dectes texanus* sampled in two fields in central Kansas.

Observations in our laboratory suggest that soybean stem diameter is only weakly correlated with the size of *D. texanus* pupae and adults that mature within them; although the largest adults tend to emerge from the largest stalks, some large stalks also yield very small individuals (A.K. Grant, unpublished data). Oviposition in smaller soybean plants might entail substantial mortality for *D. texanus*. Although very small larvae are often numerous in samples of soybean stalks, many small pupae produce non-viable adults that die prior to reproductive age in the laboratory. Nevertheless, weights of *D. texanus* pupae can vary from 5 mg to 50 mg indicating substantial genetic variance for body size. Since the smallest size classes of insects maturing in soybean still have very low apparent fitness, exploitation of soybeans by *D. texanus* may be associated with ongoing selection for reduced body size and a lower critical size threshold for adult viability.

Acknowledgments

This work was partially funded by USDA, CSREES, Northcentral Region IPM Mini-grants Program. The manuscript was improved by commentary from Dr. L. Buschman. Contribution No. 08-54-J of the Kansas State Agricultural Research Station. Voucher specimens No. 178 are located in the Kansas State University Museum of Prairie Arthropod Research.

References

Andrews, G.L. and Williams, R.L. 1988. An estimate of the effect of soybean stem borer on yields. *Tech. Bull. Miss. Agric. For. Exp.* St. No. 153 (June).

Buschman, L., and P.E. Sloderbeck. 2005. Efficacy of in-season applications of systemic insecticides to control *Dectes* stem borers in soybean. In SWREC "2005 Field Day Report." *K-State Rep. Prog.* 945: pp 53-55.

Campbell, W.V. 1980. Sampling coleopterous stem borers in soybean. In: M. Kogan and D. Herzog (eds.) Sampling methods in soybean entomology. *Springer Verlag*, pp. 357-373.

Campbell, W.V. and J.W. van Duyn. 1977. Cultural and chemical control of *Dectes texanus* on soybeans. *J. Econ. Entomol.* 70: 256-258.

Falter, JM. 1969. *Dectes* sp. (Coleoptera: Cerambycidae): A unique and potentially important pest of soybeans. *Journal of the Elisha Mitchell Scientific Society* 85: 123.

Hatchett, J.H, Daugherty, D.M, Robbins, J.C, Barry, R.M, and Houser E.C. 1975. Biology in Missouri of *Dectes texanus*, a new pest of soybean. *Annals of the Entomological Society of America* 68: 209-213.

Laster, M.L., Tupper, G.R., Hartwig, E.E. and Thom, W.O. 1981. Studies of the stem borer, *Dectes texanus*, on soybeans in Issaquena County, Mississippi, 1978. *Miss. Agric. For. Exp. St.* 6: 1-3.

Michaud, J.P. and A.K. Grant. 2005. The biology and behavior of the longhorned beetle *Dectes texanus* on sunflower and soybean. *J. Ins. Sci.* 5:25. Available online: http://www.insectscience.org/5.25/.

Michaud, J.P., A.K. Grant, and J.L. Jyoti. 2007a. Impact of the stem borer *Dectes texanus* (Coleoptera: Coccinellidae) on yield of cultivated sunflower, *Helianthus annuus*. *J. Ins. Sci.* 7.25. Available online: http:www.insectscience.org/7.25/.

Michaud, J.P., J.A. Qureshi, and A.K. Grant. 2007b. Sunflowers as a trap crop for reducing soybean losses to the stalk-borer *Dectes texanus* (Coleoptera: Cerambycidae). *Pest Man. Sci*. (in press).

Niide, T., Bowling, R.D. and Pendleton, B.B. 2006. Morphometric and mating compatibility of *Dectes trexanus texanus* (Coleoptera: Cerambycidae) from soybean and sunflower. *J.Econ. Entomol.* 99: 48-53.

Patrick, C. 1973. Observations on the biology of *Dectes texanus texanus* (Coleoptera: Cerambycidae) in Tennessee. *J. Geor. Entomol. Soc.* 8: 277-279.

Patrick, C., G. Lentz, S. Stewart and A. Thompson. Undated. *Dectes* stem borer in Tennesee. University of Tennessee Agricultural Extension Service Bull. SP503-F. Available online: http://www.utextension.utk.edu/publications/spfiles/SP503-F.pdf.

Richardson, L.G. 1975. Resistance of soybeans to a stem borer, *Dectes texanus texanus* LeConte. Ph.D. thesis, Department of Entomology, North Carolina State University, Raleigh.

SAS Institute, 2003. *The SAS system for windows Release 8.02.* SAS Institute, Cary, NC.

In: Soybean and Wheat Crops
Editors: S. Davies, G. Evans
ISBN: 978-1-60741-173-4

Chapter 8

FUSARIUM HEAD BLIGHT AND DON CONTAMINATION MANAGEMENT IN SOFT AND DURUM WHEAT CULTIVATION

Andrea Maiorano*[*], *Massimo Blandino and Amedeo Reyneri
Department of Agronomy, Forest and Land Management, University of Turin,
via Leonardo da Vinci 44, 10095 Grugliasco (TO), Italy

ABSTRACT

Fusarium Head Blight (FHB), also known as scab, is a devastating disease of enormous economic importance throughout the world that attacks all classes of barley and wheat. Every year, all of the most important cereal producers in the world are affected by this disease. *Fusarium graminearum* and *Fusarium culmorum* are considered the most pathogenic and widespread agents of FHB in wheat. Both *F. graminearum* and *F. culmorum* produce deoxynivalenol (DON, also known as vomitoxin), a mycotoxin of the trichotecenes group, one of the most widespread mycotoxins in cereals, which can be synthesised in the field in small grain cereals and in maize. This mycotoxin has a great impact on the health of animals and humans, due to their cytotoxic activity and immunosuppressive effects. As a consequence, FHB and DON contamination are responsible for serious direct and indirect economic losses. Economic losses can be attributed to yeld loss due to fungal disease, reduced technological and nutritional quality of grain, reduced crop value due to mycotoxin contamination, reduced animal productivity, human health costs. Prevention, monitoring, sampling, chemical analysis, litigation, mitigation, and research costs also need to be taken into account. Moreover, FHB and DON limits, which have been established in many parts of the world for animal feeding and human consumption, have made the worldwide cereal market much more selective.

F. graminearum and F. culmorum inoculum is usually abundant and its production, dispersal and deposition are influenced by weather conditions. Warm, humid weather,

[*] Corresponding author: Andrea Maiorano, Tel +39 011 670 8928, e-mail: andrea.maiorano@unito.it.

frequent rainfall and heavy dew favour spore germination. Temperature and water activity of the host are the fundamental factors that affect fungal growth inside grain. DON synthesis is mainly influenced by temperature and water ativity and its production seems to play a decisive role on the aggressiveness of *F. graminearum* and *F. culmorum*: it has been suggested that in the presence of DON, plant defence mechanisms are not triggered fast enough, thereby leading to an increased aggressiveness of the pathogen.

The accumulation of DON in the grain is intimately related to the development of the disease in the field. Prevention strategies through pre-harvest agronomic management can achieve the quality and safety standards required by the market and international regulations. Among the agricultural practices, FHB development and DON contamination in wheat grains are mainly affected by tillage, crop rotation, varieties and fungicide application. None of these strategies on their own are able to significantly reduce the impact of this disease. Nevertheless, careful planning of all the different management decisions canl lead to the designing of crop management systems able to reduce FHB and mycotoxin incidence and severity. In other words, it is necessary for producers to make use of the available wheat varieties that are known to be more resistant and to apply all of the good agricultural practices tuned to a determined cultivation area.

Useful tools for prevention are represented by FHB and DON predictive models: the use of a model to predict the outcome of a disease is desirable to enhance and trigger management opportunities with the aim of reaching high technological, nutritional and productivity quality and safety of the production.

Keywords: Fusarium Head Blight, deoxynivalenol, *F. graminearum* epidemiology, FHB economic impact, crop management systems, predictive models

1. INTRODUCTION

Fusarium Head Blight (FHB), also known as scab, is a devastating disease of enormous economic importance throughout the world that attacks all classes of barley and wheat. Every year, all of the most important cereal producers in the world are affected by this disease. As a consequence, FHB is responsible for serious direct and indirect economic losses.

FHB is caused by different species of the genus *Fusarium*, but two of them represent the most important species worldwide: *Fusarium graminearum* Schwabe (teleomorph *Gibberella zeae* (Schw.) Petch.) and *Fusarium culmorum* (W.G. Sm.) Sacc. (teleomorph unknown). Their temperature requirements determine their relative competitive strength and their consequent geographical predominance: *F. culmorum* tends to predominate in the cooler maritime regions of northwest Europe, where *Fusarium poae* (Peck) Wollenw., *Fusarium avenaceum* (Corda) Sacc. and *Microdochium nivale* (Fr.) Samuels & Hallet (syn.*Fusarium nivale*) also assume a greater importance (Birzele et al., 2002; Kosiak et al., 2003; Ioos et al., 2004). Other less frequently isolated species are *F. poae*, *F. cerealis* (syn. F. crookwellense), *F. equiseti* (syn. *F. scirpi*) (*G. intricans*), *F. sporotrichioides*, and *F. tricinctum*. Many other species may be sporadically encountered, including *F. acuminatum*, *F. subglutinans* (syn. *F. sacchari*), *F. solani*, F. *oxysporum*, *F. semitectum* (syn. *F. pallidoroseum*, *F. incarnatum*), *F. verticillioides* (syn.: F. *moniliforme*), and *F. proliferatum* (Bottalico and Perrone, 2002).

F. graminearum is considered the most pathogenic and widespread agent of FHB in wheat. Both *F. graminearum* and *F. culmorum* produce deoxynivalenol (DON, also known as vomitoxin), a mycotoxin of the trichotecenes group, one of the most widespread mycotoxins in cereals, which can be synthesised in the field in small grain cereals and in maize. Trichotecenes are associated with serious mycotoxicosis in humans and animals. This group of mycotoxins has a great impact on the health of animals and humans, due to their cytotoxic activity and immunosuppressive effects. Their main effects—related to their concentration in the commodity—are a reduced feed uptake, vomiting and immunosuppression (Rotter et al., 1996; Oswald et al., 2005). Therefore, FHB and DON limits, which have been established in many parts of the world for animal feeding and human consumption, have made the worldwide cereal market much more selective.

2. Occurrence and Economic Impact

2.1. Occurrence

FHB infection and DON contaminations have been reported by the most important wheat producers in the world, on all continents. The aetiological complexity of FHB is the main reason for the great diffusion of this disease and its geographical and seasonal variability. Table 1 illustrates a review of the natural occurrence of FHB and DON contamination of wheat grain in different part of the world in recent years (from 1984 to 2007). *F. graminearum* is the most prevalent species in most sites, except in England (Nicholson et al., 1993), Norway (Kosiak et al., 2003) and Latvia (Treikale et al., 2008), where the environmental conditions are more favourable to *F. avenaceum*, *F. poae* and *F. culmorum*.

Data taken in different periods in the Netherlands have shown an interesting change in the prevalence of the *Fusarium* spp. In a screening conducted in 2001 and 2002, Waalwijk et al. (2003) found *F. graminearum* to be the most prevalent species, a result that was in contrast with two previous surveys conducted in the 1980s by Daamen et al. (1991) and in the early 1990s by De Nijs et al. (1996), in which *F. culmorum* was reported to be the predominant agent for FHB. Waalwijk suggested some possible reasons for this change: i) the increase in maize production (*F. graminearum* is a major pathogen on maize and has the capacity to survive on maize stubble); ii) climatic changes might favour the propagation of *F. graminearum* over *F. culmorum*, as the former species has a higher temperature optimum; iii) the homothallic nature of *F. graminearum*, which allows the production of large masses of ascospores that can play a role in the epidemiology.

The data from the Netherlands show that FHB is a very dynamic disease, mainly due to its aetiological complexity. It is also possible that fungi may be spread from one country to another due to the increases in the global grain trade. However, as far as *Fusarium* fungi are concerned, this risk is likely to be minimal since these phytopathogens are field rather than storage organisms. On the other hand, there is now overwhelming evidence of global contamination of cereals and animal feeds with trichothecenes. Trade in wheat may, therefore, contribute to the worldwide dispersal of mycotoxins (Placinta et al., 1999).

Scab is prevalent in warm, humid regions where flowering coincides with rainy periods. In the United States, the incidence of this disease has been increasing over the last 10 years for a combination of reasons. Perhaps the most important reason concerns the increased area in which wheat is rotated with maize or other cereals. Other reasons are changes in the cropping system for soil protection purposes and changes in wheat cropping from traditional to more humid, non-traditional areas (McMullen et al., 1997; Gilchrist and Dubin, 2002).

Among the data reported in the table, the highest DON contaminations are reported in Michigan (20000 μg/kg in 2004), England (10626 μg/kg in 2003), Russia (8600 μg/kg, year not specified) and Argentina (8440 μg/kg, year not specified). The data from Africa (Kenya and Morocco) and from China show the lowest range of contamination, but these regions of the world are so large that more data would be needed to characterize their DON contamination in wheat.

2.2. Economic Impact

The economic costs of FHB and mycotoxins are very difficult to accurately determine, as so many factors are involved and all of the indirect effects are almost impossible to take into account. In general, economic losses in wheat due to FHB can be attributed to i) yield loss due to fungal disease and to ii) reduced technological and nutritional quality of wheat grain. Invasion of kernels by *Fusarium* spp. destroys the starch granules, storage proteins and cell wall, thus affecting grain quality. The embryo is usually not infected except in heavily invaded kernels. However, slightly infected kernels with apparently uninfected embryos exhibit reduced germination and vigour (Bechtel et al., 1985; Snijders, 1990). Economic losses due to DON contamination can be attributed to i) reduced crop value due to mycotoxin contamination, ii) losses in animal productivity due to mycotoxin-related contamination, iii) human health costs. Moreover, prevention, monitoring, sampling, chemical analysis, litigation, mitigation, and research costs also need to be taken into account.

Losses in quality lead to additional costs related to marketing, processing and exports. For instance, the milling industries purchase various classes of wheat for specific milling or baking purposes; therefore, high-quality grain is necessary for high prices, and grain that does not meet the minimum standards is docked to a lower grade and the producer receives a lower price (Windels, 2000). The differentiation of wheat grain quality has three main effects: 1) additional costs related to the separation of grain consignments into different lots; 2) the adaptation of all of the productive process in different production lines based on high/low quality products and their destination (industry, foods, feeds, etc.); and 3) price effects due to a reduction in high-quality grain supply.

In principle, reductions in wheat grain supply due to yield loss or to quality classification tend to increase future prices, but the beneficiaries of high price premiums for milling-quality grains are only the producers who have high-quality wheat, while a larger share of production may be discounted because of poor quality (Johnson et al., 1998).

Therefore, the economic impact of FHB and DON contamination is shared by all of the subjects in the productive chain up to the consumer: crop producers, animal producers, grain handlers, distributors, processors, consumers and society in general, because all of these costs contribute to elevate the final price (Charmley et al., 1994). Thus, the economic impact of FHB and DON is really difficult to quantify as so many factors should be taken into account, but some models have been proposed to estimate FHB and DON costs within limited contexts.

Wheat and barley losses caused by scab epidemics in the United States during the 1990s were estimated at close to 3 billion of US dollars (US$). American wheat farmers were estimated to have lost over 13.6 billion tons, evaluated at about 2.5 billion US$. Economic losses for wheat producers in Canada in the same period were estimated as 220 million US$ in Quebec and Ontario and 300 million US$ in Manitoba (Windels, 2000). In 1992, the average wheat yields in northeastern North Dakota and northwestern Minnesota were about 3.36 t/ha. In 1993, the yields lowered respectively to 1.68 t/ha and to 2.01 t/ha and from 1994 through 1998, the yields fluctuated but always remained between 1.75 and 2.29 (Windels, 2000).

In another simulation conducted for the same period (1991–1997), Johnson et al. (1998) measured the cumulative economic losses suffered by wheat producers in nine US states concerning three wheat classes (soft red winter, hard red spring, and durum), taking into account the loss of production (bushels) and changes in prices (US$/bushels). According to the analysis, during the 1991-1997 period, wheat producers in the affected regions suffered cumulative losses of 1.300 million US$ (61.8% hard red spring, 32.6% soft red winter, and 5.6% durum). Nganje et al. (2001) continued the analysis started by Johnson et al. for the 1998-2000 period, simulating the primary (production and prices) and secondary impact of losses incurred in other sectors of the economy which result from subsequent rounds of spending and re-spending within an economy. The direct combined effect of price discounts and yield reduction from FHB were estimated at 733 US$ million, while the combined direct and secondary losses were estimated at 2.300 US$ million. Income losses resulted in a reduction in farm numbers: about 2000 farms were lost in North Dakota during the 1992-1996 period, versus 500 during the previous four years. The economic cost of DON in wheat and corn was estimated in a simulation conducted by CAST (2003). In this simulation, the direct costs for yields that did not respect U.S. food and feed crop advisory limits, the direct costs for potential livestock losses, the costs of efforts to mitigate the contamination (dry storage, physical separation during milling, and testing) and human health hazard were taken into account. The simulation generated mean annual costs for DON of 637 million US$ in crop losses (wheat and corn) and 18 million US$ in feed losses. The mean simulated annual cost of livestock losses was about 2 million US$.

These simulation give an idea of the magnitude of the economic effects of FHB infection and DON contamination. They refer to contamination in some U.S. regions, but infections and contaminations have been reported in a wealth of literature from all over the world where wheat is cultivated (Table 1). Nevertheless, in order to reduce the impact of the exclusion of the food and feed industries for the DON contaminated wheat stocks, other industrial uses could be considered. Mesterházy (2004) indicated three possibilities that have greater chance: a) direct energy production, b) usage for gasoline production, c) alcohol production for industrial use or as fuel. In this way, the utilization of contaminated grain could be a feasible solution for the farmers who cannot sell their contaminated grain to the feed and food markets.

Table 1. Natural occurrence of FHB and DON contamination of wheat grain in different part of the world in recent years (from 1984 to 2007)

Continent	State	Period	FHB, species and notes	DON contamination (µg/kg) and notes	Reference
Africa	Kenya	2004	*F. graminearum* the most virulent species. Prevalent species: *Fusarium poae*, *F. chlamydosporum* and *F. oxysporum*	Range: 110–290 µg/kg	Muthomi et al. (2008)
	Morocco	2001–2005		17 samples: 7 samples contaminated. Mean of positives= 65.9 µg/kg Max=128 µg/kg	Hajjaji et al. (2006)
	Tunisia	Not specified Recent years	Presence of *F. culmorum* and *F. pseudograminearum*	All isolated strains were DON producers	Kammoun et al. (2008)
America (North)	Canada	1988–1991	Prevalent species: *F. graminearum*. Others: *F. culmorum*, *F. acuminatum*, *F. avenaceum*, *F. equiseti*, *F. poae*, *F. sporotrichioides*		Wong et al. (1992)
	Kansas	1993	29 out of 30 samples infected by *F. graminearum*	26 out of 116 samples contaminated	Trigostockli et al. (1995)
	Michigan	2004	FHB survey: average number of infected heads=77.6%; range=12%-100%	Range: 1000–20,000 µg/kg Average: 8300 µg/kg	Hart et al. (2004)
	Minnesota	1984–1986	Prevalent species: *F. graminearum*, *F. poae* Others: *F. equiseti*, *F. acuminatum*, *F. sporotrichiodes*, *F. semitectum*		Wilcoxson et al. (1988)
America (South)	Argentina	2 years (years not specified)	All samples positive to *F. graminearum*.	Range: 0–8440 µg/kg	Lori et al. (2003)

Continent	State	Period	FHB, species and notes	DON contamination (µg/kg) and notes	Reference
	Argentina	1993	*F. graminearum* prevalent species.	Range: 300–4500 µg/kg	Dalcero et al. (1997)
	Brasil	2005		National (94% contaminated, mean 332 µg/kg) and imported (Paraguay and Argentina, 88% contaminated, mean 92 µg/kg) stored wheat samples.	Calori-Dominguesi et al. (2007)
	Uruguay	2001–2002	Prevalent species: *F. graminearum* Others: *F. avenaceum, F. culmorum, F. poae, F. equiseti*		Pereira et al. (2004)
Asia	China	1989		30 samples (47% positive) Range: 7–309 µg/kg	Luo et al. (1990)
	Japan		Prevalent species: *F. graminearum, M. nivale* Others: *F. avenaceum, F. tricinctum*		Parry et al. (1995)
	Nepal	1997	Prevalent species: *F. graminearum* and *F. equiseti* Others: *G. fujikuroi MP-D, F. oxysporum, F. avenaceum F. semitectum, F. torulosum, F. acuminatum, F. chlamydosporum.*		Desjardins et al. (2000)
	Russia	1989–2002		Range: 0–8600 µg/kg (2166 samples)	Tutelyan (2004)

Table 1. Continued

	Russia	Not Specified Recent years	Widespread and dominant species: *F. poae*, *F. sporotrichioides*, *F. avenaceum*. Widespread but infrequent species: *F. tricinctum*, *F. equiseti*, *F. semitectum*. Prevalent species distribution by region: *F. graminearum* (Central and Northwestern regions), *F. acuminatum* (Central Siberia, Far East), *F. culmorum* (North-western, Central, Ural), *F. cerealis* (South, Far East), *F. sambucinum* and *G. fujikuroi* complex (South, Far East, Central)		Gagkaeva and Gavrilova (2008)
Europe	Bulgaria	1995	140 samples *Fusarium* spp and *Alterrnaria* spp	Mean=180 μg/kg Max=1800 μg/kg	Vrabcheva et al. (1996)
	Croatia	2001–2003	Prevalent species: *F. graminearum* Other: *F. culmorum*		Krstanovic et al. (2005)
	England	1991	Prevalent species: *F. avenaceum*, *F. culmorum*. Others: *F. lateritirum*, *F. poae*		Nicholson et al. (1993)
	France	2000–2002	Prevalent species: *F.graminearum*, *F. avenaceum* and *F. poae*. Others: *M. nivale*, *F. culmorum*. 749 samples		Ioos et al. (2004)
	Germany	2000–2001		Mean=24 μg/kg, Max=309 μg/kg (41 samples, Incidence 95%)	Schollenberger et al.(2006)
	Germany	1989–1993		Mean for each year from 167 to 735 μg/kg	Muller et al.(2001)
	Hungary	1991–1998		367 samples Range:70-1560 μg/kg	Rafai et al. (2000)
	Italy	2004–2006	Prevalent species: *F. graminearum* and *F. culmorum*.	Range: 25-2660 μg/kg	Maiorano et al. (2008)

Continent	State	Period	FHB, species and notes	DON contamination (µg/kg) and notes	Reference
	Italy	2000		Durum wheat: max=1000 µg/kg Soft wheat: max=330 µg/kg	Logrieco et al. (2003)
	Latvia	2006-2007	Prevalent species: *F. poae*, *F. culmorum*, *F. gibbosum*, *F. avenaceum* var. *herbarum* Other species: *F. oxysporum* var. *orthoceras*, *F. sambucinum*, *F. sporotrichioides*, *F. moniliforme*, *F. semitectum*	2006: 64 samples, 4.7% contaminated, range 28-36 µg/kg 2007: 81 samples, 27.2 contaminated, range 1.0-615.0 µg/kg	Treikale et al.(2008)
	Lithuania	1998		84 samples. Low contamination	Keblys et al.(2001)
	Norway	1994–1996	695 samples (wheat, barley and oats) Prevalent species: *F. avenaceum*, *F. poae*, *F. tricinctum* and *F. culmorum* Others: *F. graminearum*, *F. sporotrichioides*		Kosiak et al.(2003)
	Norway	1988–1996		About 5000 samples (wheat, barley, oats). 61.5% of wheat samples positive	Langseth and Elen (1997)
	Poland	1994–1995	17% of kernels with visible symptoms of scab (*F. graminearum*)	Mean=9628 µg/kg	Perkowski et al.(1997)
	Slovakia	2004-2006		139 samples, 9.3% exceeding EU food limit (1250 µg/kg)	Slikova et al.(2008)
	Spain	2003	*F. graminearum*, *F. culmorum*, *F. equiseti*		Jurado et al. (2004)
	The Czech Republic	1999–2001		Almost all samples contaminated. Only two samples exceed the EU food limit (1250 µg/kg)	Hajslova et al. (2007)

Table1. (Continued)

	The Netherlands	1974–1986	Prevalent species: *M. nivale, F. culmorum.* Other species: *F. avenaceum, F. graminearum.*		Daamen et al. (1991)
	The Netherlands	1991 and 1993	Prevalent species in 1991: *Fusarium culmorum, Fusarium avenaceum* Prevalent species in 1993: *Fusarium poae, Fusarium culmorum, Fusarium crookwellense*		De Nijs et al. (1996)
	The Netherlands	2001–2002	Prevalent species: *F. graminearum*, in contrast with Daamen et al., (1991) and De Nijs et al. (1996)		Waalwijk et al. (2003)
	The United Kingdom	2001–2005		954 samples 2001: mean=90 μg/kg, Max=5175 μg/kg 2002:mean=116 μg/kg, Max=3065 μg/kg 2003:mean=218 μg/kg, Max=10626 μg/kg	Edwards (2004b)
Oceania	New Zealand	1999–2000	Prevalent species: *F. graminearum* Other species: *F. avenaceum, F. culmorum, F. poae* FHB incidence (%): 0.0–2.4 in 1999, 0.0–3.5 in 2000	Range: 0.00-260 μg/kg in 1999, 60-3820 μg/kg in 2000	Cromey et al.(2002)
	Australia (New South Wales)	2000	Outbreak of Fusarium head blight caused by higher rainfall than average occurred during anthesis in November. *F. graminearum, F. culmorum, F. cerealis* (= *F. crookwellense*), *F. pseudograminearum* and *F. avenaceum.*		Tan et al. (2004)

3. Ecology and Epidemiology of Fusarium Head Blight in Wheat and Synthesis of Deoxynivalenol

In order to understand how Fusarium head blight invades wheat grain and synthesizes DON, it is important to consider the *Fusarium* spp. life cycle on small grain cereals

Many papers have dealt with the epidemiology of Fusarium head blight, or more specifically of *F. graminearum*, which has always been considered the most important FHB pathogen.

Atanasoff, in 1920, first described the development of FHB, and in particular of *Gibberella saubinetii* (=*Gibberella zeae*). The first attempt to describe the complete life cycle of *F. graminearum* on cereals was, however, made by Sutton (1982) who described the disease cycle of the fungus in relation to the pathosystem formed by maize, wheat and the fungus itself in which the fungus survives, spreads, develops and invades the cereal tissues, producing mycotoxin in grain. The study by Sutton, was followed by others (Snijders, 1990; Parry et al., 1995; Gilchrist and Dubin, 2002; Logrieco et al., 2003; Gilbert and Fernando, 2004; Xu and Berrie, 2005; Wagacha and Muthomi, 2007) which described the epidemiology of head blight Fusaria, and other which updated the description of the pathosystem with new knowledge about (i) the influence of the environmental and climatic conditions on the fungi development (Doohan et al., 2003; Xu, 2003), (ii) the competitive relationships between different *Fusarium* spp (Xu et al., 2007), (iii) the inoculum dispersal dynamics (Horberg, 2002; Rossi et al., 2002a; Paul et al., 2004; Maldonado-Ramirez et al., 2005), and so on. One of the most recent descriptions of the epidemiology of FHB was given by Osborne and Stein in 2007.

A brief description of the epidemiology of FHB in wheat is given in this section, through a review of the main literature.

3.1. Inoculum Sources and Inoculum Structures

The principal *Fusarium* spp. sources of inoculum are crop debris like maize stalks, grain, ear pieces and cob, and the stubble and straw of barley, wheat, and other cereals. Soil itself and contaminated seeds can be also sources of inoculum, but their role is marginal compared to the former sources. The fungi are present and survive in colonized crop residues as mycelium, and may develop saprophytically on residues on the soil surface for 2 or 3 years (Sutton, 1982; Osborne and Stein, 2007). Moreover, *Fusarium* spp can survive parasitically and saprohytically on leaves throughout the wheat growing season, which means that these leaves may contribute additional inoculum to the FHB development (Ali and Francl, 2001). In some locations, the leaves have also shown large macroconidial concentrations, suggesting that the fungus can also grow epiphytically, resulting in a higher inoculum level within the canopy (Gilbert and Fernando, 2004). In addition, noncereal residues such as soybeans, sunflower, pasture, and many gramineous weed species have been reported to be important sources of inoculum for *Fusarium* spp. (Jenkinson and Parry, 1994; Inch and Gilbert, 2003; Pereyra and Dill-Macky, 2008).

In general, the inoculum structures are spores (ascospores, macroconidia, chlamidospores) and fungal mycelia (Sutton, 1982), but macroconidia and ascospores have been reported to be the most important types of inoculum and both of them are abundantly produced on the infected residues (Khonga and Sutton, 1988; Fernando et al., 2000). *F. graminearum* has an additional epidemiological advantage because it regularly and abundantly forms perithecia of its sexual stage (*Gibberella zeae*), resulting in the production of ascospores on the colonized residues (Xu and Berrie, 2005).

3.2. Inoculum Production, Dispersal and Deposition

The production of conidia and ascospores is critically influenced by temperature, the water activity (a_w) of the host, rainfall, relative humidity and light (UV light required). Sutton (1982) reported 29°C and 25-28°C as the optimum temperature for perithecia and ascospore production, respectively. Moisture is required for formation of both macroconidia and ascospores: Xu (2003) reported that when the soil moisture content is below 30%, ascospore production is not possible. When it is greater than 80%, ascospore production is at its maximum. Macroconidia production is also dependent on the temperature o a great extend: Rossi et al. (2002b) reported that they can be produced in the 5-35°C range while the optimum temperature for *M. nivale* is 26°C, 28°C for *F. avenaceum* and 32°C for *F. graminearum* and *F. culmorum*. Macroconidia are formed in mucilage in sporodochia and rain and wetting are necessary for their liberation.

The dispersal of both macroconidia and ascospores is associated with rainfall events and high relative humidity (Paulitz, 1996; Horberg, 2002). In a study conducted in Italy, very few macroconidia were sampled in the air before rainfall, but their number progressively increased with the beginning of a rain event. In the presence of high humidity, conidia continued to be sampled at high densities for some hours after rain had ceased and they usually reached their peak under these conditions. Finally, the density of the airborne conidia rapidly decreased when the relative humidity dropped (Rossi et al., 2002a).

Turgor pressure represents the driving force behind ascospore discharge. It has been suggested that the discharge of ascospores results from the buildup of turgor pressure generated by ion fluxes and the accumulation of mannitole, which is the main simple sugar component in the fluid (the epiplasm) with which the ascospores are discharged from asci from within the ascus (Trail et al., 2002). Ascospores are discharged to distances from <1 mm to nearly 10 mm with an average between 4.6 mm to 3.9 mm. This distance is sufficient for the ascospores to surpass the laminar boundary layer of air over which they become airborne and can encounter turbulent air currents which help them ascend into the atmosphere. This mainly happens during daylight hours, when the laminar boundary layer is considered to be only millimetres thick, while during the night, it can extend to nearly 10 m above ground level. Therefore, ascospores of *G. zeae* discharged during the daylight hours have the greatest probability of becoming airborne in turbulent air currents (Schmale et al., 2005). Once ascospores become airborne, they are subjected to air turbulence: during the day, energy from the sun heats the earth's surface and there is an upward transfer of heat toward the cooler atmosphere. Ascospores can be transported in the atmosphere to significant vertical and horizontal distances over the surface of the earth by turbulent mixing masses of air. At night, the surface of the earth cools rapidly in comparison to the atmosphere and there is a downward transfer of heat, which means that at night, as turbulent mixing slows down, the ascospores may settle out of the atmosphere over the surface of the earth (Maldonado-

Ramirez et al., 2005). This could explain the apparent contradiction with the interpretation given by Paulitz (1996) concerning the results of his experiments in which higher atmospheric concentrations of *G. zeae* were reported during the night hours: the higher concentration of *G. zeae* at night might be the effect of spore deposition and not of spore release as explained by the author. On the contrary, the low number of spores detected during the daytime might be due to the presence of mixed or turbulent layers. Thus, spore deposition may be separated from spore release in both time and space (Schmale et al., 2006).

3.3. Germination, Head Infection, Growth

Warm, humid weather, frequent rainfall and heavy dew favour spore germination. Pinkish-red mycelium and conidia develop abundantly in infected spikelets and the infection can spread to adjacent spikelets or through the entire head. Infected kernels become shrivelled and discoloured with a white, pink, or light-brown scaly appearance, as a result of mycelial outgrowths from the pericarp (Logrieco et al., 2003).

Temperature, relative humidity and water activity of the host are the fundamental factors that affect fungal germination and growth. In general, *F. graminearum* mycelium grows well over a wide range of temperatures, up to 30°C, and is associated with the warmer regions of the world, whereas *F. poae* is found more frequently in temperate climates. Most of the species can be found in much of the geographical area affected by FHB, but individual species usually dominate a specific region and *F. graminearum* dominates in a majority of regions (Osborne and Stein, 2007).

Table 2. Minimum (Tmin), optimum (Topt), and maximum (Tmax) temperatures for the germination of four different fungal species that cause Fusarium head blight in wheat

Species	Tmin	Topt	Tmax
F. avenaceum	14	20–28	35
F. culmorum	16	20–26	33
F. graminearum	5–10	24–29	35
F. poae		20–25	
M. nivale	10	15–20	28

(Rossi *et al.*, 2001; Brennan *et al.*, 2003; Hope *et al.*, 2005; Ramirez *et al.*, 2006).

Rossi et al. (2001), carrying out *in vitro* experiments, studied the effect of relative humidity (RH) and temperature on the infection by *F. graminearum*, *F. culmorum*, *F. avenaceum* and *M. nivale*. They reported that when spikes were incubated at different RH the frequency of the infected glumes was very high when wetted at 100% RH and very low at all the other RH tested (95%, 85%, 75% and 65%) for all the species except for *F. culmorum* whose frequency only started to significantly increase at 75% RH. As far as temperature is concerned, they found a minimum temperature between 10.0 and 16.5°C, an optimum one between 18 and 29°C and a maximum one between 28.5 and 35.5°C (Table 4). Brennan et al. (2003) reported lower optimum temperatures for the same fungal species (differences from 2 to 8°C in the case of *F. nivale*). Ramirez et al. (2006) reported similar results concerning the optimum and maximum temperature for *F. graminearum* but a lower minimum temperature (5°C). These difference could be related to the different growth substrate that was used. Ramirez et al. (2006) also studied the effect of a_w and reported an optimum level for *F. graminearum* growth at the highest tested level: 0.995 a_w. No growth was observed at 0.900 a_w.

Once the spores have germinated, the propagation of the hyphae is more favoured in the flower parts than in the other organs. The fungus can develop both inter- and intracellularly and cause severe damages. In 2001, Bushnell provided a summary on of the way that head blight Fusaria can infect the spikelets. The stomates can be entered by *F. graminearum* (Kang and Buchenauer, 2000a; Pritsch et al., 2000). The glume, palea, lemma and the awn can all be colonized by *Fusarium* mycelia. Hyphae can enter into the floret mouth opening where it can colonize caught and retained anthers or the apical brush of the caryopsis. Another potential pathway of entry is the crevice between the palea and lemma, especially near the floret base. Whether entering from the mouth or crevice, the fungus can develop abundantly on and within interior tissues of the lemma and palea. Surface tissues of the ovary and lodicules are especially susceptible. However, the mode of penetration of *F. graminearum* into these interior tissues has not been determined yet. *F. graminearum* can grow within tissues between cells instead of entering them, establishing a biotrophic relationship with the host tissues. How and where the intercellular fungus penetrates the cells for its subsequent growth within the cells is still not known. Once established within the floret, the fungus colonizes and follows vascular tissues in the floret stalk through the rachilla or rachis into other florets (Kang and Buchenauer, 2000a; Pritsch et al., 2000; Bushnell, 2001; Wagacha and Muthomi, 2007).

Stress-related genes are known to be activated in both resistant and susceptible cultivars in response to *F. graminearum* invasion. The genes activation is observed in both the invaded and in the adjacent uncolonized spikelet, which demonstrates that this response is expressed systemically. However, the molecular and physiological responses of heads to invasion by Fusarium head blight have not been fully investigated (Pritsch et al., 2000; Pritsch et al., 2001; Kong et al., 2005).

On the other hand, Fusarium head blight may produce cell wall degrading enzymes such as cellulases, xylanases and pectinases during infection and colonization of wheat spike tissues (Kang and Buchenauer, 2000b).

3.4. Deoxynivalenol Biosynthesis and Its Role in the Pathogenesis

Deoxynivalenol (DON, vomitoxin) is a type B trichothecene, an epoxy-sesquiter-penoid (Figure 1). DON is probably the most frequently encountered mycotoxin in the FHB of wheat throughout the world (Parry et al., 1995; FAO, 2003).

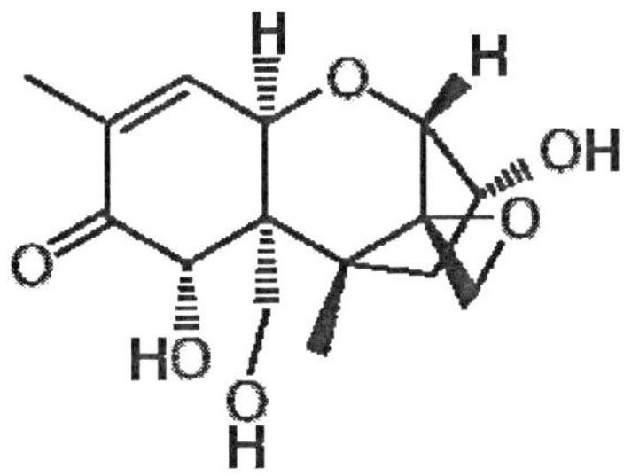

Figure 1. Chemical structure of deoxynivalenol.

Its occurrence is associated above all with *F. graminearum* and *F. culmorum*. DON is associated with serious mycotoxicosis in humans and animals. It has cytotoxic activity (protein synthesis inhibition, effects on DNA and RNA synthesis, mitochondrial function inhibition, effects on cellular membranes and on correct cellular division and apoptosis) and an immunosuppressor effect which reduce the resistance to microbial infections (Rotter et al., 1996; Minervini et al., 2004).

According to the results of in vitro experiments with *F. graminearum*, DON is synthesized in a range of temperatures between 5°C and over 30°C, with an optimum production between 25°C and 30°C, depending on the a_w. The optimum a_w for DON production has been reported to be near 1 (0.995) and the minimum requirement is between 0.900-0.930 (Hope et al., 2005; Ramirez et al., 2006). In field conditions, high FHB incidence and DON contamination are not always directly correlated and contrasting cases have been reported (Hermann et al., 1998; Höhn et al., 2002; Menniti et al., 2003; Miedaner et al., 2003), suggesting that other complex factors such as year, site, cultivar and agronomic technique are very important in DON contamination. However, DON contamination has been reported to depend to a greater extend on rainfalls around heading and anthesis and on temperatures between 10°C and 32°C around the same period (Hooker et al., 2002; Koch et al., 2006).

DON seems to play a decisive role on the aggressiveness of *F. graminearum* and *F. culmorum*. DON is a strong protein synthesis inhibitor and this may cause the inhibition of defence response genes in susceptible host-cultivars, leading to a rapid increase of the disease (Mesterhazy, 2002). Moreover, DON can induce the complete loss of chloroplast pigments at a sub-lethal concentration (Rotter et al., 1996). It has been suggested that in the presence of DON, plant defence mechanisms are not triggered fast enough, thereby leading to an increased aggressiveness of the pathogen (Wagacha and Muthomi, 2007).

4. Crop Management Systems to Reduce the Occurrence of FHB and DON Contamination in Wheat

The accumulation of DON in the grain is intimately related to the development of the disease in the field (McMullen et al., 1997). Therefore, the current strategy to control DON in cereals is to manage FHB. Most practices aimed at FHB prevention are essentially crop management practices whose goal is to reduce infection or fungal growth by toxigenic fungi (Parry et al., 1995). Among the agricultural practices, FHB development and/or DON contamination in wheat grains are mainly affected by tillage, crop rotation, varieties and fungicide application (Pirgozliev et al., 2003). None of these strategies on their own are able to significantly reduce the impact of this disease (Edwards, 2004a).

Although information is available on the basic effect of single agricultural practices on *Fusarium* infection and DON contamination in wheat, only a few studies have been made to quantify the relative importance of each of these factors compared to the others that are involved and to verify their interactions and combined effects.

Preliminary information on the relative effect of management options on FHB incidence and DON contamination could be obtained through a simplified approach to calculate the severity of the effect of the single factors, as proposed by Koch et al. (2006). In this study, the severity of the effect of the single agricultural practices was calculated as follows: the mean DON value of the treatment with the highest DON concentration is divided by the mean value of the lowest treatment.

The data obtained in the experiments conducted by Koch et al. (2006) are reported and compared in table 5 with other information obtained from the literature and with information from similar trials conducted in 2005–2007 in 4 sites in northern and central Italy. In all these studies, the effect of at least two agricultural practices on DON contamination were compared, in natural inoculated conditions.

Based on these data, the following ranking order can be obtained: susceptibility of wheat cultivar (1.6–5.6) = the preceding crop (2.0–4.3) > soil tillage (1.4–6.5) ≥ fungicide application at anthesis of wheat (1.7–2.7). These data confirm that wheat crop quality management, aiming at a low DON concentration, must start in the field and first focus on agronomic factors that influence FHB infection. Above all, conditions such as preceding host crops, especially maize, which can leave infected residues in the field and the cultivation of a susceptible cultivar contribute to increase *Fusarium* infections of wheat crops. These data confirm the results obtained in France in 7 years of trials by researchers from Arvalis (Barrier-Guillot et al., 2006; Gourdain, 2008), who obtained a grid with 7 levels of risk for DON contamination in wheat grains based on 3 combined risk factors: previous crop, tillage and varietal sensibility to FHB.

4.1. Crop Residue and Proceeding Crops

It is generally accepted that if a wheat crop follows an alternative host for *Fusarium* pathogens it is at a greater risk to FHB and DON contamination of grain. All species of *Fusarium* are able to survive through the winter as mycelia in non-decomposed residues from the previous crop, with the subsequent production of ascospores in the perithecia located on the surface of the pathogen-infested residues (Osborne and Stein, 2007). The primary reservoir of inoculum is debris from the previous crop (Bai and Shaner, 1994) and pathogens survive longer on residues that do not degrade easily, such as stem nodes (Sutton, 1982). Thus, FHB disease and DON contamination are more severe if the preceding crops are maize or sorghum, rather than wheat or barley and even less contamination is observed following other crops (Dill-Macky and Jones, 2000; Krebs et al., 2000; Champeil et al., 2004b; Maiorano et al., 2008).

Table 3. Severity of the effect of several agricultural practices on deoxynivalenol (DON) contamination in winter wheat grain on average of the other experimental factors included and with no-inoculated conditions

Country and year	Effect	Value in			Reference
		Numerator	Denominator	Severity of effect	
USA (Minnesota); 1995-1997	Preceding crop	Maize	Soybean	2.0	Dill-Macky and Jones, 2000
	Tillage	Direct sowing	Ploughing (30 cm)	1.4	
Canada (Ontario); 1996-1999	Preceding crop	Maize	Soybean	2.9	Schaafsma et al., 2001
	Tillage(*)	Minimum tillage	Ploughing (30 cm)	1.6	
	Cultivar	Highly susceptible	Moderately resistant	3.7	
Germany; 2001-2002 (Exp. 1)	Preceding crop	Winter wheat	Sugar beet	4.3	Koch et al., 2006
	Tillage(*)	Direct sowing	Ploughing (30 cm)	2.7	
	Cultivar	Highly susceptible	Moderately resistant	4.3	
Germany; 2001-2002 (Exp. 2)	Tillage (after wheat)	Direct drilling	Ploughing (30 cm)	3.4	Koch et al., 2006
	Cultivar	Highly susceptible	Moderately resistant	5.6	
	Fungicide application	Without	With	2.1	
Italy; 2002-2003	Cultivar	Highly susceptible	Moderately resistant	1.6	Blandino et al., 2006
	Fungicide application	Without	With	1.7	
Italy; 2006-2007	Tillage (after maize)	Direct sowing	Ploughing (30 cm)	6.5	Blandino, unpublished
	Cultivar	Highly susceptible	Moderately resistant	4.8	
	Fungicide application	Without	With	2.7	

* Data obtained from commercial wheet farm fields with different previous crops

This may be accounted for in two ways: first, it is related to the density of the residues left by the preceding crop, second, the nature of the preceding crop can affect the composition of the pathogen complex throughout the following year.

As clearly reported in the literature, limited soil tillage or no-till increase the frequency of head blight, whereas deep tillage, such us ploughing, decreases it (Teich, 1989; McMullen et al., 1997; Miller et al., 1998; Krebs et al., 2000; Maiorano et al., 2008), because it reduces the density of the residues on the surface of the soil and helps to decrease the production of inoculum (Teich and Hamilton, 1985) and the quantity of spores available for dispersal (Bateman et al., 1998).

The general effect of tillage practices that enhance *Fusarium* infection is highly variable: in experiments conducted in Italy, direct sowing after maize increased the DON concentration in wheat grains by a factor of 6.5 on average in all conditions. This effect was much smaller when direct sowing was applied after wheat or other crops (Teich and Hamilton, 1985; Dill-Macky and Jones, 2000). In general, the data presented clearly show that the tillage system is less important than the influence of the preceding crop and the susceptibility of the wheat cultivar. Furthermore, the relative importance of tillage practices for a low DON concentration in wheat grain is closely related to the level of FHB pressure: under highly infectious conditions, for example with grain maize as the previous crop, a highly susceptible cultivar and abundant rainfall during anthesis, minimum tillage or no-tillage could result in a considerably increased DON concentration.

4.2. Cultivar Susceptibility

Among the other factors examined in the several studies, the cultivation of a susceptible cultivar showed a clearly higher DON contamination compared to a moderately resistant one. With the exception of the experiment conducted in North Italy in 2002-2003 (Blandino et al., 2006), in all the other works, the use of a more susceptible variety increased the content of this mycotoxin in wheat grain by a factor of 4.6. These results, in each situation, reported a comparison between a moderately resistant and a susceptible variety. It is assumed that if wheat cultivars with an intermediate *Fusarium* rating instead of a moderately resistant and susceptible or highly susceptible rating are tested, a considerable decrease of the varietal effect could be expected. Furthermore, all the wheat cultivars used in these experiments were commercial and commonly present on the farm fields. This means that the choice of the less susceptible cultivars available on the markets is a key tool that both the farmer and end user could apply to reduce the risk of high DON content.

At present, no durable, fully FHB-resistant wheat cultivars exist, therefore control relies on the use of cultivars with partial resistance (Mesterhazy, 2002), but in recent studies by Mesterhazy et al. (2005) and Tóth et al. (2008), the wheat varieties that are more resistant to FHB were shown to reduce DON production to almost zero.

The development of FHB resistant varieties is one of the main goals of numerous breeding programmes across the world (Yuen and Schoneweis, 2007).

Five types of varietal resistance are currently know: they affect the penetration of the fungus into the plant (type I) (Schroeder and Christensen, 1963), the infection kinetics, with resistance to colonisation within the ear (type II) (Schroeder and Christensen, 1963), the degradation of the mycotoxins (type III) (Miller and Young, 1985; Miller et al., 1985), the consequences of infection, with tolerance to high DON concentrations (type IV) (Wang and Miller, 1988) and the resistance to grain infection, with differences in yield despite similar levels of attack (type V) (Mesterházy, 1995).

Type I and II resistances have been associated with certain morphological characteristics of wheat cultivars. Hilton et al., (1999) observed a significant negative relationship between plant height and resistance to FHB. The earliest variety seems to accumulate more deoxynivalenol that the late-flowering varieties (Jenny et al., 2000), probably due to a greater coincidence of the phase of maximum susceptibility in plants with the period most favourable for spore dispersion.

Such observations have been supported by genetic mapping of the quantitative trait loci (QTL), where the QTLs of the plant morphological traits, such as plant height and heading date, coincide with the QTLs of lower FHB severity and DON concentration (Zhu et al., 1999; Ma et al., 2000).

Several research groups have identified molecular markers linked to QTLs associated with various aspects of resistance (Buerstmayr et al., 2002; Buerstmayr et al., 2003). The most common sources of genetic resistance are derived from the Chinese variety (Zhang et al., 2004; Yang et al., 2005; Yu et al., 2008).

As far as the most recent employment of genetic engineering is concerned, Anand et al. (2003) produced transgenic wheat lines, expressing chitinase and ß-1,3-glucanase genes, that provided moderate resistance to FHB in greenhouse trials. In addition, transgenic wheat expressing the *F. sporotrichioides* Tri101 gene, which encodes a protein that detoxifies the trichothecene mycotoxins produced by *F. graminearum*, conferred partial protection against the spread of FHB in infected spikes (Okubara et al., 2002).

Combining all the desired agronomic characters, plus resistance to other important disease and insects, with a high degree of resistance to FHB is still a major challenge. The successes in enhancing resistance to FHB and DON contamination in cereals using conventional breeding, molecular markers and through transgenic approaches have been thoroughly discussed in several reviews (Dahleen et al., 2001; Kolb et al., 2001; Ruckenbauer et al., 2001; Hollins et al., 2003; Miedaner et al., 2003; Bai and Shaner, 2004; Champeil et al., 2004a; Snijders, 2004).

4.3. Fungicide Application

Adapting the crop sequence, incorporating the previous crop residues and using a resistant cultivar are precautionary measures to control FHB infection that must be taken at the beginning of wheat cultivation. During crop growth, a fungicide application may be applied at anthesis to diminish DON formation (Pirgozliev et al., 2003).

Although several studies have demonstrated that good levels of control can be achieved with fungicides in *in vitro* experiments (Matthies et al., 1999; Ramirez et al., 2004; Mullenborn et al., 2008) or in trials in which wheat is artificially inoculated (Boyacoglu et al., 1992; Mielke and Weinert, 1996; Matthies and Buchenauer, 2000; Mesterhazy et al., 2003; Haidukowski et al., 2005), the situation under natural infection condition is less clear. In the field, there is conflicting evidence as to the ability of fungicides to reduce FHB and to reduce DON contamination, due to the complex interaction between fungicide, mycotoxin production, other fungal species and weather conditions (Milus and Parsons, 1994; D'Mello et al., 1998; Simpson et al., 2001).

The outcome of the use of fungicides seems to depend on the fungal species that is present and the effect that the particular fungicide has on these species. Fungicides containing triazoles (the most important are bromuconazole, epoxiconazole, metconazole, propioconazole, tebuconazole and tetraconazole), imidazoles (Prochloraz) or triazolinthiones (prothioconazole) active ingredients which inhibit the biosynthesis of ergosterol, have proved to be the most active molecules against FHB infection and DON contamination (Jennings et al., 2000; Edwards et al., 2001; Siranidou and Buchenauer, 2001; Menniti et al., 2003; Ioos et al., 2005).

Strobilurins have instead shown poor efficacy against FHB caused by toxigenic *Fusarium* spp. (Pirgozliev et al., 2003) and often both in vitro and in field studies have revealed an increase in DON concentration following an application of these fungicides compared to unsprayed controls (D'Mello et al., 2001; Simpson et al., 2001).

Magan et al. (2002) suggested that the fungicide action could influence the different fungal species by changing the ratio between the toxigenic and non toxigenic fungi. Different authors (Bertelsen et al., 1999; Jennings et al., 2000; Simpson et al., 2001) observed that the application of azoxystrobin, unlike triazoles, has little effect on the reduction of *F. culmorum*, *F. avenaceum* and *F. graminearum*, while it guarantees a significant reduction in *M. nivale*, which is not able to synthesise DON (Edwards, 2004a). As suggested by Pirgozliev et al. (2003), the increase in DON that was observed is therefore probably due to an increase in the infection of the *Fusarium* species, following the reduction in the presence of *M. nivale*. To assure lower mycotoxin contamination in wheat grain, the application of strobilurin fungicides is only recommended in a mixture with azoles (Chala et al., 2003; Pirgozliev et al., 2003). The effect on *Fusarium* also depends on the timing and frequency of applications (Homdork et al., 2000). The best control of FHB pathogens using fungicide has been observed following applications at wheat anthesis, whereas their efficacy declined rapidly with time after or before this growth stage (Wiersma and Motteberg, 2005).

In the studies reported in Table 5, the fungicide application controlled FHB infection and DON contamination in the regions more prone to the disease and in the wheat cultivars with low levels of resistance. Thus, fungicide may be useful when weather or other cultural conditions are particularly favourable for disease development and FHB disease pressure is high.

4.4. Effect of Other Agricultural Practices

Mineral fertilization, weed control, sowing date and planting density seem to play a minor role on FHB infection and DON contamination in wheat compared to the previously reported agricultural practices (Champeil et al., 2004b).

Martin et al. (1991), Lemmens et al. (2004) and Heier et al. (2005) reported that FHB and DON contamination may be directly correlated to an increase in mineral N-input. However, the effect of mineral nutrition on head blight infection is still unclear: other works have reported that the level of the disease is similar with or without a nitrogen fertilizer application (Teich and Hamilton, 1985; Fauzi and Paulitz, 1994; Schaafsma et al., 1998) or, on the contrary, have shown that nitrogen application can limit the disease (Teich and Nelson, 1984).

Teich (1987) observed a higher FHB infection in Ontario, Canada, with the application of ammonium nitrate than when urea was applied. In other studies the application of different types of nitrogen fertilizer did not lead to any significantly different results (Lemmens et al., 2004). Potassium, phosphorus and pH did not appear to affect the DON content to any great extent (Teich and Hamilton, 1985).

The isolation of *Fusarium* species, pathogenic to winter wheat from common grasses (Cassini and 1970; Lager and Wallenhammer, 2003) and broad-leaved weeds (Jenkinson and Parry, 1994), suggests that an effective weed control can reduce the inoculum. Moreover, weeds can also favour *Fusarium* infection by modifying the microclimate and increasing heat and humidity in the canopy.

Sowing date could have an indirect effect on the FHB infection. It in fact influences the moment in which the flower appears, a moment which can sometimes coincide or not with the *Fusarium* spore release (Lacey et al., 1999). Caron (1993) reported that a late sowing date favoured the development of head blight, whereas an early sowing date favoured the development of foot rot.

Severe infections of head blight have been reported with high planting densities (Mesterházy, 1995), which could increase the humidity of the canopy (Deadman and Cooke, 1997), favouring spore germination. On the other hand, Champeil et al. (2004b) suggested that an elevated density of the canopy could be a greater obstacle to spore dispersal.

Nevertheless, at the moment, the magnitude of the effects that these agricultural practises have on FHB disease development, if indeed there is one, remain unclear and each one of them needs to be evaluated alone and in combination with the other crop techniques. Although only a few studies have been conducted on the effects of agricultural practices, it is possible to say they are probably very dependent on the climatic and soil conditions of each region. Thus, it is difficult to extrapolate the results obtained for these preventive measures from each wheat growing area and to generalize them for application in other regions.

In the future, biological control could be used to offer an additional strategy and as part of an integrated management of FHB. In vitro assays, trials in glasshouses and under field conditions have shown that some bacteria included in the genera *Bacillus* and *Pseudomonas*, some yeasts belonging to the genera *Rhodotorula*, *Sporobolomyces* and *Cryptococcus* and some fungi of the genera *Trichoderma*, are able to reduce *F. graminearum* growth (Fernandez, 1992; Schisler et al., 2000; Schisler et al., 2002; da Luz et al., 2003; Khan et al., 2004). These potential biocontrol agents could be applied at wheat anthesis in order to prevent *Fusarium* infection, when conditions for the disease exist. In a recent work under greenhouse conditions, Palazzinia et al. (2007) observed a significant reduction in DON contamination in wheat grain inoculated with bacterial strains. Unfortunately, under field conditions, the biological control effects have been shown to be too variable (McMullen et al., 2002).

4.5. Effect of Crop Management Systems

In order to optimise the control of FHB and DON contamination of grain, it could be interesting to evaluate the effect of different crop management systems.

At present, there is a lack of information about the effects of entire cropping systems on FHB.

Full-factorial experiments that take into account tillage (tilled vs not tilled), wheat cultivar (susceptible vs moderately resistant) and fungicide application (triazoles at heading vs not treated) were conducted in naturally inoculated conditions in 4 sites in Italy during the 2006-2007 period.

The results were discussed taking into account some of the possible combinations of the tested agricultural practices, representing different crop management systems classified for the expected risk in FHB infection and DON contamination (Table 6), according to the following scheme:

- HR: high risk agricultural practices, characterized by direct sowing of an untreated highly susceptible cultivar;
- R: risk agricultural practices, with direct sowing of a highly susceptible cultivar, treated with fungicide at heading;
- MR: medium-risk agricultural practices, with ploughing and cultivation of a highly susceptible, untreated cultivar;
- RI: right agricultural practices, characterized by ploughing, a moderately resistant cultivar and no fungicide application;
- CA: careful agricultural practices, in which a moderately resistant cultivar was sown after ploughing and was treated with fungicide at heading.

The crop management systems showed significant differences as far as yield, incidence and the severity of FHB, and DON contamination are concerned.

As expected, FHB severity decreased going from the higher risk crop management system to the more careful ones. Compared to the HR treatment, the reduction in the severity of FHB was 45, 43, 76 and 91% for R, MR, RI and CA, respectively. On the other hand, grain yield generally increased with the application of crop management systems that are less favourable to disease development.

The crop management system R reduced DON contamination by 60% compared to HR, while a higher reduction was observed for both the MR and RI protocols. Compared to RI, a further, not significant reduction, was observed with the fungicide application at heading (CA).

If the data of DON contamination presented by Koch et al. (2006) are extrapolated, and the agricultural practices in the same previously reported cropping systems are compared, similar results can be obtained. Compared to the HR treatment, the head blight severity reduction was 71, 80, 96 and 96% for R, MR, RI and CA, respectively.

These results underline that in years and environments with low FHB pressure the main factors that contribute to heavy *Fusarium* infections of wheat crops are the presence of residues of *Fusarium*-host crops on the soil surface and the cultivation of susceptible cultivars. Moreover, no further significant reductions were observed with fungicide applications at heading on a resistant wheat cultivar sown after ploughing.

Obst et al. (2000), in a four-year study in Germany, determined five FHB risk factors: (i) maize as previous crop, (ii) minimum tillage after maize, (iii) use of a moderate or highly susceptible wheat variety and (iv) application of strobilurin products. Again in this experiment, the combination of these risk factors showed synergic effects: a single risk factor showed a threefold relative risk of DON contamination, while four combined risk factors showed a fifty-six fold relative risk of DON contamination.

The data available at the moment concerning the comparison of different cropping systems underline that crop management strategies used to minimize DON contamination risk in wheat grains must first consider the effects of the preceding crop and cultivar. Soil tillage and fungicide application at anthesis could lead to a significant reduction in DON contamination in years and environments with higher FHB disease pressure. For this reason, these preventive measures should be taken into consideration in all the wheat growing areas most prone to FHB infection and DON contamination.

In the next few years the effects of all crop techniques on all mycotoxins must be combined to provide guidelines on Good Agricultural Practices (GAP), in order to minimise mycotoxin contamination of cereal products.

Table 4. Effect of combined agricultural practices on grain yield, FHB incidence and severity and DON contamination

Year	Treatment		Soil tillage[V]	Cultivar[W]	Fungicide treatment[X]	FHB incidence (%)[Y]	FHB severity (%)[Z]	Yield (t ha^{-1})	DON (μg kg^{-1})
2006	HR	high risk	Direct sowing	HS	untreated	39.3 a	10.2 a	6.7 c	1594 a
	R	risk	Direct sowing	HS	prochloraz at heading	18.9 c	4.9 b	7.0 bc	598 b
	MR	medium risk	Ploughing	HS	untreated	26.8 b	4.8 b	6.8 c	161 c
	RI	right	Ploughing	MR	untreated	14.3 c	1.4 c	7.3 ab	44 c
	CA	careful	Ploughing	MR	prochloraz at heading	6.1 d	0.2 d	7.5 a	13 c
2007	HR	high risk	Direct sowing	HS	untreated	68.3 a	12.3 a	4.7 c	531 a
	R	risk	Direct sowing	HS	prochloraz at heading	47.8 b	7.5 b	5.4 b	221 b
	MR	medium risk	Ploughing	HS	untreated	65.9 a	8.1 b	5.8 b	156 b
	RI	right	Ploughing	MR	untreated	49.2 b	4.3 c	6.5 a	38 c
	CA	careful	Ploughing	MR	prochloraz at heading	34.5 c	2.0 d	6.6 a	13 c

V: The previous crop was maize in each year and site.
W: HS: highly susceptible; MR: moderately resistant.
X: triazole fungicide, A.I. prochloraz (Sportak 45 EW®, BASF) was applied once at 0.59 kg (A.I.) at heading.
Y: FHB incidence was calculated as the percentage of ears with symptoms of disease at the soft dough stages (GS 85).
Z: FHB severity was calculated as the percentage of kernels per ear with symptoms of disease at the soft dough stages (GS 85).
Values, in the same column, followed by the same letter are not significantly different, according to the Student-Newman-Keuls test. The data reported in the table are the means of 4 sites in northern and central Italy.

5. Predictive Models of Fusarium Head Blight and Deoxynivalenol Contamination in Wheat

A model is a simplified representation of a system that is a limited part of the reality which contains interrelated elements. The behaviour of the system and the relations between the elements are simulated by mathematical and statistical equations and/or logical operations (France and Thornley, 1984; de Wit, 1993; Jones and Luyten, 1998). The great development and diffusion of information technologies and computer programming, above all during the 1980s, have led to the possibility of elaborating a large amount of data and developing very complex models (Barrett and Nearing, 1998). Nevertheless, the level of complexity and detail needed for a specific model depends on the objectives and questions being asked of the model and on the amount of information (data) and time that is available for model building and testing. In agriculture, models are mainly used for three purposes: i) by researchers to examine scientific hypothesis or to study and better comprehend biological/ epidemiological/agricultural systems; ii) by governmental agencies as policy analysis tools, to assist in best management solutions for problem or susceptible areas; iii) by agronomic technical services or producers as part of computerized decision support systems to evaluate optimum management practices. The use of a model to predict the outcome of a disease is desirable to enhance and trigger management opportunities with the aim of reaching high technological and/or nutritional and/or productivity quality and/or safety of the production (Boote *et al.*, 1996; Schaafsma and Hooker, 2007; Maiorano, 2008; Maiorano *et al.*, 2009). Kranz and Royle (1978) classified the epidemiological models according to their main objective in i) descriptive (models that provide generalise experimental results), ii) predictive (models that allow prediction of epidemics), and iii) conceptual (models that allow the identification of the problem by distinguishing cause from effect). The descriptive models are easy to comprehend, often require fewer inputs, and often are easier to use and apply, but they have to be calibrated to each new site and year. On the other hand, the conceptual models are better able to model genotype X environment interaction, but their complexity makes them more difficult to understand, to use and to apply, and they also require more input information (Prandini et al., 2008).

The relevant economic impact (section 1) of FHB for the main cereal producing countries in the world has stimulated many researchers to develop epidemiological models in order to simulate FHB infection and DON contamination. Most of these models have been included in computerized decision support systems mainly used to help the operator in the filed to decide weather or not applying and the correct timing of application of fungicides against FHB.

A review of the models that have been developed for FHB and for DON synthesis is presented in this section. A similar review has recently been presented by Prandini et al. (2008).

5.1. Argentina

Moschini and Fortugno (1996) developed nine empirical equations, in the region of Pergamino, Argentina, associating mean head blight incidence data with temperature and moisture variables. The models were obtained with linear regression techniques by fitting a 12-year dataset (1978-1990). The two equations showing the best results were validated with data from Pergamino for the years 1991, 1992 and 1993 and subsequently also in the northern Argentinean Pampas Region (from 1993 to 1995) by Moschini et al. (2001). The meteorological variables (temperature, humidity, rainfall) and the time segment of observation (beginning eight days prior to the heading date and finishing after the accumulation of 530 degree days) where chosen for their agreement with the fungal environmental requirements. The authors also incorporated a factor for cultivar susceptibility in the model for its validation in the Pampas Region in 1993-1995, thus improving the fit between the observed and predicted head blight incidence.

Both models developed in Pergamino satisfactorily predicted head blight incidence in both Pergamino (Moschini and Fortugno, 1996) and in the Pampas region (Moschini et al., 2001) showing that these models could be successfully used to predict FHB incidence in those regions.

5.2. Argentina, Brazil, Uruguay

Fernandes et al. (2004) used a linked process based modelling approach to explain FHB epidemics that developed at three sites in South America: Pergamino (Argentina), La Estanzuela (Uruguay), and Passo Fundo (Brazil). In order to model FHB in wheat, they used a wheat development model (Cropsim-Ceres 2002 included in DSSAT 4.0) and an FHB model which calculate three parameters, i) proportion of susceptible tissue, ii) infection rate, iii) spore cloud density. The final risk is calculated from the summation of the partial indices. The rates and rules in the model are influenced by weather variables such as daily mean temperature, daily mean relative humidity, daily solar radiation, and consecutive rainy days.

In Passo Fundo, Brazil, the model evaluation with disease data from 5 years of epidemics, showed that the risk estimated by the model explained over 95% of variation in disease field epidemics. Details on development of the model development still have to be published.

5.3. Belgium

Detrixhe et al. (2003) developed an agrometeorological model that is able to estimate the risk of FHB infection at a regional scale (1km x 1km grid) in winter wheat. Moisture on plant surfaces has a considerable influence on fungal pathogen development. Therefore, estimation of leaf wetness and duration is an important agrometeorological parameter that needs to be taken into account in the development of FHB in wheat. This model is based on a leaf wetness duration calculation module (Oger et al., 2002), which estimates the duration of surface wetness from data provided by a weather station network (for standard meteorological data) and a weather radar which cover the whole Belgian territory (for precipitation data). Weather data (temperature, relative humidity, wind speed, short-wave and long-wave radiation) are collected over a period that starts 8 days before wheat flowering and ends 7 days after and these data are interpolated to a grid size of 1 km x 1 km, which cover the entire Belgian territory. A first evaluation of the model was carried out in 2002 with the collection of 43 wheat samples and their analyses for *Fusarium* spp. incidence. The first evaluations highlighted some instrumental limitations concerning the risk assessment in regions located at more than 120 km from the weather radar and the necessity to include FHB parameters like crop rotation, cultivar characteristics and field disease history in the model. Nevertheless, the use of spatial interpolation of meteorological data and of the estimation of leaf wetness makes this model very interesting and original. Further calibration/validation tests are in progress to optimise the model. The institute (Agricultural Research Center, Gemloux, Belgium) is also still working on the question of leaf wheat duration (Oger, 2008, *personal communication*).

5.4. Canada

DONcast is a prediction model that has been used successfully in Ontario, Canada, as a forecasting tool mainly to help with fungicide spray decisions at heading and for grain marketing decisions (Hooker et al., 2002). Growers and crop advisors in Ontario have used it since 2000 and it continues to evolve with the inclusion of environmental and agronomic diversities. It was developed from data collected from over 750 farms across Ontario, since 1996, using multiple regression techniques which allowed the most important climatic variables associated with the increase or decreases of DON to be indentified. The model is based on three equations, related to three different critical periods around wheat heading: i) from 4 to 7 days before heading, ii) from 3 to 6 days after heading, and iii) from 7 to 10 days after heading. The first critical period (i) corresponds to inoculum production: in this period, rainfall amounts of>5 mm per day trigger and increase the DON potential while daily minimum air temperatures of less than 10°C limit the DON potential. Similarly, weather variables in critical periods ii and iii correspond to infection during flowering and fungal growth. Here, the number of rainy days and days with relative humidity over 75% at 11.00 h increase the DON potential; daily maximum temperatures over 32°C and average temperatures of less than 12°C instead limit the DON potential. Daily weather data are converted to binary values using a set of criteria for each variable within each critical period. The binary values are summed within each weather variable and critical period and the complex summations are plugged into empirical equations to forecast the concentration of DON in wheat grain at harvest. Validation analysis has shown a greater accuracy than 80% in determining whether a fungicide application is able to reduce DON. In 2004, a web-based interactive model was developed for industry in Ontario. This model allows input of field-specific weather and agronomic variables to be used as input for more accurate predictions. The model is available on the website www.ontarioweathernetwork.com/DONcast.cfm.

Since 2000, DONcast has been validated and calibrated not only in Canada, but also in other regions including the United States, Uruguay and France (Hooker and Schaafsma, 2004; Schaafsma and Hooker, 2006; Schaafsma et al., 2006; Schaafsma and Hooker, 2007).

5.5. The Czech Republic

Data about DON contamination in wheat grain, weather conditions during the growing season and cultivation practices from two field experiments conducted in 2002-2003 were used for the development of a neural network model designed for DON contamination prediction in The Czech Republic (Klem et al., 2007). Using the data from field experiments, Klem et al. trained neural networks to predict DON content, on the basis of weather data, as a continuous input variable and preceding crop as a categorical input variable. The neural network works with five input variables: the categorial variable (preceding crop) and continuous variables (average April temperature, sum of April precipitation, average temperature five days prior to anthesis, sum of precipitation five days prior to anthesis).

5.6. Italy

Rossi et al. (2003) developed a dynamic simulation model for the risk of FHB in wheat. The model calculates a daily infection risk based on sporulation, spore dispersal and infection of the host tissue of the four main species that cause the disease (*G. zeae*, *F. culmorum*, *Gibberella. avenacea*, *Monographella nivalis*). Spore yield and dispersal are calculated as functions of temperature, rainfall and relative humidity, while the main factors affecting the infection rate are temperature, wetness and the host growth stage. The model also calculates the risk for mycotoxin production by *G. zeae* and *F. culmorum* in the infected head tissue. The model was validated with data from different wheat growing areas in northern Italy from 2002 to 2004 using different soft and durum wheat cultivars. The validation showed a general agreement between the model simulations and actual FHB epidemics and DON contamination under different epidemiological conditions.

The interesting aspect of this model is that it was developed using the System Analysis principles (Leffelaar, 1993) and unlike the others, this is the first known attempt to build a mechanistic model of FHB development and DON contamination in wheat.

5.7. Switzerland

Musa et al. (2007) developed the internet-based decision support system FusaProg. The model included in FusaProg takes into account the effects of cropping factors, previous crops, soil and straw management, as well as the susceptibility of the wheat varieties to FHB. These factors are used as driving variables and are combined with the prevailing weather conditions and growth stage in order to predict the deoxynivalenol content of a specific wheat plot before harvest. Hourly measured and forecasted data from 60 stations of a Swiss weather service and private stations are transferred to a server where the data are automatically analyzed for periods conducive for *F. graminearum* infection, according to weather rules based on three parameters: daily mean temperature, sum of precipitation and daily mean of the relative humidity. Weather conditions over the last three days as well as the weather forecast are taken into account for the DON risk assessment. The FusaProg model assesses the risk of DON contamination in three steps: i) a specific field is classified according to the previous crop, the straw management, and the seedbed tillage; ii) the susceptibility of the variety and the growth stage of the host is evaluated; iii) weather data are taken into account. In 2006, the systems was evaluated by Swiss cantonal plant protection officers and in 2007, the system was made available to Swiss wheat producers on the internet site http://www.fusaprog.ch/.

5.8. The United States

In the United States, the Fusarium Head Blight Risk Assessment Tool was developed by a cooperative research project at five universities. Using historical records of the weather and observations of disease severity in field plots, logistic regression models were developed to predict the probability of the mean disease severity exceeding 10%. Several empirical models were developed that used weather data for seven days prior to flowering, seven or 10 days starting from flowering, or both pre- and post-flowering time windows. The currently-used models, which were 80% accurate in predicting the data used in model development, utilize only pre-flowering environmental data. The implicit assumption underlying the models is that scab epidemics are determined (at least in part) by inoculum availability at flowering and that the weather immediately preceding flowering determines the magnitude of the available inoculum. In 2004, Penn State and Ohio State Universities deployed the predictive system in a Web-based format for evaluation purposes in 23 states. Separate models were used for winter wheat with a low level of corn residue, winter wheat with a high level of corn residue, and spring wheat production systems. Scab risk maps were produced at a 20-km resolution using temperature and relative humidity data. The model is available on the Internet site http://www.wheatscab.psu.edu/riskTool.html

6. Conclusions and Future Outlook on the Prevention of FHB and DON Contamination in Wheat

In these sections it has been illustrated that Fusarium head blight and *Fusarium* toxin contamination of soft and durum wheat depends on many factors. This means that a unique solution to this problem cannot exist.

Thus, the producers' objective must be to find technical strategies that can minimize the risk of DON contamination as much as possible in wheat grain, flour, semolina and in foodstuffs, taking into account that a total absence of the toxin is almost impossible in the field.

In order to reduce the risk of contamination, it is necessary for producers to adopt a strategy that integrates and modulates all of the preventive measures presented in section 4. In other words, it is necessary for producers to make use of the available wheat varieties that are known to be more resistant and to apply all of the good agricultural practices tuned to a determined cultivation area.

Even though this kind of strategy could appear to be easy to apply, the use of this type of approach is only just a beginning. In fact, until now, soft and durum wheat production has been dominated by strategies aimed at maximising the productivity and the technological quality for the transformation industry. These aspects will doubtlessly continue to guide the decisions of the producers and of the whole food and feed production chain, but now they will also have to pay attention to the safety aspects, which are considered a fundamental pre-requisite in cereal production. Thus, all of the agricultural practices aimed at the optimization of productivity and technological quality should now be inserted into production strategies aimed at the reduction of the presence of contaminants like DON, not the other way around.

In this context, the higher the probability of running into high DON contamination, the tighter the agricultural practice constraint necessary to reduce the risk of FHB and consequently the more the other cultivation possibilities will need to be reduced to improve other technical aspects.

In an environment with a high DON contamination risk, the agronomic constraint to prevent this toxin could become so tight that the production destined to specific products (such as baby food) becomes economically inconvenient, or that the premium prices necessary to compensate the losses or the higher costs for more expensive agricultural practices result to be no longer sustainable for the chain production. In this perspective, the expansion of durum wheat cultivation towards cooler areas—as observed, for example, in Italy in recent years—could be more hampered by the uncertainty about DON contamination than productivity or technological quality difficulties (e.g., protein content, flour indexes) that this cultivation could encounter in new environments.

In the future, should stricter limits be established for the most widespread mycotoxins—e.g., DON—wheat cultivation will have to abandon some cultivation areas unless more effective solutions to contrast the toxigenic fungi are found. To this end, it would be necessary to have i) more resistant wheat varieties with the same productivity and quality characteristics, ii) more effective and innovative fungicides, and iii) competitor microorganisms against the toxigenic ones, or those that can destroy the toxic molecules. Nevertheless, at the moment none of these solutions appear to be solely able to solve the problem, if the capability of the fungal species to develop resistances or adaptive strategies is also taken into account.

Thus, we come back to the initial affirmation: FHB development and DON synthesis depend on many factors and, consequently, the elimination of the contamination is not a realistic objective. Only an integrated strategy is realistic and will be able to answer the need for reducing the presence of this contaminant as much as possible.

References

Ali, S.,Francl, L., 2001. Progression of Fusarium species on wheat leaves from seedling to adult stages in North Dakota. In: S.M. Canty, J. Lewis, L. Siler and R.W. Ward (Editors), 2001 National Fusarium Head Blight Forum. Michigan State University, Erlanger, KY, pp. 99.

Anand, A., Zhou, T., Trick, H.N., Gill, B.S., Bockus, W.W.,Muthukrishnan, S., 2003. Greenhouse and field testing of transgenic wheat plants stably expressing genes for thaumatin-like protein, chitinase and glucanase against Fusarium graminearum. Journal of Experimental Botany, 54(384): 1101-1111.

Atanasoff, D., 1920. Fusarium blight (scab) of wheat and other cereals. J. Agric. Res, 20(1): 1-32.

Bai, G.H.,Shaner, G., 1994. Scab of Wheat - Prospects for Control. Plant Disease, 78(8): 760-766.

Bai, G.H.,Shaner, G., 2004. Management and resistance in wheat and barley to Fusarium head blight. Annual Review of Phytopathology, 42: 135-161.

Barrett, J.R.,Nearing, M.A., 1998. Humanization of decision support using information from simulation. In: R.M. Peart and R.B. Curry (Editors), Agricultural systems modeling and simulation. Marcel Dekker, Inc., Gainesville, Florida (USA), pp. 1-18.

Barrier-Guillot, B., Delambre, M., Morel, A., Maumene, C., Gouet, H., Grosjean, F.,Leuillet, M., 2006. Identification of agronomic factors that influence the level of deoxynivalenol (DON) in wheat grown in France. In: H. Njapau, S. Trujillo, H.P. van Egmond and D.L. Park (Editors), Mycotoxins and phycotoxins. Advances in determination, toxicology and exposure management. Wageningen Academic Publishers, The Netherlands, pp. 239-247.

Bateman, G.L., Murray, G., Gutteridge, R.J.,Coskun, H., 1998. Effects of method of straw disposal and depth of cultivation on populations of Fusarium spp in soil and on brown foot rot in continuous winter wheat. Annals of Applied Biology, 132(1): 35-47.

Bechtel, D.B., Kaleidau, L.A., Gaines, R.L.,Seitz, L.M., 1985. The Effects of Fusarium Graminearum Infection on Wheat Kernels. Cereal Chemistry, 62: 191-197.

Bertelsen, J.R., de-Neergaard, E.,Smedegaaed-Petersen, V., 1999. Reason for improved yield when using azoxystrobin in winter wheat, 16th Danish Plant Protection Conference. Crop Protection in Organic Farming, Pests and Diseases, Tjele, Denmark.

Birzele, B., Meier, A., Hindorf, H., Krämer, J.,Dehne, H.W., 2002. Epidemiology of Fusarium Infection and Deoxynivalenol Content in Winter Wheat in the Rhineland, Germany. European Journal of Plant Pathology, 108(7): 667-673.

Blandino, M., Minelli, L.,Reyneri, A., 2006. Strategies for the chemical control of Fusarium head blight: Effect on yield, alveographic parameters and deoxynivalenol contamination in winter wheat grain. European Journal of Agronomy, 25(3): 193-201.

Boote, K.J., Jones, J.W.,Pickering, N.B., 1996. Potential Uses and Limitations of Crop Models. Agron J, 88(5): 704-716.

Bottalico, A.,Perrone, G., 2002. Toxigenic Fusarium species and Mycotoxins Associated with Head Blight in Small-Grain Cereals in Europe. European Journal of Plant Pathology, 108(7): 611-624.

Boyacoglu, D., Hettiarachy, N.S.,Stack, R.W., 1992. Effect of three systemic fungicides on deoxynivalenol (vomitoxin) production by *Fusarium graminearum* in wheat. Canadian Journal of Plant Science, 72: 93-101.

Brennan, J.M., Fagan, B., van Maanen, A., Cooke, B.M.,Doohan, F.M., 2003. Studies on in vitro growth and pathogenicity of European Fusarium fungi. European Journal of Plant Pathology, 109(6): 577-587.

Buerstmayr, H., Lemmens, M., Hartl, L., Doldi, L., Steiner, B., Stierschneider, M.,Ruckenbauer, P., 2002. Molecular mapping of QTLs for Fusarium head blight resistance in spring wheat. I. Resistance to fungal spread (Type II resistance). TAG Theoretical and Applied Genetics, 104(1): 84-91.

Buerstmayr, H., Steiner, B., Hartl, L., Griesser, M., Angerer, N., Lengauer, D., Miedaner, T., Schneider, B.,Lemmens, M., 2003. Molecular mapping of QTLs for Fusarium head blight resistance in spring wheat. II. Resistance to fungal penetration and spread. Theoretical and Applied Genetics, 107(3): 503-508.

Bushnell, W.R., 2001. What is known about infection pathways in Fusarium Head Blight? In: S.M. Canty, J. Lewis, L. Siler and R.W. Ward (Editors), 2001 National Fusarium Head Blight Forum. Michigan State University, Erlanger, KY, pp. 105.

Calori-Dominguesi, M.A., de Almeida, R.R., Tomiwaka, M.M., Gallo, C.R., da Gloria, E.M.,Dias, C.T.S., 2007. Occurrence of deoxynivalenol in national and imported wheat used in Brazil. Ciencia E Tecnologia De Alimentos, 27(1): 181-185.

Caron, D., 1993. Les Fusarioses. In: ITCF (Editor), Maladies des blés et orges, pp. 30-39.

Cassini, R.,, E., Dijon, pp. 271-279., 1970. Facteurs favorables ou défavorables au development des fusarioses et septorioses du blé, Proceedings of the Meeting of sections Cereals and Physiology, Eucarpia, Dijon, pp. 271-279.

CAST, 2003. Mycotoxins: risk in plant, animal, and human systems, Task Force Report 139. Council for Agricultural Science and Technology, Ames, IA, USA, 199 pp.

Chala, A., Weinert, J.,Wolf, G.A., 2003. An integrated approach to the evaluation of the efficacy of fungicides against Fusarium culmorum, the cause of head blight of wheat. Journal of Phytopathology, 151(11-12): 673-678.

Champeil, A., Dore, T.,Fourbet, J.F., 2004a. Fusarium head blight: epidemiological origin of the effects of cultural practices on head blight attacks and the production of mycotoxins by Fusarium in wheat grains. Plant Science, 166(6): 1389-1415.

Champeil, A., Fourbet, J.F., Dore, T.,Rossignol, L., 2004b. Influence of cropping system on Fusarium head blight and mycotoxin levels in winter wheat. Crop Protection, 23(6): 531-537.

Charmley, L.L., Rosenburg, A.,Trenholm, H.L., 1994. Factors responsible for economic losses due to *Fusarium* mycotoxin contamination of grains, foods, and feedstuffs. In: J.D. Miller and H.L. Trenholm (Editors), Mycotoxins in grain: Compounds other than aflatoxin. American Association of Cereal Chemists, St. Paul, MN.

Cromey, M.G., Shorter, S.C., Lauren, D.R.,Sinclair, K.I., 2002. Cultivar and crop management influences on fusarium head blight and mycotoxins in spring wheat (*Triticum aestivum*) in New Zealand. New Zealand Journal of Crop and Horticultural Science, 30: 235-247.

D'Mello, J.P.F., Macdonald, A.M.C.,Rinna, R., 2001. Effects of azoxystrobin on mycotoxin production in a carbendazim-resistant strain of Fusarium sporotrichioides. Phytoparasitica, 29(5): 431-440.

D'Mello, J.P.F., Macdonald, A.M.C., D., P.,Dijksma, W.T.P., 1998. 3-Acetyl deoxynivalenol and esterase production in a fungicide-insensitive strain of *Fusarium culmorum*. Mycotoxin Research, 14: 8-9.

da Luz, W.C., Stockwell, C.A.,Bergstrom, G.C., 2003. Biological control of Fusarium graminearum. In: K.J.L.a.W.R. Bushnell (Editor), Fusarium Head Blight of Wheat and Barley. Amer. Phytopathol. Soc. Press, St. Paul, MN, pp. 381–394.

Daamen, R.A., Langerak, C.J.,Stol, W., 1991. Surveys of cereal diseases and pests in the Netherlands. 3. Monographella nivalis and Fusarium spp. in winter wheat fields and seed lots. European Journal of Plant Pathology, 97(2): 105-114.

Dahleen, L.S., Okubara, P.A.,Blechl, A.E., 2001. Transgenic approaches to combat fusarium head blight in wheat and barley. Crop Science, 41(3): 628-637.

Dalcero, A., Torres, A., Etcheverry, M., Chulze, S.,Varsavsky, E., 1997. Occurrence of deoxynivalenol and Fusarium graminearum in Argentinian wheat. Food Additives and Contaminants, 14(1): 11-14.

de Nijs, M., Soentoro, P., Asch, E.D.-v., Kamphuis, H., Rombouts, F.M., Notermans, S.,H.W, 1996. Fungal Infection and Presence of Deoxynivalenol and Zearalenone in Cereals Grown in The Netherlands. Journal of Food Protection, 59: 772-777.

de Wit, C.T., 1993. Philosophy and terminology. In: P.A. Leffelaar (Editor), On systems analysis and simulation of ecological processes. Kluwer Academic Publishers, Wageningen, pp. 3-4.

Deadman, M.L.,Cooke, B.M., 1997. Wheat disease and cereal-clover bi-crops. Manage, Focus 2: 15-19.

Desjardins, A.E., Manandhar, G., Plattner, R.D., Maragos, C.M., Shrestha, K.,McCormick, S.P., 2000. Occurrence of Fusarium species and mycotoxins in nepalese maize and wheat and the effect of traditional processing methods on mycotoxin levels. Journal of Agricultural and Food Chemistry, 48(4): 1377-1383.

Detrixhe, P., Chandelier, A.,Cavelier, M., 2003. Development of an agrometeorological model integrating leaf wetness duration estimation to assess the risk of head blight infection in wheat. Aspects of Applied Biology 68: 199-204.

Dill-Macky, R.,Jones, R.K., 2000. The effect of previous crop residues and tillage on Fusarium head blight of wheat. Plant Disease, 84(1): 71-76.

Doohan, F.M., Brennan, J.,Cooke, B.M., 2003. Influence of climatic factors on Fusarium species pathogenic to cereals. European Journal of Plant Pathology, 109(7): 755-768.

Edwards, S.G., 2004a. Influence of agricultural practices on fusarium infection of cereals and subsequent contamination of grain by trichothecene mycotoxins. Toxicology Letters, 153(1): 29-35.

Edwards, S.G., 2004b. Investigation of Fusarium mycotoxins in UK wheat production. In: S.M. Canty, T. Boring, K. Versdahl, J. Warwell and R.W. Ward (Editors), 2nd International Symposium on Fusarium Head Blight. Michigan State University, Orlando, FL, USA, pp. 398-400.

Edwards, S.G., Pirgozliev, S.R., Hare, M.C.,Jenkinson, P., 2001. Quantification of Trichothecene-Producing Fusarium Species in Harvested Grain by Competitive PCR To Determine Efficacies of Fungicides against Fusarium Head Blight of Winter Wheat. Appl. Environ. Microbiol., 67(4): 1575-1580.

FAO, 2003. Manual on the application of the HACCP system in mycotoxin prevention and control. FAO Food and Nutrition Papers, 73. FAO, Rome, 122 pp.

Fauzi, M.T.,Paulitz, T.C., 1994. The Effect of Plant-Growth Regulators and Nitrogen on Fusarium Head Blight of the Spring Wheat Cultivar Max. Plant Disease, 78(3): 289-292.

Fernandes, J.M., Cunha, G.R., Del Ponte, E., Pavan, W., Pires, J.L., Baethgen, W., Gimenez, A., Magrin, G.,Travasso, M.I., 2004. Modeling fusarium head blight in wheat under climate change using linked process-based models. In: S.M. Canty, T. Boring, K. Versdahl, J. Wardwell and R. Ward (Editors), 2nd International Symposium on Fusarium Head Blight. Michigan State University, Orlando, FL, USA, pp. 441-444.

Fernandez, M.R., 1992. The Effect of Trichoderma-Harzianum on Fungal Pathogens Infesting Wheat and Black Oat Straw. Soil Biology & Biochemistry, 24(10): 1031-1034.

Fernando, W.G.D., Miller, J.D., Seaman, W.L., Seifert, K.A.,Paulitz, T.C., 2000. Daily and seasonal dynamics of airborne spores of *Fusarium graminearum* and other *Fusarium* sepcies sampled over wheat plots. Canadian Journal of Botany, 78: 497-505.

France, J.,Thornley, J.H.M., 1984. Role of mathematical models in agriculture and agricultural research. In: J. France and J.H.M. Thornley (Editors), Mathematical models in agriculture. Butterworths, Hurley, Maidenhead, UK, pp. 1-14.

Gagkaeva, T.Y.,Gavrilova, O.P., 2008. Current distribution of *Fusarium* fungi on small grain cereals in Russia. Journal of Plant Pathology - Abstracts of presentations of the X International Fusarium and Fusarium Genomics Workshop 2008, Alghero, Sardinia (Italy), August 30-September 2, 2008, 90(3S): S3.47-S3.48.

Gilbert, J.,Fernando, W.G.D., 2004. Epidemiology and biological control of Gibberella zeae Fusarium graminearum. Canadian Journal of Plant Pathology-Revue Canadienne De Phytopathologie, 26(4): 464-472.

Gilchrist, L.,Dubin, H.J., 2002. Fusarium Head Blight. In: B.C. Curtis, S. Rajaram and H. Gómez Macpherson (Editors), Bread Wheat. Improvement and Production. .

Gourdain, E., 2008. Maîtriser le risque sur les cultures de blés: quels outils pour quelles utilisations? In: I.d.v. ARVALIS (Editor), Maîtriser le risque fusariotoxines. ARVALIS, Paris, Salons de l'Aveyron, pp. 27-40.

Haidukowski, M., Pascale, M., Perrone, G., Pancaldi, D., Campagna, C.,Visconti, A., 2005. Effect of fungicides on the development of Fusarium head blight, yield and deoxynivalenol accumulation in wheat inoculated under field conditions with Fusarium graminearum and Fusarium culmorum. Journal of the Science of Food and Agriculture, 85(2): 191-198.

Hajjaji, A., El Otmani, M., Bouya, D., Bouseta, A., Mathieu, F., Collin, S.,Lebrihi, A., 2006. Occurrence of mycotoxins (ochratoxin A, deoxynivalenol) and toxigenic fungi in Moroccan wheat grains: impact of ecological factors on the growth and ochratoxin A production. Molecular Nutrition & Food Research, 50(6): 494-499.

Hajslova, J., Lancova, K., Sehnalova, M., Krplova, A., Zachariasova, M., Moravcova, H., Nedelnik, J., Markova, J.,Ehrenbergerova, J., 2007. Occurrence of trichothecene mycotoxins in cereals harvested in the Czech Republic. Czech Journal of Food Sciences, 25(6): 339-350.

Hart, P., Munn, B., Ward, R.,Siler, L., 2004. Fusarium head blight in Michigan in 2004. In: S.M. Canty, T. Boring, K. Versdahl, J. Warwell and R.W. Ward (Editors), 2nd International Symposium on Fusarium Head Blight. Michigan State University, Orlando, FL, USA, pp. 453-457.

Heier, T., Jain, S.K., Kogel, K.H.,Pons-Kuhnemann, J., 2005. Influence of N-fertilization and fungicide strategies on Fusarium head blight severity and mycotoxin content in winter wheat. Journal of Phytopathology, 153(9): 551-557.

Hermann, W., Kuebler, E.,Aufhammer, W., 1998. Aehrenbefall mit fusarien und toxingehalt im korngut bei verschiedenen wintergetreidearten. Pflanzenbauwiss. Ger. J. Agron., 6(1-8).

Hilton, A.J., Jenkinson, P., Hollins, T.W.,Parry, D.W., 1999. Relationship between cultivar height and severity of Fusarium ear blight in wheat. Plant Pathology, 48(2): 202-208.

Höhn, D., Rosenberger, A., Kaul, H.-P., Kübler, E.,Aufhammer, W., 2002. Infection with Fusarium head blight (F. culmorum) and deoxynivalenol accumulation of winter wheat as affected by varied maturation. Pflanzenbauwissenschaften, 6(1): 1-8.

Hollins, T.W., Ruckenbauer, P.,De Jong, H., 2003. Progress towards wheat varieties with resistance to fusarium head blight. Food Control, 14(4): 239-244.

Homdork, S., Fehrmann, H.,Beck, R., 2000. Effects of field application of tebuconazole on yield, yield components and the mycotoxin content of Fusarium-infected wheat grain. Journal of Phytopathology, 148(1): 1-6.

Hooker, D.C.,Schaafsma, A.W., 2004. The DONcast model: predicting deoxynivalenol (DON) in wheat. In: S.M. Canty, T. Boring, J. Wardwell and R. Ward (Editors), Proceedings of the Second International Symposium on Fusarium Head Blight. Michigan State University, Orlando, FL, USA, pp. 458.

Hooker, D.C., Schaafsma, A.W.,Tamburic-Ilincic, L., 2002. Using weather variables pre- and post-heading to predict deoxynivalenol content in winter wheat. Plant Disease, 86(6): 611-619.

Hope, R., Aldred, D.,Magan, N., 2005. Comparison of environmental profiles for growth and deoxynivalenol production by *Fusarium culmorum* and *F. graminearum* on wheat grain. Letters in Applied Microbiology, 40(4): 295-300.

Horberg, H.M., 2002. Patterns of splash dispersed conidia of Fusarium poae and Fusarium culmorum. European Journal of Plant Pathology, 108(1): 73-80.

Inch, S.,Gilbert, J., 2003. The incidence of Fusarium species recovered from inflorescences of wild grasses in southern Manitoba. Canadian Journal of Plant Pathology-Revue Canadienne De Phytopathologie, 25(4): 379-383.

Ioos, R., Belhadj, A.,Menez, M., 2004. Occurrence and distribution of Microdochium nivale and Fusarium species isolated from barley, durum and soft wheat grains in France from 2000 to 2002. Mycopathologia, 158(3): 351-362.

Ioos, R., Belhadj, A., Menez, M.,Faure, A., 2005. The effects of fungicides on Fusarium spp. and Microdochium nivale and their associated trichothecene mycotoxins in French naturally-infected cereal grains. Crop Protection, 24(10): 894-902.

Jenkinson, P.,Parry, D.W., 1994. Isolation of fusarium species from common broad-leaved weeds and their pathogenicity to winter wheat. Mycological Research, 98: 776-780.

Jennings, P., Turner, J.A.,Nicholson, P., 2000. Overview of Fusarium ear blight in the UK—effect of fungicide treatment on disease control and mycotoxin production, BCPC Crop Protection Conference—Pest and Disease. Farnham: British Crop Protection Council, pp. 707–712.

Jenny, E., Hecker, A., Kessler, P., Külling, C.,H.-R., F., 2000. Getreidefusariosen: Sortenresistenz und Toxingehalte. Agrarforschung, 7(270-273).

Johnson, D.D., Flaskerud, G.K., Taylor, R.D.,Satyanarayana, V., 1998. Economic impacts of fusarium head blight in wheat, North Dakota State University, Department of Agribusiness and Applied Economics, Fargo.

Jones, J.W.,Luyten, J.C., 1998. Simulation of biological processes. In: R.M. Peart and R.B. Curry (Editors), Agricultural systems modeling and simulation. Marcel Dekker, Inc., Gainesville, Florida (USA), pp. 19-62.

Jurado, M., Vàzquez, C., Lòpez-Errasquin, E., Patiño, B.,Gonzàlez-Jaén, M.T., 2004. Analysis of the occurrence of *Fusarium* species in spanish cereals by PCR assay. In: S.M. Canty, T. Boring, K. Versdahl, J. Warwell and R.W. Ward (Editors), 2nd International Symposium on Fusarium Head Blight. Michigan State University, Orlando, FL, USA, pp. 460-464.

Kammoun, G.L., Barreau, C., Richard-Forget, F., Gargouri, S.,Hajlaoui, M.R., 2008. Ocurrence of trichothecene mycotoxins in Tunisian *Fusarium* species isolated from wheat. Journal of Plant Pathology - Abstracts of presentations of the X International Fusarium and Fusarium Genomics Workshop 2008, Alghero, Sardinia (Italy), August 30-September 2, 2008, 90(3S): S3.54.

Kang, Z.,Buchenauer, H., 2000a. Cytology and ultrastructure of the infection of wheat spikes by Fusarium culmorum. Mycological Research, 104(9): 1083-1093.

Kang, Z.,Buchenauer, H., 2000b. Ultrastructural and Cytochemical Studies on Cellulose, Xylan and Pectin Degradation in Wheat Spikes Infected by <i>Fusarium culmorum</i>. Journal of Phytopathology, 148(5): 263-275.

Keblys, M., Flaoyen, A.,Langseth, W., 2001. The occurrence of type A and B trichothecenes in Lithuanian cereals. Acta Agriculturae Scandinavica Section B-Soil and Plant Science, 50(3-4): 155-160.

Khan, N.I., Schisler, D.A., Boehm, M.J., Lipps, P.E.,Slininger, P.J., 2004. Field testing of antagonists of Fusarium head blight incited by Gibberella zeae. Biological Control, 29(2): 245-255.

Khonga, E.B.,Sutton, J.C., 1988. Inoculum production and survival of Gibberella zeae in maize and wheat residues. Can. J. Plant Pathol.(10): 232-239.

Klem, K., Vanova, M., Hajslova, J., Lancova, K.,Sehnalova, M., 2007. A neural network model for prediction of deoxynivalenol content in wheat grain based on weather data and preceding crop. Plant Soil and Environment, 53(10): 421-429.

Koch, H.-J., Pringas, C.,Maerlaender, B., 2006. Evaluation of environmental and management effects on Fusarium head blight infection and deoxynivalenol concentration in the grain of winter wheat. European Journal of Agronomy, 24(4): 357-366.

Kolb, F.L., Bai, G.H., Muehlbauer, G.J., Anderson, J.A., Smith, K.P.,Fedak, G., 2001. Host plant resistance genes for fusarium head blight: Mapping and manipulation with molecular markers. Crop Science, 41(3): 611-619.

Kong, L., Anderson, J.M.,Ohm, H.W., 2005. Induction of wheat defense and stress-related genes in response to *Fusarium graminearum*. Genome, 48: 29-40.

Kosiak, B., Torp, M., Skjerve, E.,Thrane, U., 2003. The Prevalence and Distribution of *Fusarium* species in Norwegian Cereals: a Survey. Acta Agriculturae Scandinavica, Section B - Plant Soil Science, 53(4): 168 - 176.

Kranz, J.,Royle, D.J., 1978. Perspective in mathematical modelling of plant disease epidemics. In: P.R. Scott and A. Bainbridge (Editors), Plant Disease Epidemiology. Blackwell Scientific Pubblications, Oxford, pp. 111-120.

Krebs, H., Streit, B.,Forrer, H.-R., 2000. Effect of tillage and preceding crops on Fusarium infection and deoxynivalenol content of wheat. In: T. Alfoldi (Editor), 13th International IFOAM Scientific Conference, *The World Grows Organic* IOS, Amsterdam, Basel, Convention Center, pp. 28-31.

Krstanovic, V., Klapec, T., Velic, N.,Milakovic, Z., 2005. Contamination of malt barley and wheat by Fusarium graminearum and Fusarium culmorum from the crop years 2001-2003 in eastern Croatia. Microbiological Research, 160(4): 353-359.

Lacey, J., Bateman, G.L.,Mirocha, C.J., 1999. Effects of infection time and moisture on development of ear blight and deoxynivalenol production by Fusarium spp. in wheat. Annals of Applied Biology, 134(3): 277-283.

Lager, J.,Wallenhammer, A.C., 2003. Crop loss from soil-borne pathogens in white clover stands assessed by chemical treatments. Z. Pflanzenk. Pflanzen, 110: 120-128.

Langseth, W.,Elen, O., 1997. The occurrence of deoxynivalenol in Norwegian cereals - Differences between years and districts, 1988-1996. Acta Agriculturae Scandinavica Section B-Soil and Plant Science, 47(3): 176-184.

Leffelaar, P.A., 1993. On systems analysis ans simulation of ecological processes. Kluwer Academic Publlisher, Dordrecht, The Netherlands, 294 pp.

Lemmens, M., Haim, K., Lew, H.,Ruckenbauer, P., 2004. The effect of nitrogen fertilization on Fusarium head blight development and deoxynivalenol contamination in wheat. Journal of Phytopathology, 152(1): 1-8.

Logrieco, A., Bottalico, A., Mule, G., Moretti, A.,Perrone, G., 2003. Epidemiology of toxigenic fungi and their associated mycotoxins for some Mediterranean crops. European Journal of Plant Pathology, 109(7): 645-667.

Lori, G.A., Sisterna, M.N., Haidukowski, M.,Rizzo, I., 2003. Fusarium graminearum and deoxynivalenol contamination in the durum wheat area of Argentina. Microbiological Research, 158(1): 29-35.

Luo, Y., Yoshizawa, T.,Katayama, T., 1990. Comparative study on the natural occurrence of Fusarium mycotoxins (trichothecenes and zearalenone) in corn and wheat from high- and low-risk areas for human esophageal cancer in China. Applied and Environmental Microbiology, 56(12): 3723-3726.

Ma, Z.P., Steffenson, B.J., Prom, L.K.,Lapitan, N.L.V., 2000. Mapping of quantitative trait loci for Fusarium head blight resistance in barley. Phytopathology, 90(10): 1079-1088.

Magan, N., Hope, R., Colleate, A.,Baxter, E.S., 2002. Relationship between growth and mycotoxin production by Fusarium species, biocides and environment. European Journal of Plant Pathology, 108(7): 685-690.

Maiorano, A., 2008. Control of fusarium-toxins in cereal grain: effect of the agronomic practices and development of decision support systems. PhD Thesis., University of Turin, Turin (Italy), 103 pp.

Maiorano, A., Blandino, M., Reyneri, A.,Vanara, F., 2008. Effects of maize residues on the Fusarium spp. infection and deoxynivalenol (DON) contamination of wheat grain. Crop Protection, 27(2): 182-188.

Maiorano, A., Reyneri, A., Sacco, D., Magni, A.,Ramponi, C., 2009. A dynamic risk assessment model (FUMAgrain) of fumonisin synthesis by Fusarium verticillioides in maize grain in Italy. Crop Protection, 28(3): 243-256.

Maldonado-Ramirez, S.L., Schmale Iii, D.G., Shields, E.J.,Bergstrom, G.C., 2005. The relative abundance of viable spores of Gibberella zeae in the planetary boundary layer suggests the role of long-distance transport in regional epidemics of Fusarium head blight. Agricultural and Forest Meteorology, 132(1-2): 20-27.

Martin, R.A., Macleod, J.A.,Caldwell, C., 1991. Influences of Production Inputs on Incidence of Infection by Fusarium Species on Cereal Seed. Plant Disease, 75(8): 784-788.

Matthies, A.,Buchenauer, H., 2000. Effect of tebuconazole (Folicur R) and prochloraz (Sportak R) treatments on Fusarium head scab development, yield and deoxynivalenol (DON) content in grains of wheat following artificial inoculation with *Fusarium culmorum*. Z. Pflanz. Pflanzenschutz., 1: 33-52.

Matthies, A., Walker, F.,Buchenauer, H., 1999. Interference of selected fungicides, plant growth retardants as well as piperonyl butoxide and 1-aminobenzotriazole in trichothecene production of Fusarium graminearum (strain 4528) in vitro. Zeitschrift Fur Pflanzenkrankheiten Und Pflanzenschutz-Journal of Plant Diseases and Protection, 106(2): 198-212.

McMullen, M., Jones, R.,Gallenberg, D., 1997. Scab of Wheat and Barley: A Re-emerging Disease of Devastating Impact. Plant Disease, 81(12): 1340-1348.

McMullen, M., Lukach, J., McKay, K.,Schatz, B., 2002. Fungicide and biological agent effects on Fusarium head blight across two wheat classes. J. Appl. Genet., 43A: 223-226.

Menniti, A.M., Pancaldi, D., Maccaferri, M.,Casalini, L., 2003. Effect of Fungicides on Fusarium Head Blight and Deoxynivalenol Content in Durum Wheat Grain. European Journal of Plant Pathology, 109(2): 109-115.

Mesterhazy, A., 2002. Role of deoxynivalenol in aggressiveness of Fusarium graminearum and F-culmorum and in resistance to Fusarium head blight. European Journal of Plant Pathology, 108(7): 675-684.

Mesterházy, A., 1995. Types and components of resistance to *Fusarium* head blight of wheat. Plant Breeding, 114(5): 377-386.

Mesterházy, A., 2004. Possible ways to utilize mycotoxin contaminated grain. In: S.M. Canty, T. Boring, K. Versdahl, J. Warwell and R.W. Ward (Editors), 2nd International Symposium on Fusarium Head Blight. Michigan State University, Orlando, FL, USA, pp. 413-414.

Mesterhazy, A., Bartok, T., Kaszonyi, G., Varga, M., Toth, B.,Varga, J., 2005. Common resistance to different Fusarium spp. causing Fusarium head blight in wheat. European Journal of Plant Pathology, 112(3): 267-281.

Mesterhazy, A., Bartok, T.,Lamper, C., 2003. Influence of wheat cultivar, species of Fusarium, and isolate aggressiveness on the efficacy of fungicides for control of Fusarium head blight. Plant Disease, 87(9): 1107-1115.

Miedaner, T., Schneider, B.,Geiger, H.H., 2003. Deoxynivalenol (DON) content and Fusarium head blight resistance in segregating populations of winter rye and winter wheat. Crop Science, 43(2): 519-526.

Mielke, H.,Weinert, J., 1996. Investigations on the effect of various fungicides on the pathogen of partial head blight (*Fusarium culmorum* (W. G. Sm.) Sacc.). Nachr bl Dtsch Pflanzenschutzd, 48: 93-95.

Miller, J.D., Culley, J., Fraser, K., Hubbard, S., Meloche, F., Ouellet, T., Seaman, W.L., Seifert, K.A., Turkington, K.,Voldeng, H., 1998. Effect of tillage practice on fusarium head blight of wheat. Canadian Journal of Plant Pathology-Revue Canadienne De Phytopathologie, 20(1): 95-103.

Miller, J.D.,Young, J.C., 1985. Deoxynivalenol in an experimental *F. graminearum* infection of wheat. Canadian Journal of Plant Pathology, 7: 132-135.

Miller, J.D., Young, J.C.,Sampson, D.R., 1985. Deoxynivalenol and Fusarium head blight resistance in spring cereals. Journal of Phytopathology, 113: 359-367.

Milus, E.A.,Parsons, C.E., 1994. Evaluation of foliar fngicide for controlling Fusarium Head Blight of wheat. Plant Disease, 78: 697-699.

Minervini, F., Fornelli, F.,Flynn, K.M., 2004. Toxicity and apoptosis induced by the mycotoxins nivalenol, deoxynivalenol and fumonisin B-1 in a human erythroleukemia cell line. Toxicology in Vitro, 18(1): 21-28.

Moschini, R.C.,Fortugno, C., 1996. Predicting wheat head blight incidence using models based on meteorological factors in Pergamino, Argentina. European Journal of Plant Pathology, 102(3): 211-218.

Moschini, R.C., Pioli, R., Carmona, M.,Sacchi, O., 2001. Empirical Predictions of Wheat Head Blight in the Northern Argentinean Pampas Region. Crop Sci, 41(5): 1541-1545.

Mullenborn, C., Steiner, U., Ludwig, M.,Oerke, E.C., 2008. Effect of fungicides on the complex of Fusarium species and saprophytic fungi colonizing wheat kernels. European Journal of Plant Pathology, 120(2): 157-166.

Muller, H.M., Reimann, J., Schumacher, U.,Schwadorf, K., 2001. Further survey of the occurrence of Fusarium toxins in wheat grown in southwest Germany. Archives of Animal Nutrition-Archiv Fur Tierernahrung, 54(2): 173-182.

Musa, T., Hecker, A., Vogelgsang, S.,Forrer, H.R., 2007. Forecasting of Fusarium head blight and deoxynivalenol content in winter wheat with FusaProg. EPPO Bulletin, 37(2): 283-289.

Muthomi, J.W., Ndung'u, J.K., Gathumbi, J.K., Mutitu, E.W.,Wagacha, J.M., 2008. The occurrence of Fusarium species and mycotoxins in Kenyan wheat. Crop Protection, 27(8): 1215-1219.

Nganje, W.E., Johnson, D.D., Wilson, W.W., Leistritz, F.L., Bangsund, D.A.,Tiapo, N.M., 2001. Economic impacts of Fusarium Head Blight in wheat and barley: 1998-2000, North Dakota State University, Department of Agribusiness and Applied Economics, Agribusiness & Applied Economics Report, Fargo, ND.

Nicholson, P., Jenkinson, P., Rezanoor, H.N.,Parry, D.W., 1993. Restriction fragment length polymorphism analysis of variation in Fusarium species causing ear blight of cereals. Plant Pathology, 42(6): 905-914.

Obst, A., Lepschy, J., Beck, R., Bauer, G.,Bechtel, A., 2000. The risk of toxins by Fusarium graminearum in wheat — interactions between weather and agronomic factors. Mycotoxin Research, 16(0): 16-20.

Oger, R., Buffet, D.,Kestemont, M.H., 2002. Estimation of leaf wetness duration using standard weather station data and radar images to assess the risk of wheat contamination by mycotoxins at regional scale in Belgium.

Okubara, Okubara, P., Blechl, Blechl, A., McCormick, McCormick, S., Alexander, Alexander, N., Dill, M., Dill-Macky, R., Hohn,Hohn, T., 2002. Engineering deoxynivalenol metabolism in wheat through the expression of a fungal trichothecene acetyltransferase gene. TAG Theoretical and Applied Genetics, 106(1): 74-83.

Osborne, L.E.,Stein, J.M., 2007. Epidemiology of Fusarium head blight on small-grain cereals. International Journal of Food Microbiology, 119(1-2): 103-108.

Oswald, I.P., Marin, D.E., Bouhet, S., Pinton, P., Taranu, I.,Accensi, F., 2005. Immunotoxicological risk of mycotoxins for domestic animals. Food Additives and Contaminants, 22(4): 354-360.

Palazzinia, J.M., Ramirez, M.L., Torres, A.M.,Chulze, S.N., 2007. Potential biocontrol agents for Fusarium head blight and deoxynivalenol production in wheat. Crop Protection, 26(11): 1702-1710.

Parry, D.W., Jenkinson, P.,Mcleod, L., 1995. Fusarium Ear Blight (Scab) in Small-Grain Cereals - a Review. Plant Pathology, 44(2): 207-238.

Paul, P.A., El-Allaf, S.M., Lipps, P.E.,Madden, L.V., 2004. Rain Splash Dispersal of Gibberella zeae Within Wheat Canopies in Ohio. Phytopathology, 94(12): 1342-1349.

Paulitz, T.C., 1996. Diurnal release of ascospores by Gibberella zeae in inoculated wheat plots. Plant Disease, 80(6): 674-678.

Pereyra, S.A.,Dill-Macky, R., 2008. Colonization of the Residues of Diverse Plant Species by Gibberella zeae and Their Contribution to Fusarium Head Blight Inoculum. Plant Disease, 92(5): 800-807.

Pereyra, S.A., Dill-Macky, R.,Castro, M., 2004. *Fusarium* species present in wheat and barley grains in Uruguay. In: S.M. Canty, T. Boring, K. Versdahl, J. Warwell and R.W. Ward (Editors), 2nd International Symposium on Fusarium Head Blight. Michigan State University, Orlando, FL, USA, pp. 488.

Perkowski, J., Stachowiak, J., Kiecana, I., Golinski, P.,Chelkowski, J., 1997. Natural occurrence of Fusarium mycotoxins in Polish cereals. Cereal Research Communications, 25(3): 379-380.

Pirgozliev, S.R., Edwards, S.G., Hare, M.C.,Jenkinson, P., 2003. Strategies for the control of Fusarium head blight in cereals. European Journal of Plant Pathology, 109(7): 731-742.

Placinta, C.M., D'Mello, J.P.F.,Macdonald, A.M.C., 1999. A review of worldwide contamination of cereal grains and animal feed with Fusarium mycotoxins. Animal Feed Science and Technology, 78(1-2): 21-37.

Prandini, A., Sigolo, S., Filippi, L., Battilani, P.,Piva, G., 2008. Review of predictive models for Fusarium head blight and related mycotoxin contamination in wheat. Food and Chemical Toxicology, In Press, Corrected Proof.

Pritsch, C., Muehlbauer, G.J., Bushnell, W.R., Somers, D.A.,Vance, C.P., 2000. Fungal Development and Induction of Defense Response Genes During Early Infection of Wheat Spikes by Fusarium graminearum. Molecular Plant-Microbe Interactions, 13(2): 159-169.

Pritsch, C., Vance, C.P., Bushnell, W.R., Somers, D.A., Hohn, T.M.,Muehlbauer, G.J., 2001. Systemic expression of defense response genes in wheat spikes as a response to Fusarium graminearum infection. Physiological and Molecular Plant Pathology, 58(1): 1-12.

Rafai, P., Bata, r., Jakab, L.,Vanyi, A., 2000. Evaluation of mycotoxin-contaminated cereals for their use in animal feeds in Hungary. Food Additives & Contaminants: Part A, 17(9): 799 - 808.

Ramirez, M.L., Chulze, S.,Magan, N., 2004. Impact of environmental factors and fungicides on growth and deoxinivalenol production by Fusarium graminearum isolates from Argentinian wheat. Crop Protection, 23(2): 117-125.

Ramirez, M.L., Chulze, S.,Magan, N., 2006. Temperature and water activity effects on growth and temporal deoxynivalenol production by two Argentinean strains of Fusarium graminearum on irradiated wheat grain. International Journal of Food Microbiology, 106(3): 291-296.

Rossi, V., Giosuè, S., Pattori, E., Spanna, F.,Del Vecchio, A., 2003. A model estimating the risk of *Fusarium* head blight on wheat. EPPO/OEPP Bulletin, 33(3): 421-425.

Rossi, V., Languasco, L., Pattori, E.,Giosue, S., 2002a. Dynamics of airborne Fusarium macroconidia in wheat fields naturally affected by head blight. Journal of Plant Pathology, 84(1): 53-64.

Rossi, V., Pattori, E., Ravanetti, A.,Giosue, S., 2002b. Effect of constant and fluctuating temperature regimes on sporulation of four fungi causing head blight of wheat. Journal of Plant Pathology, 84(2): 95-105.

Rossi, V., Ravanetti, A., Pattori, E.,Giosuè, S., 2001. Influence of temperature and humidity on the infection of wheat spikes by some fungi causing fusarium head blight. Journal of Plant Pathology, 83(3): 189-198.

Rotter, B.A., Prelusky, D.B.,Pestka, J.J., 1996. Toxicology of deoxynivalenol (vomitoxin). Journal of Toxicology and Environmental Health, 48(1): 1-34.

Ruckenbauer, P., Buerstmayr, H.,Lemmens, M., 2001. Present strategies in resistance breeding against scab (Fusarium spp.). Euphytica, 119(1-2): 121-127.

Schaafsma, A.W.,Hooker, D.C., 2006. Applications in forecasting deoxynivalenol in wheat using DONcast. In: D. Barug, D. Bhatnagar, H.P. van Egmond, J.W. van der Kamp, W.A. van Osenbruggen and A. Visconti (Editors), The mycotoxin factbook. Food & feed topics. Wageningen Academic Publishers, The Netherlands, pp. 211-222.

Schaafsma, A.W.,Hooker, D.C., 2007. Climatic models to predict occurrence of Fusarium toxins in wheat and maize. International Journal of Food Microbiology, 119(1-2): 116-125.

Schaafsma, A.W., Hooker, D.C., Pineiro, M., Diaz de Ackermann, M., Pereyra, S.,Castaño, J.P., 2006. Pre-harvest forecasting of deoxynivalenol for regulatory action in wheat grain in Uruguay using readily available weather inputs. In: H. Njapau, S. Trujillo, H.P. van Egmond and D.L. Park (Editors), Mycotoxins and Phycotoxins. Advances in determination, toxicology and exposure management. Wageningen Academic Publishers, The Netherlands, pp. 227-238.

Schaafsma, A.W., Tamburic-Illincic, L.,Miller, J.D., 1998. The effect of agronomic practice on the accumulation of deoxynivalenol (DON) in winter wheat fields in Ontario, 1996-1997. In: M.S. University (Editor), National Fusarium Head Blight Forum, USA, pp. 7-9.

Schisler, D.A., Khan, N.I.,Boehm, M.J., 2002. Biological control of Fusarium head blight of wheat and deoxynivalenol levels in grain via use of microbial antagonists. Mycotoxins and Food Safety, 504: 53-69.

Schisler, D.A., Slininger, P.J., Hanson, L.E.,Loria, R., 2000. Potato cultivar, pathogen isolate and antagonist cultivation medium influence the efficacy and ranking off bacterial antagonists of Fusarium dry rot. Biocontrol Science and Technology, 10(3): 267-279.

Schmale, D.G., Arntsen, Q.A.,Bergstrom, G.C., 2005. The forcible discharge distance of ascospores of Gibberelia zeae. Canadian Journal of Plant Pathology-Revue Canadienne De Phytopathologie, 27(3): 376-382.

Schmale, D.G., Shields, E.J.,Bergstrom, G.C., 2006. Night-time spore deposition of the fusarium head blight pathogen, Gibberella zeae, in rotational wheat fields. Canadian Journal of Plant Pathology-Revue Canadienne De Phytopathologie, 28(1): 100-108.

Schollenberger, M., Muller, H.M., Rufle, M., Suchy, S., Plank, S.,Drochner, W., 2006. Natural occurrence of 16 Fusarium toxins in grains and feedstuffs of plant origin from germany. Mycopathologia, 161(1): 43-52.

Schroeder, H.W.,Christensen, J.J., 1963. Factors aff ecting resistance of wheat to scab caused by *Gibberella zeae*. Phytopathology, 53: 831-838.

Simpson, D.R., Weston, G.E., Turner, J.A., Jennings, P.,Nicholson, P., 2001. Differential control of head blight pathogens of wheat by fungicides and consequences for mycotoxin contamination of grain. European Journal of Plant Pathology, 107(4): 421-431.

Siranidou, E.,Buchenauer, H., 2001. Chemical control of Fusarium head blight on wheat. Zeitschrift Fur Pflanzenkrankheiten Und Pflanzenschutz-Journal of Plant Diseases and Protection, 108(3): 231-243.

Slikova, S., Sudyova, V.,Gregova, E., 2008. Deoxynivalenol in wheat from the growing areas of Slovakia. Cereal Research Communications, 36(2): 279-287.

Snijders, C.H.A., 1990. Fusarium head blight and mycotoxin contamination of wheat, a review. Netherland Journal of Plant Pathology, 96: 187-198.

Snijders, C.H.A., 2004. Resistance in wheat to Fusarium infection and trichothecene formation. Toxicology Letters, 153(1): 37-46.

Sutton, J.C., 1982. Epidemiology of wheat head blight and maize ear rot caused by *Fusarium graminearum*. Canadian Journal of Plant Pathology, 4: 195-209.

Tan, M.K., Simpfendorfer, S., Backhouse, D.,Murray, G.M., 2004. Occurrence of Fusarium head blight (FHB) in southern NSW in 2000: identification of causal fungi and determination of putative chemotype of Fusarium graminearum isolates by PCR. Australasian Plant Pathology, 33(3): 385-392.

Teich, A.H., 1987. Less wheat scab with urea than with ammonium nitrate fertilisers. Cereal Research Communications, 15: 35-38.

Teich, A.H., 1989. Epidemiology of wheat (Triticum aestivum L.) scab caused by *Fusarium* spp. In: J. Chelkowsky (Editor), Fusarium: mycotoxins, taxonomy and pathogenicity. Elsevier, Amsterdam, pp. 269-282.

Teich, A.H.,Hamilton, J.R., 1985. Effect of Cultural Practices, Soil Phosphorus, Potassium, and pH on the Incidence of Fusarium Head Blight and Deoxynivalenol Levels in Wheat. Appl. Environ. Microbiol., 49(6): 1429-1431.

Teich, A.H.,Nelson, K., 1984. Survey of Fusarium head blight and possible effects of cultural practices in wheat fields in Lambton County in 1983. Can. Plant Dis. Surv., 64(1): 11-13.

Toth, B., Kaszonyi, G., Bartok, T., Varga, J.,Mesterhazy, A., 2008. Common resistance of wheat to members of the Fusarium graminearum species complex and F-culmorum. Plant Breeding, 127(1): 1-8.

Trail, F., Xu, H., Loranger, R.,Gadoury, D., 2002. Physiological and environmental aspects of ascospore discharge in Gibberella zeae (anamorph Fusarium graminearum). Mycologia, 94(2): 181-189.

Treikale, O., Priekule, I., Pugacheva, J.,Lazareva, L., 2008. Fusarium species in winter wheat in Latvia: distribution and toxigenic effect. Journal of Plant Pathology - Abstracts of presentations of the X International Fusarium and Fusarium Genomics Workshop 2008, Alghero, Sardinia (Italy), August 30-September 2, 2008, 90(3S): S3.59.

Trigo-Stockli, D.M., Curran, S.P.,Pedersen, J.R., 1995. Distribution and occurrence of mycotoxins in 1993 kansas wheat. Cereal Chemistry, 72(5): 470-474.

Tutelyan, V.A., 2004. Deoxynivalenol in cereals in Russia. Toxicology Letters, 153(1): 173-179.

Vrabcheva, T., Gessler, R., Usleber, E.,Martlbauer, E., 1996. First survey on the natural occurrence of Fusarium mycotoxins in Bulgarian wheat. Mycopathologia, 136(1): 47-52.

Waalwijk, C., Kastelein, P., de Vries, I., Kerényi, Z., van der Lee, T., Hesselink, T., Köhl, J.,Kema, G., 2003. Major Changes in Fusarium spp. in Wheat in the Netherlands. European Journal of Plant Pathology, 109(7): 743-754.

Wagacha, J.M.,Muthomi, J.W., 2007. Fusarium culmorum: Infection process, mechanisms of mycotoxin production and their role in pathogenesis in wheat. Crop Protection, 26(7): 877-885.

Wang, Y.Z.,Miller, J.D., 1988. Effects of *Fusarium graminearum* Metabolites on Wheat Tissue in Relation to Fusarium Head Blight Resistance. Journal of Phytopathology, 122(2): 118-125.

Wiersma, J.J.,Motteberg, C.D., 2005. Evaluation of five fungicide application timings for control of leaf-spot diseases and fusarium head blight in hard red spring wheat. The Canadian Journal of Plant Pathology, 27: 25-37.

Wilcoxson, R.D., Kommedahl, T., Ozmon, E.A.,Windels, C.E., 1988. Occurrence of Fusarium Species in Scabby Wheat from Minnesota and their Pathogenicity to Wheat. Phytopathology, 78: 586-589.

Windels, C.E., 2000. Economic and Social Impacts of Fusarium Head Blight: Changing Farms and Rural Communities in the Northern Great Plains. Phytopathology, 90(1): 17-21.

Wong, L.S.L., Tekauz, A., Leisle, D., Abramson, D.,McKenzie, R.I.H., 1992. Prevalence, distribution, and importance of fusarium head blight in wheat in Manitoba. Canadian Journal of Plant Pathology-Revue Canadienne De Phytopathologie, 14(3): 233-238.

Xu, X.M., 2003. Effects of environmental conditions on the development of Fusarium ear blight. European Journal of Plant Pathology, 109(7): 683-689.

Xu, X.M.,Berrie, A.M., 2005. Epidemiology of mycotoxigenic fungi associated with Fusarium ear blight and apple blue mould: A review. Food Additives & Contaminants: Part A, 22(4): 290 - 301.

Xu, X.M., Nicholson, P.,Ritieni, A., 2007. Effects of fungal interactions among Fusarium head blight pathogens on disease development and mycotoxin accumulation. International Journal of Food Microbiology, 119(1-2): 67-71.

Yang, J., Bai, G.,Shaner, G., 2005. Novel quantitative trait loci (QTL) for Fusarium head blight resistance in wheat cultivar Chokwang. TAG Theoretical and Applied Genetics, 111(8): 1571-1579.

Yu, J.B., Bai, G.H., Zhou, W.C., Dong, Y.H.,Kolb, F.L., 2008. Quantitative trait loci for Fusarium head blight resistance in a recombinant inbred population of Wangshuibai/Wheaton. Phytopathology, 98(1): 87-94.

Yuen, G.Y.,Schoneweis, S.D., 2007. Strategies for managing Fusarium head blight and deoxynivalenol accumulation in wheat. International Journal of Food Microbiology, 119(1-2): 126-130.

Zhang, X., Zhou, M., Ren, L., Bai, G., Ma, H., Scholten, O.E., Guo, P.,Lu, W., 2004. Molecular characterization of Fusarium head blight resistance from wheat variety Wangshuibai. Euphytica, 139(1): 59-64.

Zhu, H., Gilchrist, L., Hayes, P., Kleinhofs, A., Kudrna, D., Liu, Z., Prom, L., Steffenson, B., Toojinda, T.,Vivar, H., 1999. Does function follow form? Principal QTLs for Fusarium head blight (FHB) resistance are coincident with QTLs for inflorescence traits and plant height in a doubled-haploid population of barley. Theoretical and Applied Genetics, 99(7-8): 1221-1232.

In: Soybean and Wheat Crops
Editors: S. Davies, G. Evans

ISBN: 978-1-60741-173-4

Chapter 9

GROWING WHEAT FOR HIGH ALCOHOL YIELD – HOMOGENEOUS AND HETEROGENEOUS APPROACHES

J. S. Swanston and A. C. Newton
Scottish Crop Research Institute, Invergowrie, Dundee, DD2 5DA, UK

ABSTRACT

In the 1980s, Scottish grain distillers began to change from imported maize to home-grown wheat as their preferred adjunct, requiring an annual intake in excess of 0.5 million tonnes. For ease of processing, soft-milling varieties were required, and low grain protein contents were desirable. However, as this was a localised market, requiring a small proportion of the UK wheat harvest, it received little attention from wheat breeders until European interest developed in the use of wheat-based fuel ethanol as a partial petrol replacement. A greater number of varieties with potentially high alcohol yields are now being entered into national trials, but breeders face problems in early generation selection for alcohol yield, as rapid testing procedures are still being developed, including the use of Near Infra-Red Spectroscopy (NIR). Research to locate genetic factors responsible for alcohol yield, on wheat chromosomes, is also at an early stage, although this should facilitate the use of DNA-based selection systems in future breeding programmes. Changes in the quantity and timing of nitrogen fertiliser may also be necessary, as grain nitrogen content has a significant and negative effect on alcohol yield and reduced inputs are also desirable to enhance the energy balance associated with fuel ethanol production. However, these have to be achieved without a deleterious effect on grain yield. As a number of current varieties have good alcohol yield potential, but may have agronomic weaknesses, an alternative approach for cultivation is in the form of varietal mixtures. Complex mixtures, i.e. those with four or more components, have also been shown to increase grain yield and to reduce the spread of disease and thus the need for prophylactic spraying of fungicides. Mixtures are also likely to provide greater stability, across sites and seasons, for both yield and quality.

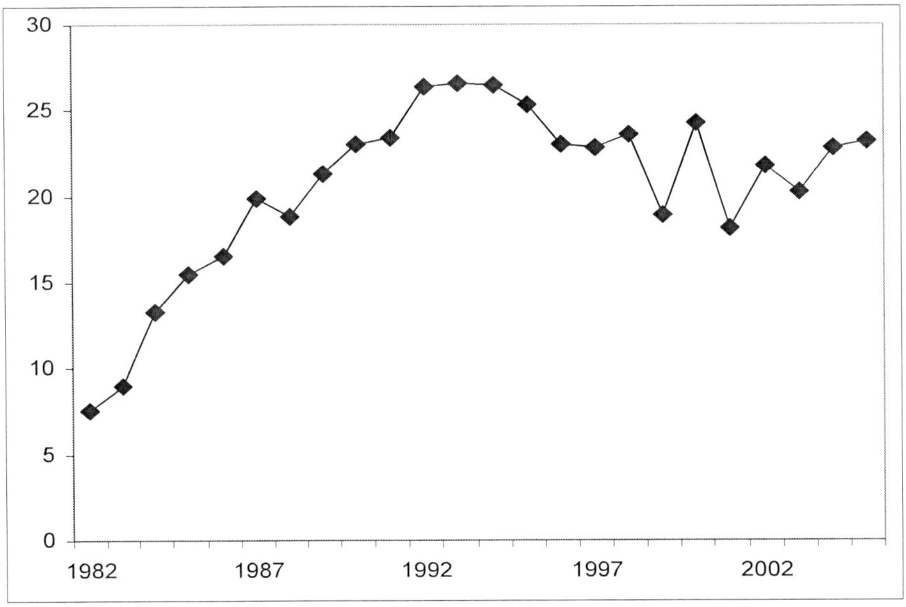

Figure 1. Wheat as a percentage of the Scottish Cereal Acreage 1981-2005.

INTRODUCTION

There are two types of whisky distilled in Scotland [6]. For malt whisky, the sugars for fermentation derive entirely from the enzymic activity of malted barley on its own reserves of starch, whereas, in grain distilling, only a small proportion of malted barley, usually with high levels of starch-degrading enzyme activity, is added to the mash. The rest of the cereal is in the form of unmalted grain, cooked, under pressure, at high temperature to gelatinise the starch. From the 1860s, the preferred adjunct was maize [78] due to its superior alcohol yield per tonne of grain [77], but, during the 1980s distillers began to switch to wheat, largely for economic reasons [49]. This change occurred quickly and, within 10 years, had already created a demand of 450,000 tonnes per annum [72]. The opportunity to replace an imported crop with a home-grown one had a major impact on wheat growing in Scotland, with a three- to four-fold increase in the area under cultivation during the 1980s (Figure 1). Wheat is now the second most widely-cultivated cereal in Scotland, after spring barley, occupying 20-25% of the area sown to cereals [55].

However, the spectrum of varieties differs considerably from that required for breadmaking. Distillers experienced processing problems if hard wheat was used [72] and also noted a strong negative association between alcohol yield and grain protein content [49], [73], so have restricted their intake to samples from soft wheat varieties, preferably of low protein (nitrogen) content. One variety in particular, Riband, was used widely, due to its very high alcohol yield [9], [69], but, for many years, distillers did not have many varieties to choose from. Only four varieties were acceptable for grain distilling until 2003, when two soft wheat varieties were added to the UK Recommended List [24], [69]. The absence of any wheat breeding programme, specifically targeted at the distilling industry, was considered, in part, to stem from lack of understanding of the genetic factors contributing to alcohol yield and the absence of appropriate testing procedures [66]. The distilling industry has devised a laboratory test, based on a scaled-down version of the industrial process [1], but this is not applicable to the very large numbers of lines generated in a breeding programme.

From an annual UK crop of around 16 million tonnes [4] the proportion required for distilling constitutes around 4% and, within a European context, is inconsequential, hence the minimal interest of wheat breeders in this market. However, ethanol derived from plant-based materials, frequently referred to as bioethanol, is now used as a partial petrol replacement in a number of countries, creating increased demand [65]. Unmodified engines can accommodate up to 10% petrol replacement [52]. With engine modifications, vehicles can run on fuels with blends of 85% ethanol:15% petrol and such fuels are readily available in Brazil [43] and, increasingly, in the USA. In the USA, bioethanol has been produced since the 1970s [52], through a distillation process that uses maize as a feedstock. Interest in Europe has centred on the use of wheat, which can give both the grain and the alcohol yields required for a positive energy balance, when the energy released by the fuel produced is compared with that expended in growing and processing the crop [48].

The production of Scotch whisky is covered by legislation which constrains or eliminates the use of certain additives [46]. In particular the enzymes required for starch degradation must come only from malted barley. However, as the wheat used in grain distilling primarily acts only as a substrate for such enzymes, Swanston and Newton [65] suggested that the suitability of wheat for alcohol production was likely to be independent of the source of the enzymes. Initial commercial use of wheat appeared to confirm this view [58] as the preferred varieties for fuel ethanol production were the same as those used for grain distilling. As the market for soft wheat with high alcohol yield potential seemed to be growing, there was increasing interest in varietal production and agronomic practise that would permit such potential to be fully realised. In this chapter we consider the development of selection tools that can facilitate the breeding process, but also the most effective means of deploying and cultivating the products of these breeding programmes.

Selection for Distilling Quality

As noted in the previous section, the distilling industry has devised a testing procedure that can be applied to samples of 30g [1], but includes a fermentation stage that requires 68 h, so is unsuited to rapid screening of large populations. It is, however, a useful procedure for testing new varieties and is applied to all soft wheat candidates for recommendation in the UK. This permits the distilling industry to identify the varieties that they are willing to accept and suitability for distilling is included as a characteristic on UK recommended lists produced by HGCA (www.hgca.com), for England and Wales and SAC (www.sac.co.uk), for Scotland. Results are obtained from samples grown at a range of sites across the UK [70] and these can cover a fairly wide range of grain protein contents and alcohol yields. This enables the mean performance of individual varieties to be assessed, although genotype x environment interaction can cause changes in the relative performance of varieties between both sites and seasons and this is illustrated in the comparison of two distilling varieties in Figure 2. Some varieties may perform less well in certain areas and this is usually reflected in regional, rather than national, recommendations.

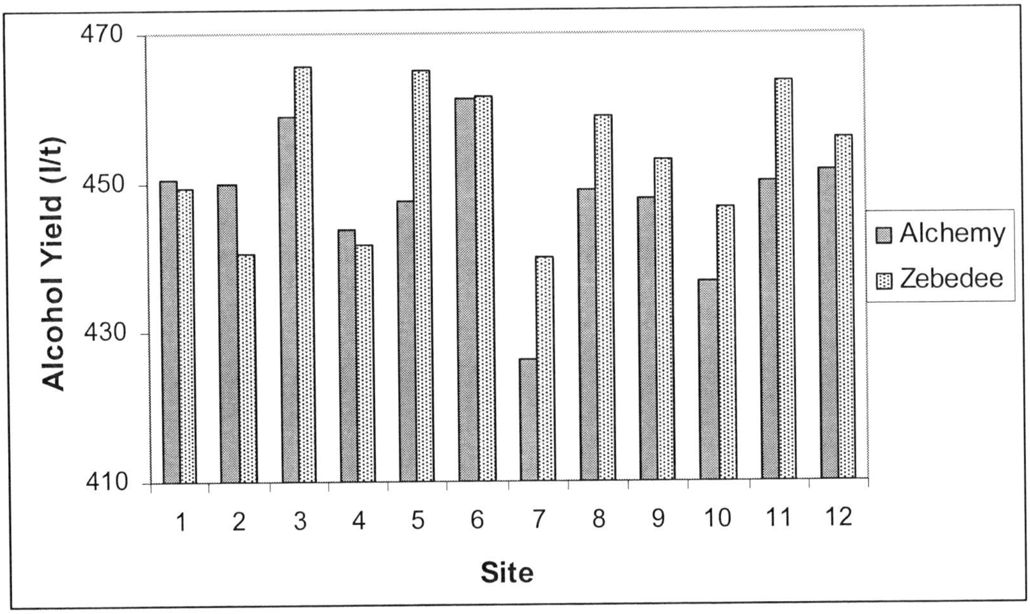

Figure 2. Alcohol yields from 2 varieties grown at 12 sites over 2 seasons (sites 1-6 harvested in 2005 and sites 7-12 in 2006).

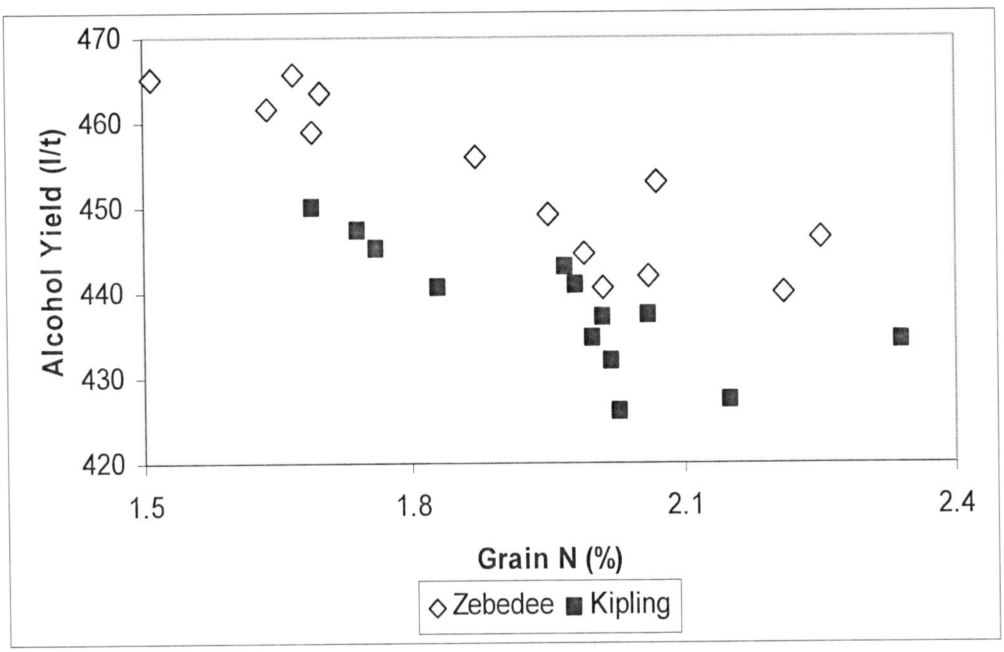

Figure 3. Alcohol yields in the varieties Zebedee and Kipling, classed as good and poor distilling quality, respectively [61], over a range of sites harvested between 2005 and 2007.

Selection at early stages in a breeding programme is rather more problematic. Several authors have observed a strong negative correlation between alcohol yield and grain nitrogen content [49], [70], [73], but the latter is unsuited as a means of selecting the best genotypes due to the strong influence of environment on its expression [70]. In addition, Kindred et al. [31] noted that a good distilling wheat had a higher starch content and alcohol yield than a poor one at any given level of grain nitrogen. That data was obtained from a trial grown at a single site, but with a range of N fertiliser regimes. Similar results were shown when good and poor distilling varieties were compared over a range of sites and several seasons (Figure 3). Thus while grain nitrogen content contributes to the phenotypic expression of alcohol yield, other factors determine the differences between varieties in genetic potential.

The extent of grain filling may also be a key contributor to alcohol yield in at least some of the main distilling varieties. Agu et al. [3] suggested that grain samples with a higher proportion of small grain tended to give lower yields of alcohol. In that work, samples were assessed by determining the proportions of grain retained by a series of sieves, indicating that grain width, rather than length was the determining factor. This supported the findings of Swanston et al. [70], who proposed an equation to predict alcohol yield from grain nitrogen, thousand grain weight (TGW) and the ratio of grain length to grain width. Length : width ratio (L:W), in which high values are indicative of narrow grain, had a negative effect on alcohol yield. Swanston et al. [70] also showed that L:W was primarily a varietal trait, while TGW was subject to both varietal and environmental variation. However, while high alcohol yields were associated with large grain size in the variety Riband [31], other varieties such as Claire [66] and Glasgow [70] could give good alcohol yields despite having relatively small grain, so grain dimensions did not always give an accurate indication of alcohol yield when compared across varieties.

Plant breeding has, traditionally, utilised phenotypic selection i.e. breeders have relied on characters which they could observe or measure to distinguish between desirable and undesirable progeny. Such characteristics needed to be sufficiently heritable for selection to be meaningful, but it was not necessary for their underlying genetic control to be fully understood. In general, however, it was easier to select, in early generations, for characters controlled by single genes, such as reduced height resulting from the inclusion of dwarfing or semi-dwarfing genes. Characters such as yield and processing quality, which had more complex genetic control, were assessed on populations that had already been reduced by earlier-generation selection. In addition, to ensure that the contribution of the genotype could be distinguished in the presence of environmental variation and genotype x environment interaction, it was necessary to carry out replicated trials over an appropriate range of sites and seasons. The advent of molecular breeding, based on selection of DNA sequences, therefore offered 2 major advantages. Firstly, it enabled direct selection of genetic potential, so could therefore be applied to unreplicated samples not grown together in trial and, secondly, it could be used in early generations of breeding programmes. This offered the potential to prioritise characters for selection on the basis of economic importance rather than ease of phenotypic assessment.

Initially, selection was based on anonymous DNA sequences which were linked either to individual genes or to genetic factors that contributed to traits controlled by more than one gene. The latter were termed quantitative trait loci (QTLs). A review of selection for a range of characters in wheat, including yield, disease resistance and quality parameters, using such molecular markers, is given by Gupta et al. [23]. These studies tended to reflect priorities within breeding programmes, as they were carried out to facilitate selection for breadmaking varieties. However, characters such as grain texture and protein content and composition, for which genetic factors were located by Nelson et al [39], could also be relevant to selection for distilling quality. Charmet et al. [10] also detected QTLs associated with the rate of N accumulation in the grain and consequent effects on the composition of protein, particularly the quantities of low molecular weight glutenins and gliadins. Selecting for differences in gliadin storage would be particularly useful in distilling wheat [71] as low gliadin content in the grain enhances alcohol yield.

Alcohol is produced through fermentation of the products of carbohydrate (essentially starch) breakdown and Smith et al. [61] provided a formula to calculate the theoretical alcohol yield from the starch content of the grain. Differences in starch content appear to be mainly genotypic [70], so selection between breeding lines should be feasible. However, several authors show that the association between starch content and alcohol yield, while positive, is not particularly strong [49], [70]. These data were generally derived from different varieties grown together in trials, where fertiliser applications would have been the same to each plot, so higher yields would have diluted grain nitrogen contents. Starch and nitrogen contents are inversely related, but the precise nature of this relationship varies with variety [31]. While starch content is, therefore, likely to be a major component of alcohol yield, achieving sufficiently similar grain N levels within trial plots to enable meaningful selection, may be problematic. Direct selection for alcohol yield would be preferable, both within breeding programmes and in detection and location of QTLs. Sylvester-Bradley and Kindred [71] noted the development of an NIR calibration, to predict alcohol yield in wheat, as a potentially valuable means of phenotype assessment.

Near Infra-Red (NIR) spectroscopy provides rapid, reproducible results, with little by way of sample preparation, so is widely used, by maltsters and other grain processors, for assessment of intake samples [7]. The original analysers, developed in the 1970s, measured the amount of reflectance and required samples to be finely milled [79], but later machines utilised the very-near infrared spectrum (800 – 1100 nm). These measured the amount of transmission through samples of whole grain [79]. Routine analyses are predominantly for nitrogen and moisture [30], but extending the use of NIR calibration, to predict processing performance of raw materials for the malting and brewing industries has been investigated for many years. Initial studies on hot water extract in barley, for example, began in the 1970s [37]. The choice of calibration samples, to represent the complete range that will be covered by samples for future assessment and estimates of precision of the reference method, against which the NIR spectra will be calibrated, are key steps in the process [7]. With an NIR calibration able to predict around 80% of the variation in alcohol yield [71], this appears to be a very promising development, as a further advantage of NIR calibrations is ease of transfer between machines [30]. This would make it possible to put the same type of equipment and software into both a breeder's and a distiller's laboratory, thus enabling breeders to target selection towards the industry's specifications.

Growing Wheat for Distilling and Bioethanol Production

The response of wheat, in terms of grain yield, to increased application of nitrogen (N) fertiliser has been studied fairly extensively, to enable optimal applications to be calculated. Factors such as the timing of fertiliser application [13], the source of N used [14] and the effect of the previous crop [74] have also been assessed, though essentially for their effect on improving breadmaking quality. For distilling wheat, the key objective is to maximise alcohol yield per hectare [61]. While grain yield is thus highly important, an increase of 1% in grain protein content will reduce alcohol yield by about 6 litres/tonne [31], so a balance has to be achieved. As an example, studies on the variety Istabraq [71] suggested that 184kg of N fertiliser would have been required to maximise alcohol yield compared to 236kg/ha to maximise grain yield, in the site and season in which the study was carried out. This illustrates the potential for optimising resource use and profitability and the importance of developing both the crop genetics and agronomy with alcohol production as a specific objective.

The economic optimum for N fertiliser application may, however, be lower at times of relatively low grain prices and high fertiliser costs [61] and the environmental impact of N fertiliser application is also coming increasingly under scrutiny. Richards [48] assessed the energy required to grow and process wheat for bioethanol production, in comparison to the energy released on combustion of the fuel, i.e. the energy balance. In growing the crop, the largest single contribution to energy consumption was the production of inorganic N fertiliser (Figure 4), so reduction of fertiliser application would be a key element in enhancing the energy balance for bioethanol production from wheat [65]. Additionally, Sylvester-Bradley and Kindred [71] noted that most (around 75%) of the greenhouse gas emissions associated with growing a wheat crop result from the use of N fertiliser – half from the use of fossil fuels in its manufacture and the other half from emissions of nitrous oxide in the field. Limited reductions in N application could be achieved without large adverse effects on grain yield [28], while Vaidyanathan et al. [74] noted that the optimal N application for wheat was reduced considerably when wheat followed rape or legumes, in a rotation, rather than another cereal. More precise monitoring of the growing crop, in addition to the soil, combined with variable rate fertiliser application [47] would improve the efficiency of N utilisation and could also reduce fertiliser application. Swanston and Newton [65] calculated that an overall 25% reduction in N fertiliser would equate to an energy saving approaching 2000Mj/ha.

The other major contributor to energy consumption, in growing a crop, is vehicle fuel use. Swanston and Newton [65] pointed out that fertiliser and fuel, together, accounted for 75% of the energy requirement. One field operation requiring a significant amount of fuel, i.e. 24% of total consumption [48], is ploughing, so reduced or minimum tillage (min-till) appears to be a possible means of significantly reducing fuel use. A further advantage of minimum tillage, particularly in some environments, can be reduced soil erosion [56], but, in the absence of ploughing, requirements for weed control and, therefore, herbicide use could increase [65]. Herbicides, along with other protectant chemicals, do not make a large contribution to energy consumption [48] or greenhouse gas emissions [71] in growing a crop, but may raise concerns about toxicity [75]. Comprehensive, prophylactic spraying regimes are not likely to affect the safety of an end-product, especially when a crop is being cultivated for non-food use, but may not be perceived as appropriate, if environmentally-benign claims are being made for the end-products. However, Taylor and Roscrow [73] argued that high alcohol yields per hectare were strongly dependent on high grain yield, as this diluted grain nitrogen content, thus justifying increased levels of all inputs, including crop protection chemicals, where this enhanced yield.

Bringhurst et al. [9] noted the decline of the favoured distilling variety Riband as being due to reduced agronomic viability. In a major crop such as wheat, where there are strong, competitive breeding programmes, older varieties are readily outclassed in grain yield potential, while disease resistance genes may be overcome by new races of a pathogen. Growers will, therefore, improve their margins by seeking varieties that can give higher yields, or possibly have lower input requirements, although prophylactic spraying regimes, to preclude disease development, are commonplace. Varieties, previously favoured by processors, may be lost unless a premium price can be guaranteed. In addition, newer varieties, which have the potential to meet industry specifications, may not be recommended for growing in some areas due to specific weaknesses. These could include factors such as a tendency for lodging or pre-harvest sprouting, which will be exacerbated under particular environmental conditions. The varieties that have replaced Riband, for agronomic reasons, have proved inferior in distilling quality [9]. New varieties, with alcohol yield potential, have been developed and Smith et al. [61] note the most promising of these as Zebedee and, in particular, Glasgow, but Glasgow has not received recommendation for growing in Scotland. It has a tendency to be weak-strawed [61] so, in the event of early N fertiliser application, which would be advantageous both for yield and for distilling quality, could be prone to lodging.

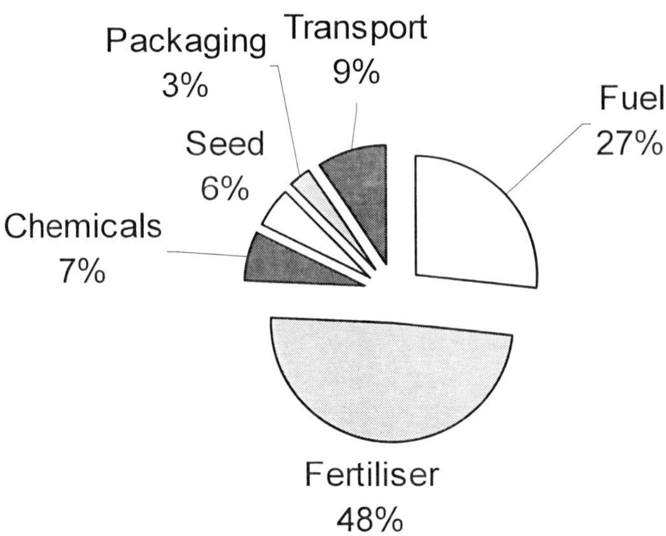

Figure 4. Relative contributions of various factors to the energy required in growing a wheat crop (data from Richards, 2000).

Utilisation of Varietal Mixtures

One conclusion from the previous section could be that it is difficult to find individual varieties that will achieve all the requirements for growers and end-users. An alternative would be to grow a crop comprised of two or more varieties which compensate for each other's deficiencies and exploit synergies. Mixtures have been cultivated over significant areas in a number of cereal species, e.g. Gacek et al, [19] reported an annual cultivation of 60,000 ha of spring barley mixtures in Poland, while, in 2000, 18% of the winter wheat area of Washington State and 13% in Oregon was sown to mixtures [38]. The major motivation for cultivating mixtures has been to slow the spread of disease and thus reduce or eliminate the requirement for fungicide application. Mixtures, for example, have been employed, over an area exceeding 3,000ha in China [81], to counter the problem of rice blast, caused by the fungus *Magnaporthe grisea*, with the initial results generating such interest that an increase to 40,000ha was projected for the following year. However, mixtures were also deployed very successfully in the former German Democratic Republic, to grow barley for malting and brewing in the 1980s. While this had a very significant effect on reducing the incidence of powdery mildew and removing the need for fungicide applications [80], the mixture components were also selected on the basis of their malting quality, thus ensuring that commercial objectives were achieved.

Commercial utilisation of mixtures has, however, met with some resistance, especially within the malting and brewing industries. This may reflect a persistent view that growers are the only real beneficiaries of mixtures [67], but the industry has also cited reduced quality and, particularly, heterogeneity as a major problem [53]. This view places emphasis on variety as the major source of variation between samples received for processing and may underestimate the effects of environment and genotype x environment interaction [68]. This can be illustrated from research in spring barley, but will be equally applicable in other crops such as wheat.

Friedt et al. [18] noted the importance of the Czechoslovakian short-strawed mutant Diamant as a precursor of modern European malting varieties. It is therefore likely that a considerable number of varieties derive from a narrow gene-pool and will be broadly similar in malting potential. In contrast with this narrow varietal variation, considerable variation in nitrogen can exist within a single field due to its topography which is reflected in the range of nitrogen values obtained from individual grains [2]. Modern malting plants operate on a very large scale [76] and require to access grain from a large range of sites, as do grain distilleries using wheat. Prior to malting, barley grain is separated on the basis of variety and nitrogen content, but, at any given nitrogen level, grain samples from the same variety can differ considerably in grain texture [11] or malting potential [34]. Similarly, Agu et al. [3] suggested that differences in grain filling, in wheat, may cause variation in the relationship between grain nitrogen and alcohol yield. In reality, there is thus a degree of heterogeneity in every sample of grain going through commercial processing, but it is only when it is sufficient to cause problems, e.g. in the brewhouse [15], that heterogeneity is seen as problematic.

A mixture derived from varieties with similar ancestry may be no more heterogeneous than any of its components and Newton et al. [42] were able to demonstrate this with a mixture of three winter barley malting varieties. A subsequent comparison was made between four malting varieties and all four possible three-component mixtures, grown at three sites in eastern Scotland, with the extent and evenness of water uptake and cell-wall breakdown (modification) determined by tests on individual grain [68]. It was demonstrated that samples of different varieties, grown at the same site, could be more similar in patterns of modification than samples of the same variety, grown at different sites. The mixtures also gave greater consistency between sites in malting performance than all but one of the individual varieties. This resulted from a reduced effect of site and absence of the change in ranking order that led to a significant genotype x site interaction, for malting quality characters, in the individual varieties [68]. Wheat varieties are also subject to genotype x site interaction for alcohol yield, as has already been shown (Fig 2).

Where barley mixture components differed significantly in quality, problems with heterogeneity did occur [42]. A similar situation may be observed after processing in wheat e.g. in breadmaking, where Sammons and Baenzinger [50] suggested that the quality of a mixture was very similar to the mean of its components, so inclusion of a component with very low quality would adversely affect the performance of the mixture. However, where differences between components were less extreme, mixtures comparable to the higher quality component could be identified [26]. Similarly, Manthey and Fehrmann [33] noted no problems from the use of mixtures. Osman [44], working with organically-grown wheat, detected a small advantage, in loaf volume, for a mixture compared to the mean of its components. Most of these data were obtained from experimental plots, but mixtures and monocultures have also been compared on a field scale [36]. Inclusion of one component with slightly lower quality, within a four-component mixture, did not cause the mixture to have significantly lower baking quality that the best component.

Wheat mixtures have also been accepted for grain distilling, on the conditions that they are grown from accredited seed and achieve acceptable specific weights [69], and Swanston et al. [66] noted that a mixture of three components was included in a commercial batch of grain that passed through the entire distilling process without problems. A major difference between wheat and barley for distilling is the absence of a malting stage and this appears to accommodate some variation in quality amongst the components. The variety Deben gave the highest grain yields of four varieties included in a field trial [66], but its alcohol yield was significantly lower than that of the other three cultivars, Riband, Claire and Consort. A mixture, comprised of Claire, Consort and Deben gave both grain yields and alcohol yields comparable with the best individual variety and alcohol yields significantly higher than those of Deben [66].

Mixtures provide the opportunity to extend the commercial lifespan of older, high-quality varieties or to exploit newer varieties with good processing attributes, but agronomic weaknesses that have precluded their recommendation [69]. The average value for alcohol yield per hectare, obtained from such mixtures, is likely to exceed that achieved by high yielding, but poorer distilling quality varieties. In addition, varietal mixtures tend to give more stable grain yields between locations and seasons [17] and, as a consequence, should show less fluctuation in nitrogen dilution effects and, thus, grain N levels. Wheat mixtures should, therefore, be potentially attractive to the grain distiller, while the emergence of several more varieties with good or moderate distilling quality, in the last few years, also provides the opportunity to create more complex mixtures.

Newton et al. [40] demonstrated that the efficacy of varietal mixtures, in reducing the spread of fungal disease and increasing yield, improves with complexity, i.e. a greater number of mixture components. This was observed, particularly, in winter barley, where mixtures with 5 or 6 components outyielded the mean of the components, when grown separately, by up to 15%. Yield increases were not simply a result of lower disease levels, since similar results were obtained when plots were given a full fungicide treatment. Armour et al. [5] suggested that high yields occur when the grain filling period coincides with the time of peak solar radiation, as photosynthesis will contribute around 75% of grain yield [27]. Potential yield will be lost by premature senescence of the canopy through drought, pest or disease [62], so high-input systems utilise fungicides [22] and, if necessary, irrigation [45] to retain green-leaf area.

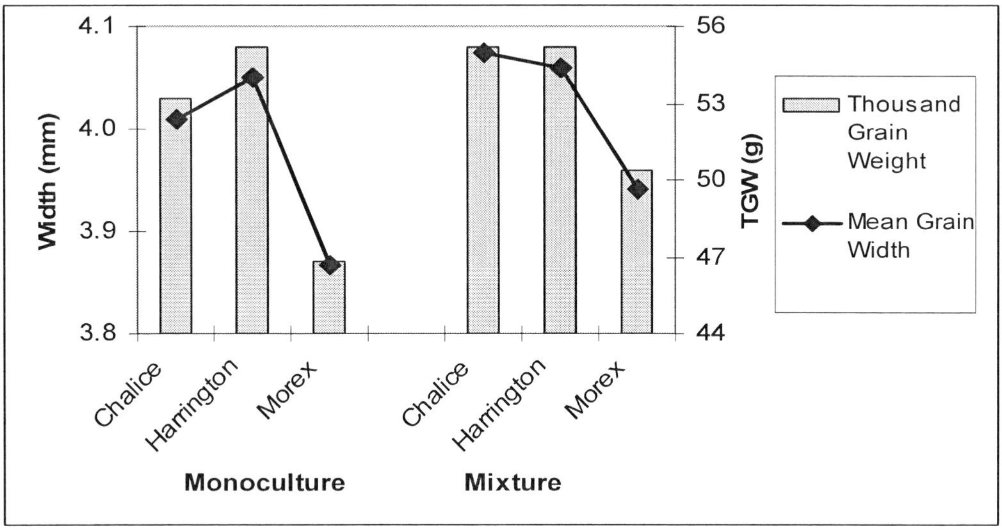

Figure 5. Grain dimensions of 3 barley varieties grown in monoculture or as components in a 3-way mixture.

Delaying canopy senescence is important in monoculture, as all the genotypes utilise the same resources simultaneously [51], whereas within mixtures, a degree of temporal and spatial variation may create an increase in resource availability. Essah and Stoskopf [16] noted a yield advantage in barley mixtures where components contrasted in phenotype, particularly where maturity differences enabled resources to be exploited over a slightly longer period, while Sarandon and Sarandon [51] demonstrated an increase in total biomass, in a wheat mixture, compared to that of the components. Both of these studies were carried out on two-component mixtures and the potential to modify the canopy, exploiting spatial variation, should be enhanced by greater complexity. Disease control will also be improved and while spraying may not be eliminated, the need for prophylactic treatment should be reduced [65]. In deriving a scheme for optimal bioethanol production from wheat, Swanston and Newton [65] suggested the use of mixtures with at least five components.

There is, however, a requirement for a better understanding of the range of interactions between mixture components. Sarandon and Sarandon [51] describe two possible interactions, i.e. the competition for the same resources by plants that are phenotypically very similar and the complementary utilisation of wider resources by plants with a degree of dissimilarity in physical appearance or maturity. However, growers have noted a reduction in the expected height differential between 6-row and 2-row barleys, when they were grown as mixtures [63] and Swanston and Newton [64] also observed convergence in the expression of certain traits among contrasting mixture components. The 6-row variety Morex, grown as a component in a mixture, with two 2-row varieties, produced fatter grain, with a higher thousand grain weight, than when grown, in monoculture, in the same trial (Figure 5). Recent work [41] has shown that certain varieties, or combinations of varieties, appear to exert a much stronger influence on the performance of mixtures than do others. Conversely, some varieties appear to respond differently within mixtures, compared to monocultures [41] and this may, at least in part, explain the greater stability of mixtures across sites [68]. Statistical techniques have been developed, to identify components that operate in the most synergistic way [32], [41] and a future trialling system, which assessed new varieties for their performance within mixtures, could be highly beneficial within arable crops.

Biomass and Second-generation Biofuels

Richards [48] noted that the energy balance, associated with the use of wheat as a fuel source, improved greatly when energy was derived from the whole crop. Straw can be burned to produce heat and/or electricity and its potential use in this area has been considered for more than a decade. Culshaw [12] noted that the low sulphur and high calcium contents of straw, compared to coal, reduced emissions of acid gases on combustion. A number of small, combined heat and power plants have been constructed in Denmark. These were straw-burning, with a typical annual consumption of 40,000 tonnes and a capacity of 9 MW of electricity and 20Mj of heat [8]. This could supply the requirements of two small towns, with a combined population of 14,000 inhabitants [8]. Culshaw [12] also noted the potential for electricity generation, citing the building of a 31 MW capacity power station, in eastern England, with a projected annual requirement of 180,000 tonnes of straw.

Straw has also been considered as a potential feedstock for bioethanol production. Shepherd [58] suggested that cellulose-based feedstocks were, potentially, the least expensive, but, technologically, the most complex, in terms of extracting the sugars for fermentation. Keller et al. [29] indicated that a priority was to reduce the severity of acid hydrolysis required in the initial stage of cellulose breakdown. A further problem is that ligocellulosic feedstocks, such as straw, contain hemicellulose in addition to cellulose and microorganisms capable of converting the full range of 5- in addition to 6-carbon sugars, occurring as breakdown products, are required [56]. Schubert [54] reviewed many of the current research and development projects in these areas, which included improved enzymes for cellulose breakdown and a range of micro-organisms for fermentation, including thermophylic bacteria as well as pathogenic species, such as *E.Coli*, which have been genetically manipulated.

If wheat is to be widely utilised for fuel, in addition to potable alcohol, conversion of cellulose and hemicellulose is likely to be necessary. Shepherd [58] suggested that a 5% replacement of petrol in the UK would be possible if a quantity of wheat grain, equivalent to that currently exported as surplus to domestic requirements, was converted into ethanol. However Sylvester-Bradley and Kindred [71] noted the concerns being expressed about the diversion of food crops into non-food uses, especially at times when global grain reserves are low and, as a consequence, food prices are escalating. Since the late 1940s, wheat breeders have been extremely successful in increasing grain yields [59], but this has been achieved by lowering the harvest index i.e. the contribution made by above-ground biomass, other than grain, to whole-crop yield. To design crops that can meet both food and non-food use, it may now be necessary to increase or improve total biomass, preferably without a negative effect on grain yield, although this might not be essential if there was an increase in the total energetically-positive and usable biomass.

Whether wheat straw can be improved as a substrate for ethanol production is not clear. The brown-midrib mutant in maize is associated with low lignin content and higher digestibility, under a wide range of environmental conditions [57] and this should also make it more amenable to other forms of processing. However, similar phenotypes are not presently available in wheat. The potential opportunities to increase biomass could include the use of F1 hybrids or transformation. A number of hybrid winter barleys, all 6-row feed types, have been cultivated in the UK in recent years, as they offer an advantage in grain yield, but there have been no similar developments in hybrid wheat. This may be due to the difficulty in obtaining significant yield advantage, as conventional varieties should be able to achieve 15t/ha under optimal conditions [5]. Additionally, any differences in grain quality between the parents, use to create a hybrid, could be exacerbated if random assortment of genetic factors led to transgressive segregation in the F2 grain, which F1 hybrid plants would produce, as this could be problematic for processing.

A number of authors have made claims for increases in biomass, in different species, through genetic modification. Two examples, where genes from a cereal species have been inserted into another crop [21], [60], are given in Table 1. Sylvester-Bradley and Kindred [71] noted the potential value of the alanine aminotrasferase gene from barley as it appeared to express under low N conditions, in GM Canola [21], so could therefore be used to enhance biomass production in a low input situation. At present, however, these claims for biomass increase have been made for plants grown under controlled experimental conditions and there is little evidence from field trials, as yet, to conclusively demonstrate the value of a GM approach to increasing biomass. Good et al. [21], however, suggested that seed yields in GM Canola, comparable with those of the parent variety could be obtained, but with a reduction of 40% in fertiliser application.

Table 1. Examples of 2 genes from cereals, which have been reported to increase biomass, when used to genetically modify another species

Crop	Enzyme	Source	Biomass Increase	Reference
Wheat	ADP-glucose pyrophosphorylase	Maize (Sh2 modified)	31%	Smidansky et al., 2002 [60]
Canola (Oil Seed Rape)	Alanine aminotransferase	Barley (AlaAT cDNA)	30-75% Low N only	Good et al., 2007 [21]

The use of varietal mixtures to improve biomass has been successful in energy crops such as willow [35], with the additional advantage of controlling rust without the need for fungicide application. There is limited information from cereals, although Swanston and Newton [65] hypothesised that the advantages in grain yield observed in complex mixtures would be mirrored by gains in biomass yield. Sarandon and Sarandon [51], however, noted an increase in biomass, within a mixture, even in the absence of an increase in grain yield. A considerable amount of further research in this area will, therefore, be required, particularly as there does not appear to have been any studies on the effect that increasing component number would have on biomass content.

Conclusion

The cultivation of wheat, for potable alcohol production, will continue to be important in Scotland although, at the present time, the extent to which wheat will be used for fuel alcohol, within the UK, is unclear. Development of biofuel production, especially from biomass, would have implications for the plant breeding industry, since, at present, there is no selection of biomass, either for its content or its suitability for processing, within UK cereal breeding. The use of fuel alcohol derived from grain remains controversial [71], but utilising wheat for this purpose could require 3m tonnes per annum within the UK, initially [58]. Without this market, it is difficult to see distilling quality being a major target for wheat breeders.

As wheat, however, remains the major human food crop, on a global scale, there will be a continuing requirement to enhance its productivity, to meet demands from a population predicted to reach 10 billion during the 21st century. The possibility of exploiting biotechnology to increase yield [60] is likely to be explored, and the potential to enhance nitrogen uptake under low N conditions [21], [71] has been briefly considered earlier in this chapter. A more far-reaching application of genetic modification would be to develop wheat varieties capable of association with nitrogen-fixing bacteria, thus greatly reducing the need for inorganic fertilsers. Recent work has enhanced the understanding of root nodule formation in legumes, permitting genetically modified plants to form nodules in the absence of *Rhizobia* [20]. This raises the possibility [25] of transferring the capacity for nodule formation, an essential aspect of the symbiotic relationship with nitrogen-fixing bacteria, to non-leguminous plants, but such developments are unlikely in the immediate future.

However, there will be both economic and environmental pressures to grow crops with reduced inputs, particularly of N fertilisers, due to their contribution to energy use [48], [65] and greenhouse gas emissions [71]. This is likely to encourage more effective targeting of fertiliser application and the cultivation of crops that can maximise exploitation of above- and below-ground resources. Additionally, if fertiliser prices continue to increase, maximising yield, irrespective of input costs, will become an unattractive option, as growers seek an economic optimum for inputs [61]. It is also widely predicted that current changes in climate may give rise to greater fluctuation in weather and increased likelihood of more extreme conditions. Under these circumstances, stability in performance may become as important, if not more so, than yield potential.

Future plant breeding targets will require to address such issues and this will have implications for national testing systems under which recommendation are often based, currently, on yield obtained under comprehensive fertiliser and pesticide regimes. It may also be appropriate to re-consider the way that varieties are deployed in agriculture. As homogeneity appears to be much less of an issue for wheat quality markets than it is for malting barley, varietal mixtures may be grown more widely than at present. Mixtures have been shown to give greater yield stability across sites and seasons [17] and also to reduce the effects of environment and genotype x environment interaction on some components of quality [68]. However, as the contributions of certain varieties, or combinations of varieties, are clearly more effective in mixtures than those of others [41], devising a testing regime, that would assess the potential of new varieties, as possible mixture components, would be a useful objective. In addition, for markets such as distilling, where the stimulus for future varietal production remains unclear, utilising mixtures, to prolong the commercial life-span of existing, high quality varieties, may become increasingly important.

REFERENCES

[1] Agu, RC; Bringhurst, TA; Brosnan, JM. Production of grain whisky and ethanol From wheat, maize and other cereals. *Journal of the Institute of Brewing,* 2006 112, 314-323.

[2] Agu, RC; Palmer, GH. Patterns of nitrogen distribution in barley grains grown in the field. *Journal of the Institute of Brewing,* 2003 109, 110-113.

[3] Agu, RC; Swanston, JS; Bringhurst, TA; Brosnan, JM; Jack, FR; Smith, PL.The influence of nitrogen uptake on the quality of distilling wheat cultivars. In: *Production, Technology and Innovation, Proceedings of the Second Worldwide Distilled Spirits Conference, Edinburgh,* Ed by Bryce J, Piggott J and Stewart GG., 2008, 67-74.

[4] Angus, WJ. United Kingdom wheat pool. In: *The world wheat book: A history of wheat breeding,* Ed by Bonjean AP and Angus WJ, Paris, Lavoisier Publishing, 2001, 103-126.

[5] Armour, T; Jamieson, PD; Nicholls, A; Zyskowski, R. Breaking the 15 t/ha wheat yield barrier. In: *New directions for a diverse planet: Proceedings of the 4th International Crop Science Congress, Brisbane, Australia,* 2004, Poster papers 2.7.3., www.cropscience.org.au.

[6] Bathgate, G. Cereals in Scotch whisky production. In: *Cereal Science and Technology,* Ed by Palmer GH, Aberdeen, University Press, 1989, 243-278

[7] Beaumont, VH. New characterization tools for whiskey raw materials. In: *Production, Technology and Innovation, Proceedings of the Second Worldwide Distilled Spirits Conference, Edinburgh,* Ed by Bryce J, Piggott J and Stewart GG., 2008, 29-35.

[8] Blake, A. Danes all set for green energy. *Farmers Weekly,* 7 June 2002, 60.

[9] Bringhurst, TA; Agu, RC; Brosnan, JM; Fotheringham, AL. Wheat for Scotch whisky production: broadening the horizon. In: *Production, Technology and Innovation, Proceedings of the Second Worldwide Distilled Spirits Conference, Edinburgh,* Ed By Bryce J, Piggott J and Stewart GG., 2008, 51-58.

[10] Charmet, G; Robert, N; Branlard, G; Linossier, L; Martre, P; Triboni, E. Genetic analysis of dry matter and nitrogen accumulation and protein composition in wheat kernels. *Theoretical and Applied Genetics,* 2005 111, 540–550,

[11] Cowe, IA; Cuthbertson, DC; Swanston, JS. The effect of moisture and nitrogen levels on milling energy of barley. *Journal of the Institute of Brewing,* 1989 95, 423- 425.

[12] Culshaw, D. Straw as a fuel. In *Cereals: Novel uses and processes,* Ed by Campbell GM, Webb C and McKee SL., New York, Plenum Press, 1997, 153-158.

[13] Dampney, PMR. The effect of applications of nitrogen during stem extension and grain filling on the quality of wheat grain used for breadmaking, *Aspects of Applied Biology,* 1987 15, 239-248.

[14] Dampney, PMR; Salmon, S. The effect of rate and timing of late nitrogen applications to breadmaking wheats as ammonium nitrate or foliar urea-N, and the effect of foliar sulphur application I. Effect on yield, grain quality and recovery of nitrogen in grain. *Aspects of Applied Biology,* 1990 25, 229–241

[15] Edney, MJ. Barley. In: *Cereal Grain Quality*, Ed by Henry RJ and Kettlewell PS., London, Chapman and Hall, 1996, 113-151.

[16] Essah, SYC; Stoskopf, NC. Mixture performance of phenotypicaly contrasting barley cultivars. *Canadian Journal of Plant Science*, 2002 82, 1-6.

[17] Finckh, MR; Gacek, ES; Goyeau, H; Lannou, C; Merz, U; Mundt, CC; Munk, L; Nadziak, J; Newton, AC; de Vallavieille-Pope, C; Wolfe, MS. Cereal variety and species mixtures in practice, with emphasis on disease resistance. *Agronomie*, 2000 20, 813-837.

[18] Friedt, W; Werner, K; Ordon, F. Genetic progress as reflected in highly successful and productive modern barley cultivars. In: *Proceedings of the Eighth International Barley Genetics Symposium, Adelaide, Volume 1*, 2000, 271-279.

[19] Gacek, ES; Czembor, HJ; Nadziak, J. Disease restriction, grain yield and its stability in winter barley cultivar mixtures. In: *Proceedings of the Third Workshop on Integrated Control of Cereal Mildews across Europe*, 1996, 193-196.

[20] Gleason, C; Chaudhuri, S; Yang, T; Muñoz, A; Poovaiah, BW; Oldroyd, GED. Nodulation independent of rhizobia induced by a calcium-activated kinase lacking autoinhibition. *Nature* 2006 441, 1149-1152.

[21] Good, AG; Johnson, SJ; De Pauw, M; Carroll, RT; Savidov, N; Vidmar, J; Lu, Z; Taylor, G; Stroeher, V. Engineering nitrogen use efficiency with alanine aminotransferase. *Canadian Journal of Botany*, 2007 85, 252-262.

[22] Gooding, MJ; Dimmock, JPRE ; France, J; Jones, SA. Green leaf area decline of wheat flag leaves: the influence of fungicides and relationships with mean grain weight and grain yield. *Annals of Applied Biology*, 2000 136, 77-84.

[23] Gupta, PK; Mir, RR; Mohan, A; Kumar, J. Wheat genomics: present status and future prospects. *International Journal of Plant Genomics* Volume 2008, Article ID 896451, 2008, 36 pages (ijpg@hindawi.com).

[24] HGCA. Recommended lists 2003/04 for cereals and oilseeds. London, HGCA, 2003.

[25] Hopkin, M. Crops could make their own fertilizers. *Nature News*, 28th June 2006, (http://www.nature.com/news).

[26] Jackson, LF; Wennig, RW. Use of cultivar blends to improve grain yield and quality and reduce disease and lodging. *Field Crops Research* 1997 52, 261-269.

[27] Jamieson, PD; Semenov, MA; Brooking, IR; Francis GS. *Sirius*: a mechanistic model of wheat response to environmental variation. *European Journal of Agronomy,* 1998 8, 161-179.

[28] Johnson, P. The agronomic and economic effects of reducing nitrogen fertiliser use. In:*Reducing agrochemical use on the arable farm: The TALISMAN and SCARAB projects*, Ed by Young JEB, Griffin MJ, Alford DV and Ogilvy SE., London, Defra, 2001, 51-61.

[29] Keller, FA; Hamilton, JE; Nguyen, QA. 2003. Microbial pretreatment of biomass-Potential for reducing severity of thermochemical biomass pretreatment. *Applied Biochemistry and Biotechnology,* 2003 105, 27–41.

[30] Lockyer, P; Wardlaw, A. The use of NIR spectroscopy to quantify various malted barley analytes. In: *Production, Technology and Innovation, Proceedings of the Second Worldwide Distilled Spirits Conference, Edinburgh*, Ed by Bryce J, Piggott J and Stewart GG., 2008, 47-50.

[31] Kindred, DR; Verhoeven, TM; Weightman, RM; Swanston, JS; Agu, RC; Brosnan, JM; Sylvester-Bradley, R. Effects of variety and fertiliser nitrogen on alcohol yield, grain yield, starch and protein content, and protein composition of winter wheat. *Journal of Cereal Science,* 2008 48, 46-57.

[32] Knott, EA; Mundt, CC. Mixing ability analysis of wheat cultivar mixtures under diseased and non diseased conditions. *Theoretical and Applied Genetics,* 1990 80, 313–320.

[33] Manthey, R; Fehrmann, H. Effect of cultivar mixtures in wheat on fungal diseases, yield and profitability. *Crop Protection*, 1993 12, 63-68.

[34] Maule, AP. Use of comparamill milling energy tests in the evaluation of malting barley purchase samples. *Proceedings of the Third Aviemore Conference on Malting, Brewing and Distilling*, Ed by Campbell I., London, Institute of Brewing, 1991, 501-502.

[35] McCracken, AR; Dawson, WM; Bowden, G. Yield responses of willow (Salix) grown in mixtures in short rotation coppice (SRC). *Biomass & Energy* 2001 21, 311-319.

[36] Mille, B; Belhaj Fraj, M; Manod, H; de Vallavieille-Pope, C. Assessing four-way mixtures of winter wheat cultivars from the performances of their two-way and individual components. *European Journal of Plant Pathology*, 2006 114, 163-173.

[37] Morgan, AG; Gothard, PG. Rapid prediction of malt hot water extract by near infra-red reflectance spectroscopy studies on barley. *Journal of the Institute of Brewing*, 1979 85, 339-341.

[38] Mundt, CC. Performance of wheat cultivars and cultivar mixtures in the presence of *Cephalosporium* stripe. *Crop Protection*, 2002 21, 93-99.

[39] Nelson, JC; Andreescu, C; Breseghello, F; Finney, PL; Gualberto, DG; Bergman, CJ; Pena, RJ; Reine Perretant, M; Leroys, P. Quantitative trait locus analysis of wheat quality traits. *Euphytica* 2006 149, 145-159.

[40] Newton, AC; Ellis, RP; Hackett, CA; Guy, DC. The effect of component number on *Rhynchosporium secalis* infection and yield in mixtures of winter barley cultivars. *Plant Pathology* 1997 46, 930–938.

[41] Newton, AC; Hackett, CA; Swanston, JS. Analysing the contribution of component cultivars and cultivar combinations to malting quality, yield and disease in complex mixtures. *Journal of the Science of Food and Agriculture*, 2008 88, 2142-2152.

[42] Newton, AC; Swanston, JS; Guy, DC; Ellis, RP. The effect of cultivar mixtures on malting quality in winter barley. *Journal of the Institute of Brewing*, 1998 104, 41- 45.

[43] Newton, R. Biofuels are the future. *Chemistry and Industry*, 2003 11, 14-15.

[44] Osman A, 2006. The effect of growing cultivar mixtures on baking quality of organic spring wheat. In: *Cereal diversity: Implications for production and products, Proceedings of COST SUSVAR workshop, La Besse, France*, 2006, 17-22.

[45] Panozzo, JF; Eagles, HA. Rate and duration of grain filling and grain nitrogen accumulation of wheat cultivars grown in different environments. *Australian Journal of Agricultural Research* 1999 50, 1007-15.

[46] Piggott, JR. Whiskies. In: *Fermented Beverage Production 2nd Edition*, Ed by Lea AGH and Piggott JR, New York, Kluwer Academic/Plenum Publishers, 2003, 239-262.

[47] Raun, WR; Solie, JB; Johnson, GV; Stone, ML; Mullen, RW; Freeman, KW; Thomason, WE; Lukina, EV. Improving nitrogen use efficiency in cereal grain production with optical sensing and variable rate application. *Agronomy Journal*, 2002 94, 815 - 820.

[48] Richards, IR. Energy balances in the growth of oilseed rape for biodiesel and wheat for bioethanol. *Levington Agriculture Report for the British Association for Biofuels and Oils. www.biodiesel.co.uk/levington.htm,* 2000, Accessed 15 August 2002.

[49] Riffkin, HL; Bringhurst, TA; McDonald, AML; Hands, E. Quality requirements of wheat for distilling. *Aspects of Applied Biology*, 1990 25, 29-40.

[50] Sammons, DJ; Baenzinger, PS. Performance of four winter wheat cultivars in blended populations. *Field Crops Research* 1985 10, 135-142.

[51] Sarandon, SJ; Sarandon, R. Mixture of cultivars: pilot field trial of an ecological alternative to improve production or quality of wheat (*Triticum aestivum*). *Journal of Applied Ecology*, 1995 32, 288-294.

[52] Scheller, WA. Gasohol: The U.S. experience, in *Cereals: A Renewable Resource, Theory and Practise*, Ed by Pomeranz Y and Munck L., St. Paul, Minnesota, American Association of Cereal Chemists, 1981, 633-649.

[53] Schildbach, G. The pros and cons of mixed variety cultivation of quality malting barley. *Brauwelt* 1991 131, 420-424.

[54] Schubert, C. Can biofuels finally take center stage? *Nature Biotechnology*, 2006 7, 777-784.

[55] Scottish Government Statistics, 2006, http://cci.scot.nhs.uk/Publications/2006/02

[56] Sheehan, J; Aden, A; Paustian, K; Killian, K; Brenner, J; Walsh, M; Nelson, R. Energy and environmental aspects of using corn stover for fuel ethanol. *Journal of Industrial Ecology* 2003 7(3–4), 117–146.

[57] Sheldrick, RD. The quality of 'brown midrib-3'mutant maize grown for forage under field conditions in southern England. *Grass and Forage Science*, 1979 34, 283-291.

[58] Shepherd, M. Cereals for bioethanol. *Presentation at conference 'Be Ready for Biofuels'*, Sponsored by United Oilseeds and HL Hutchison Ltd., Peterborough, UK, Nov.27 2002.

[59] Silvey, V. The contribution of new varieties to cereal yields in England and Wales between 1947 and 1983. *Journal of the National Institute of Agricultural Botany*, 1986 17, 155-168.

[60] Smidansky, ED; Clancy, M; Meyer, FD; Lanning, SP; Blake, NK; Talbert, LE; Giroux, MJ. Enhanced ADP-glucose pyrophosphorylase activity in wheat endosperm increases seed yield. *Proceedings of the National Academy of Sciences* 2002 99, 1724-1729.

[61] Smith, TC; Kindred, DR; Brosnan, JM; Weightman, RM; Shepherd, M; Sylvester-Bradley, R.Wheat as a feedstock for alcohol production. HGCA Research Review No. 61. Home-Grown Cereals Authority, Caledonia House, 223, Pentonville Road, London, N1 9HY, 2006.

[62] Spiertz, JHJ; Vos, J. (1985). Grain growth of wheat and its limitation by carbohydrate and nitrogen. In *Wheat Growth and Modelling*. Ed by Day W and Atkin RK., New York, Plenum Press, 1985, Vol. 86, 129-141.

[63] Swallow, A. Blends push yield limits. *Farmers Weekly 2nd August*, 2002, 58.

[64] Swanston, JS; Newton, AC. Do components of barley variety mixtures converge for malting quality attributes? *Proceedings of the Ninth International Barley Genetics Symposium, Brno, Poster Presentations*, 2004 505-510 (on CD-ROM).

[65] Swanston, JS; Newton, AC. Mixtures of UK wheat as an efficient and environmentally-friendly source for bioethanol. *Journal of Industrial Ecology*, 2005 9, 109–126

[66] Swanston, JS; Newton, AC; Brosnan, JM; Fotheringham, A; Glasgow, E. Determining the spirit yield of wheat varieties and variety mixtures. *Journal of Cereal Science*, 2005 42, 127-134.

[67] Swanston, JS; Newton, AC; Hoad, SP; Spoor, W. Barleys grown as cultivar mixtures compared with blends made before and after malting, for effects on malting performance. *Journal of the Institute of Brewing*, 2005 111, 144-152.

[68] Swanston, JS; Newton, AC; Hoad, SP; Spoor, W. Variation across environments in patterns of water uptake and endosperm modification in barley varieties and variety mixtures. *Journal of the Science of Food and Agriculture*, 2006 86, 826-833.

[69] Swanston, JS; Newton, AC; Smith, PL. Grain quality characters in complex mixtures of soft wheat and acknowledgement by farmers and distillers. In: *Cereal diversity: Implications for production and products, Proceedings of COST SUSVAR workshop, La Besse, France*, 2006, 29-32.

[70] Swanston, JS; Smith, PL; Gillespie, TL; Brosnan, JM; Bringhurst, TA; Agu, RC. Associations between grain characteristics and alcohol yield among soft wheat varieties. *Journal of the Science of Food and Agriculture*, 2007 87, 676-683.

[71] Sylvester-Bradley, R; Kindred, DR. Developing and growing wheat for the biofuels Market. In: *Arable cropping in a changing climate. Proceedings of HGCA conference – 23 and 24 January 2008*

[72] Taylor, BR; Cranstoun, DAS; Roscrow, JC. The quality of winter wheat varieties for distilling from Scottish sites. *Aspects of Applied Biology*, 1993 36, 481-484.

[73] Taylor, BR; Roscrow, JC. Factors affecting the quality of wheat grain for distilling in Northern Scotland. *Aspects of Applied Biology*, 1990 25, 183-191.

[74] Vaidyanathan, LV; Sylvester-Bradley, R; Bloom, TM; Murray, AWA. Effects of previous cropping and applied nitrogen on grain nitrogen content in winter wheat. *Aspects of Applied Biology*, 1987 15, 227-237.

[75] Van den Broek, R; Treffers, D-J; Meeusen, M; van Wijk, A; Nieuwlaar, E; Turkenburg, W. Green energy or organic food? A life-cycle assessment comparing two uses of set-aside land. *Journal of Industrial Ecology*, 2002 5(3), 65–87.

[76] Wainwright, T. Britain's biggest malting site officially opened. *Ferment*, 1998 11, 331-334.

[77] Walker, EW. Grain spirit - which cereal? In *Proceedings of the Second Aviemore Conference on Malting, Brewing and Distilling*, Ed by Campbell I and Priest FG., London, Institute of Brewing, 1986, 375-380.

[78] Weir, RB. Distilling and agriculture 1870-1939. *The Agricultural History Review*, 1984 32, 49-62.

[79] Wilson, PR. Analysis of whole cereal grains by near infrared transmission. *Aspects of Applied Biology* 1993 36, 477-480.

[80] Wolfe, MS. Maintaining the value of our varieties. *Proceedings of the Sixth International Barley Genetics Symposium, Helsingborg*, Vol. 2, 1992, 1055-1067.

[81] Zhu, Y; Chen, H; Fan, J; Wang, Y; Li, Y; Chen, J; Fan, JX; Yang, S; Hu, L; Leung, H; Mew, TW; Teng, PS; Wang, Z; Mundt, CC. Genetic diversity and disease control in rice. *Nature,* 2000 406, 718-722.

In: Soybean and Wheat Crops
Editors: S. Davies, G. Evans
ISBN: 978-1-60741-173-4

Chapter 10

GENETIC IMPROVEMENT OF WHEAT YIELD POTENTIAL AND ADAPTATION IN CHINA

Zhonghu He[1,2,*] and Xiaoke Zhang[3]
[1]Institute of Crop Science, National Wheat Improvement Center/The National Key Facility for Crop Gene Resources and Genetic Improvement, Chinese Academy of Agricultural Sciences (CAAS), 12 Zhongguancun South Street, Beijing 100081, China
[2]CIMMYT China Office, C/O CAAS, Zhongguancun South Street, Beijing 100081, China
[3]College of Agronomy, Northwest Sci-Tech University of Agriculture and Forestry, Yangling, Shaanxi 712100, China

ABSTRACT

Genetic improvement of yield potential has always been an important objective in China. Averaged annual genetic gain in grain yield ranged from 13.96 kg/ha/year to 72.11 kg/ha/year or from 0.31% to 1.23% annually in different wheat regions. The genetic improvement in grain yield was primarily attributed to increased grain weight per spike, reduced plant height, and increased harvest index. Three dwarfing genes and 1B/1R translocation have been successfully used in wheat production. *Rht-D1b* (45.5%) and *Rht 8* (46.8%) were more frequent, followed by *Rht-B1b* (24.5%). The frequencies of *Rht-B1b* and *Rht-D1b* increased, from 8.6% to 32.2% and 36.2% to 53.4%, respectively, whereas the frequency of *Rht8* has remained constant over time, when compared with cultivars released before and after 1990. From the late 1970s to the early 1990s, wheat breeding in autumn-sown wheat regions focused on the utilization of the 1B/1R translocation. The dominant *Vrn-D1* allele showed the highest frequency in Chinese wheat cultivars (37.8%), followed by the dominant *Vrn-A1*, *Vrn-B1*, and *Vrn-B3* alleles. All cultivars released in the Northern Winter Wheat Zone were winter type. Winter

* Corresponding author. z.he@cgiar.org, Phone number: +86-10-62170333, Fax: +86-10-68918547

(53.0%), spring (36.1%) and early heading (10.9%) cultivars were grown in the Yellow and Huai River Valleys Facultative Wheat Zone. Most of the spring genotypes from this zone carried only the dominant *Vrn-D1* allele, which was also predominant (64.1%) in the Middle and Lower Yangtze Valleys Autumn-Sown Spring Wheat Zone and Southwestern Autumn-Sown Spring Wheat Zone. The average frequency of the photoperiod-insensitive *Ppd-D1a* allele was 66.0%, with the frequencies of 38.6% and 90.6% in landraces and improved cultivars, respectively. Therefore, in addition to utilization of dwarfing genes and the 1B/1R translocation, dominant vernalization and photoperiod genes for early maturity have also contributed to yield improvement of Chinese wheat. The future challenge of wheat breeding is to continually raise grain yield, or to both maintain the genetic gain in grain yield and improve grain quality, without increasing inputs for the wheat based double cropping system.

INTRODUCTION

China is the largest wheat producer and consumer in the world and wheat ranks as the third leading crop in China after rice and maize.

Chinese wheat area has been divided into ten major agro-ecological zones (Figure 1), based on wheat types, varietal reactions to temperature, photoperiod, moisture, biotic and abiotic stress, and wheat growing seasons (Jin, 1986, 1997; He et al., 2001). At present, autumn-sown wheats account for about 90% of production and acreage and include zones I (4%), II (60%), III (13%), IV (minor area of production) and V (10%). Spring-sown wheats represent 7% of the wheat acreage in China and are grown in zones VI, VII, and VIII. Zones IX and X cover less than 3% of the total wheat area and include both spring- and fall-sown wheats. Although wheat is grown in 30 of the 31 provinces, more than 90% is produced in 13 provinces; of these, more than 70% of Chinese wheat is produced in five provinces, i.e., Henan, Shandong, Hebei, Anhui, and Jiangsu.

Great progress has been achieved in wheat production during the last 57 years, average yield has increased 1.9% annually from 660 kg/ha in 1950 to 4487 kg/ha, and the production has increased more than six times, from less than 20 million tones in 1950 to 105 million tones in 2007. Many factors have contributed to the significant increase of average yield, including adoption of improved cultivars, extension of high-yielding cultivation technologies, increased use of fertilizers and irrigation, expansion of farm mechanization, and improvement of rural policy. More than 2000 wheat cultivars have been released, and 59 outstanding cultivars, each covering annually an acreage of 670,000 hectares, had made significant contribution to China wheat production. It has been recorded that farmers have replaced their wheat cultivars six to eight times in the major wheat areas (He et al., 2001; Zhuang, 2003). The objective of this chapter is to review the advances in genetic improvement of wheat grain yield, and adaptation in China accomplished by breeders from 1950s to the present.

I: Northern Winter Wheat Zone; II: Yellow and Huai River Valleys Facultative Wheat Zone; III: Middle and Low Yangtze Valleys Autumn-Sown Spring Wheat Zone; IV: Southern Autumn-Sown Spring Wheat Zone; V: Southwestern Autumn-Sown Spring Wheat Zone; VI: Northeastern Spring Wheat Zone; VII: Northern Spring Wheat Zone; VIII: Northwestern Spring Wheat Zone; IX: Qinghai-Tibetan Plateau Spring-Winter Wheat Zone; X: Xingjiang Winter-Spring Wheat Zone

Figure 1. Wheat zones of China (Jin, 1986; He et al., 2001).

GENETIC IMPROVEMENT OF GRAIN YIELD AND YIELD COMPONENTS

Autumn-sown region in China, including Zones I, II, III and V, is the most important region, sharing more than 90% of wheat production. During the 2001-2002 and 2002-2003 crop seasons, six yield potential trials were conducted in Zone I located in Beijing, Zone II located in Shijiazhuang of Hebei province, Jinan of Shandong province and Zhengzhou of Henan province, Zone III located in Nanjing of Jiangsu province, Zone V located in Chengdu of Sichuan province, respectively (Zhou et al., 2007a and b). Each trial consisted of the leading cultivars from the 1949 to 2000 in each province. Only cultivars sharing more than 20% of wheat area in the respective province were included in the trials. The mean values of grain yield and annual genetic grain were different in diverse wheat regions/provinces (Table 1). It showed that mean of grain yield ranged from 4.56 t/ha to 7.02 t/ha, average annual genetic gain in grain yield ranged from 13.96 kg/ha/year to 72.11 kg/ha/year or from 0.31% to

1.23% annually in different provinces. The most significant increase in grain yield occurred in the early 1980s in Zones I, II and III.

On average, spikes/m^2 decreased from Beijing (741), to Hebei (643), Henan (591), Shandong (568), Jiangsu (448) and Sichuan (411), whereas kernels per spike and kernel weight per spike increased from north to south, but thousand kernal weight (TKW) was higher in Shandong and Henan than other provinces (Table 2). However, in terms of genetic improvement for yield components, notable variation for was found in all six provinces (Table 3). There was no clear common trend across the trials in terms of changes for kernels per spike and spikes number, with the exception of kernel weight per spike and TKW.

Table 1. Average wheat grain yield and its genetic improvement in autumn-sown zones from 1949 to 2000

Zone	No. of cultivars	Province/Location	Mean of yield (t/ha)	Annual genetic gain in grain yield	
				%	kg /ha/year
I	10	Beijing	5.39	1.23	64.27
II	11	Hebei/Shijiazhuang	6.84	0.48	32.07
	15	Shandong/Jinan	6.55	0.48	32.09
	11	Henan/Zhengzhou	7.02	1.05	72.11
III	11	Jiangsu/Nanjing	4.56	0.31	13.96
V	15	Sichuan/Chengdu	5.94	0.74	40.96

Table 2. Yield components of leading cultivars released in Zones I, II, III and V from 1949 to 2000

Province	Beijing	Hebei	Shandong	Henan	Jiangsu	Sichuan
Spikes/m^2	741	643	568	591	448	411
Kernels per spike	20.8	29.5	33.0	31.2	40.0	41.5
Thousand kernal weight (g)	35.5	36.7	43.3	40.3	37.2	38.7
Kernel weight per spike (g)	0.74	1.09	1.19	1.15	1.23	1.61

Table 3. Genetic gain (%) of yield components in leading cultivars released in Zones I, II, III and V

Zone	Province	Spikes/m^2	Kernels per spike	TKW	Kernel weight per spike
I	Beijing	0.65*	0.60	1.30**	1.79**
II	Hebei	-.079*	0.99**	0.06	1.00**
	Shandong	-0.74*	0.54*	0.35	0.78**
	Henan	0.59	-0.10	0.51	0.54
III	Jiangsu	0.51	0.14	0.32	0.58
V	Sichuan	-0.11	0.20	0.65**	0.87**

* and ** indicate significance at $p = 0.05$ and 0.01, respectively.

In Zone I, cultivars released in Beijing from 1949 to 2000 were characterized by a significant increase in TKW (1.30%, $P<0.01$), decreased spikes/m^2 (-0.65%, $P < 0.05$), and increased kernels/spike (0.60%, $P > 0.05$) and kernel weight per spike (1.79%, $P < 0.01$).

In Zone II, cultivars developed in Hebei province was characterized by a significant increase in kernels per spike (0.99%, $P < 0.01$) and kernel weight per spike (1.00%, $P < 0.01$), reduction of spike/m^2 (-0.79%, $P < 0.05$), and a slight increase in TKW (0.06%, $P > 0.05$). Cultivars sown in Shandong province was characterized by significantly reduced spikes/m^2 (-0.74%, $P < 0.05$), increased kernels per spike (0.54%, $P < 0.05$) and kernel weight per spike (0.78%, $P < 0.01$), and slightly increased TKW (0.35%, $P > 0.05$). Cultivars planted in Henan province was only characterized by significantly increased spikes/m^2 (0.59%, $P < 0.05$), increased TKW (0.51%, $P >0.05$) and kernel weight per spike (0.54%, $P >0.05$), and slightly decreased kernels per spike (-0.10%, $P>0.05$).

In Zone III, spikes/m^2 (-0.51%) was reduced, kernels per spike (0.14%), TKW (0.32%), and kernel weight per spike (0.58%) were slightly increased, but all of them are not significantly at 5% probability level. In Zone V, it was characterized with a significant increase of TKW (0.65%, $P<0.01$) and kernel weight per spike (0.87%, $P<0.01$). Kernels per spike (0.20%) was slightly increased, and spikes/m^2 (-0.11%) was slightly reduced, but they were not significant at $P = 0.05$.

The simultaneous increase of TKW and kernel per spike observed in Beijing, Hebei, Shandong, Jiangsu and Sichuan, or the shift of negative relationships between them observed in Henan, resulted in an increased kernel weight per spike, and thus led to the increased grain yield. An increase in the number of kernels per spikelet, rather than the numbers of spikelets, seemed to play a more important role in increasing kernels per spike based on observations in Shandong. This is further supported by the significant positive correlations between grain yield and kernel weight per spike ranging from 0.73 to 0.95 ($P < 0.05$) observed in five trials excluding Jiangsu (Table 4).

Increased grain yield can also be achieved in different ways in the same location. For example, Lumai 7 in Shandong province was characterized by average number of spikes/m^2 (553), greater kernels per spike (36.0) and relatively small grain size (40.8g), whereas Jimai 19 had lower number of spikes/m^2 (484), but greater kernels per spike (38.8) and grain size (45.4g). Lumai 22, called a large spiked wheat cultivar by farmers, is yet another example. It had less spikes/m^2 (399), but more kernels per spike (45.0) and big kernel size (TKW 47.9 g), i.e., large and heavy spike, 1.81g, although it is generally believed that a certain number of spikes is needed for stable yield performance across environments (Zhuang, 2003).

Table 4. Correlation coefficients between grain yield and kernel weight per spike

Province	Beijing	Hebei	Shandong	Henan	Jiangsu	Sichuan
r	0.95**	0.77**	0.73**	0.75*	0.53	0.91**

* and ** indicate significance at $p = 0.05$ and 0.01, respectively.

Changes in Plant Height, Harvest Index, Biomass and Heading Date

Plant height in Chinese wheat cultivars was significantly reduced (Zhou et al., 2007a and b). The annual genetic gain in plant height was -0.83% ($P < 0.01$) in Beijing of Zone I, and -1.33% ($P < 0.01$), -0.52% ($P < 0.05$) and -1.51% ($P < 0.01$) in Heibei, Shandong and Henan provinces of Zone II, respectively. The leading cultivars such as Jingdong 8 in Zone I are still relatively tall (103 cm). The new cultivars, such as CA 9722 and Nongda 3219 with height of 85 cm, are not widely cultivated. All cultivars released after the 1980s in Zone II are 75 to 85 cm tall. Plant height was significantly reduced by the annual genetic gain of -0.76% ($P<0.01$), from 122 cm of Mentana to around 90 cm of current cultivars in Zone III. A significant reduction in plant height (-0.69%, $P<0.01$) was gained in Zone V. A negative association between grain yield and plant height was observed in all six trials, i.e, Beijing (-0.76, $P < 0.05$), Hebei (-0.55, $P > 0.05$), Shandong (-0.29, $P > 0.05$), Henan (-0.83, $P < 0.01$), Jiangsu (-0.49, $P > 0.05$) and Sichuan (-0.76, $P < 0.01$) (Table 5).

Harvest index (HI) increased in all four zones: Zone I (0.92%, $P < 0.01$, Beijing), Zone II (0.42%, $P < 0.01$, Hebei; 0.46%, $P < 0.01$, Jinan; 0.50%, $P < 0.05$, Henan), Zone III (0.19, $P > 0.05$, Jiangsu) and Zone V (0.63, $P < 0.01$, Sichuan). Significantly positive associations between grain yield and HI were observed in five trials of Zones I, II and V, ranging from 0.67 to 0.96 ($P < 0.05$) (Table 5). Slightly positive associations between grain yield and HI (0.40, $P > 0.05$) were observed in trial of Zone III. This gave an indication that further increases in HI may continue to contribute to grain yield improvement.

Only slight changes in biomass were observed in Zones I (0.21%, $P > 0.05$, Beijing), III (0.16%, $P > 0.05$, Jiangsu), V (0.11%, $P > 0.05$, Sichuan) and some regions of Zone II (0.05%, $P > 0.05$, Hebei; -0.01%, $P > 0.05$, Shandong), but cultivars released from Henan province was an exception, with the annual genetic gain in biomass of 0.50% ($P < 0.01$). Significantly positive correlations between grain yield and biomass were observed in Zones I, II, III and V, ranging from 0.62 to 0.79 ($P < 0.05$). This indicates the importance of improving, or at least maintaining, current biomass in future cultivar development.

Heading date was significantly decreased, with the annual genetic gains of -0.12% ($P < 0.01$) and -0.22% ($P < 0.01$) in Beijing and Henan, respectively. Only slight changes in heading date in Jiangsu (0.03%, $P > 0.05$), Sichuang (-0.17%, $P > 0.05$), Hebei (-0.04%, $P > 0.05$) and Shandong provinces were observed. Significantly negative correlations between grain yield and heading date were observed in Beijing (Zone I) and Sichuan (Zone V) (Table 5).

Table 5. Correlation coefficients between grain yield (GY) and plant height (PH), harvest index (HI) and heading date (HD) in autumn-sown wheat cultivars

Location	Beijing	Hebei	Shandong	Henan	Jiangsu	Sichuang
GY-PH	-0.76*	-0.55	-0.29	-0.83**	-0.49	-0.76**
GY-HI	0.94**	0.86**	0.67**	0.76**	0.40	0.96**
GY-HD	-0.81**	-0.35	0.02	-0.53	-0.42	-0.71**
HI-PH	-0.84**	-0.88**	-0.71**	-0.79**	-0.59	-0.80**

* and ** indicate significance at p =0.05 and 0.01, respectively.

Table 6. Frequency (%) of *Rht-B1b*, *Rht-D1b* and *Rht8* genes in Chinese cultivars in various regions

Gene	Zone I	Zone II	Zone III	Zone V	Total
Rht-B1b	22.9	27.6	45.0	9.1	24.5
Rht-D1b	34.3	60.3	20.0	39.4	45.5
Rht8	40.0	49.1	30.0	48.5	46.8

Numbers of cultivars tested in Zone I, II, III and V were 35, 116, 20 and 33, respectively.

UTILIZATION OF DWARFING GENES

The improvement of grain yield and HI, largely due to a reduction in plant height (*r* between HI and plant height ranged from -0.59 to -0.88) (Table 5), are closely associated with incorporation of dwarfing genes into the leading cultivars (Zhang et al., 2006; Zhou et al., 2007a and b). Reduced height genes *Rht-B1b*, *Rht-D1b* and *Rht 8* of 220 cultivars from four main autumn-sown wheat regions in China were detected by molecular markers (Zhang et al., 2006). They include landmark landraces, leading cultivars and core parents used in various breeding programs from 1950s to the present, and introductions used both in production and breeding, and therefore provide valuable information on varietal replacements and evolution in diverse environments (Table 6). Overall, *Rht-D1b* and *Rht 8* showed similar frequency in Chinese wheats, with 45.5% and 46.8%, respectively. The presence of *Rht-B1b* is much lower, with a frequency of 24.5%. However, notable difference was observed in various regions. For example, *Rht-B1b* is quite high in Zone III with 45.0%, and it is only with 9.1% in Zone V. *Rht-D1b* showed high frequency in Zone II, with 60.3%, while *Rht 8* has high frequency in Zones I, II, and V.

Twenty-one genotypes with only *Rht 8* gene were presented in cultivars released before 1970s in Zone II or current cultivars in Zones I, III, and V. This is agreeable with general observation that plant height is much shorter in Zone II in comparison with three other regions (Zhuang, 2003). The distribution of both *Rht-B1b* and *Rht-D1b* genes shows a regional pattern or close linked with various institutional breeding activities. Forty varieties (six from Beijing, two from Hebei, five from Shandong, 19 from Henan, two from Shaanxi, one from Anhui, one from Hubei, three from Sichuan, and one from Yunnan) contain both *Rht-D1b* and *Rht 8* genes. Twenty-eight genotypes (two from Beijing, five from Hebei, one from Shandong, three from Henan, six from Shaanxi, five from Shanxi, four from Jiangsu, one from Anhui, and one from Yunnan) have the combination of *Rht-B1b* and *Rht 8* genes. Only two Chinese varieties, i.e., CA9532 from Beijing and Shannong 1355 from Shandong Province, and Suwon 86 from Korea carry both *Rht-B1b* and *Rht-D1b*. Zhoumai 11 from Henan Province has three dwarfing genes, *Rht-B1b, Rht-D1b,* and *Rht 8.* This might give an indication that combination of *Rht-B1b* or *Rht-D1b* with *Rht 8* could meet the needs of wheat production in most autumn-sown regions.

None of the above three dwarfing genes were identified in 30 genotypes, most of them are tall landraces or cultivars released before 1970s, however, cultivars including Beijing 837, Yuandong 9428, Jing 411, and Chuanmai 107 have plant height around 90-95cm, and Henong 326 and Xinong 291 have plant height of about 85cm in Beijing. This indicates that other dwarfing genes might be presented, but could not be identified at present. This has agreed with breeder experiences in China and CIMMYT when crossing Chinese wheats with known CIMMYT germplasm.

A notable change in the frequencies of three dwarfing genes was observed when comparing cultivars released before and after 1990 (Table 7). Overall, the frequencies of *Rht-B1b* and *Rht-D1b* have increased, from 8.6 to 32.2%, 36.2 to 53.4%, respectively, while the frequency of *Rht 8* remains constant over time. Before 1990, *Rht-B1b* was only presented in Zone II with frequency of 15.2%, *Rht-D1b* was not presented in Zone III, and its frequency has increased significantly in all four zones. It is also interesting to observe that frequency of *Rht 8* has reduced in Zone III and V, but significantly increased from 20.0 to 48.0% in Zone I. This is agreement with breeding experience in China since both *Rht-B1b* and *Rht-D1b* has strong effect on reducing plant height in comparison with *Rht 8* (Zhuang, 2003). Our study indicated that plant height has reduced from ca 110 to 80 cm in Beijing, from ca 110 to 75-85 in the Provinces of Hebei, Henan, Shandong, and Shaanxi, from ca 120 to 90 cm in Jiangsu and Sichuan Provinces, respectively.

Based on pedigree analysis, *Rht-B1b* in Chinese wheat is derived from two sources, viz.., Norin 10 and the Italian introduction St2422/464. The identity of *Rht-B1b* in these two sources still needs to be confirmed. Suwon 86 carrying both *Rht-B1b* and *Rht-D1b*, and Chinese cultivars, Huixianhong and Yaobaomai, are the primary sources of *Rht-D1b* in Chinese wheats. It is likely that in Youbaomai derives from an unknown introduction. Italian instructions such as Funo, Abbondanza, Lovrin 10 and Chinese landraces are the major sources of *Rht 8*.

Table 7. Frequency (%) of *Rht-B1b*, *Rht-D1b* and *Rht8* in varieties released before and after 1990

Gene	Zone I	Zone II	Zone III	Zone V	Total
Rht-B1b	0[a]/32.0[b]	15.2/32.5	0/60.0	0/13.0	8.6/32.2
Rht-D1b	20.0/40.0	51.5/63.9	0/26.7	20.0/47.8	36.2/53.4
Rht 8	20.0/48.0	39.4/53.0	60.0/20.0	60.0/43.5	41.4/47.3

[a] and [b] are percentage of various genes in varieties released before and after 1990, respectively.

Table 8. Distribution of 1B/1R translocation in Chinese wheat cultivars among difference regions

Zone	I	II	III	V	Total
No. of cultivars tested	3[a]/22[b]	18/80	9/10	5/15	35/127
No. of cultivars with 1B/1R translocation	0/13	0/33	0/2	0/3	0/51
Frequency of 1B/1R translocation (%)	0/59.1	0/41.3	0/20.0	0/20.0	0/40.2

[a] and [b] are number (percentage) of 1B/1R translocation in cultivars released before and after 1980, respectively.

UTILIZATION OF 1B/1R TRANSLOCATION

In total, 162 predominant cultivars in recent 30 years from Chinese four main autumn-sown wheat Zones (I, II, III and V) were examined by SDS-PAGE and SCAR for the presence of 1B/1R translocation (Zhou et al., 2004). Overall, frequency of cultivars with 1B/1R translocation in four autumn-sown regions was 31.5%, but those in diverse zones were different. The 1B/1R translocation was highest frequency in Zone I (52.0%), followed by Zone II (33.7%), and the lowest was found in Zones III (10.5%) and V (15.0%). However, cultivars with the 1B/1R translocation released before 1980 were absent in China. On average, 40.2% of Chinese wheats released since 1980 contained 1B/1R, but its frequency also differed in different zones, which was 59.1% for Zone I, 41.3% for Zone II and 20.0% for Zone III and V (Table 8). From the late 1970s to the early 1990s, wheat breeding in autumn-sown wheat regions focused on the utilization of the 1B/1R translocation from introductions such as Lovrin 10, Lovrin 13, and Neuzucht, which showed resistance to rusts, powdery mildew, and to heat stress after anthesis (He et al., 2001). Fengkang 8 in Beijing, Jimai 24 in Hebei, Lumai 7 in Shandong, Yumai 13 in Henan, Een 1 in Hubei and Miannong 4 in Sichuan, all have 1B/1R, and were developed in 1980, 1982, 1981, 1987, 1982 and 1986, respectively. The current leading cultivars in Beijing and Hebei continue to have 1B/1R. The low frequencies of 1B/1R in Shandong and Henan are largely due to the intensive selection for processing quality; thus good quality cultivars such as Jinan 17, Jimai 19, and Yumai 34, which lack 1B/1R, are the primary cultivars being grown.

DISTRIBUTION OF GROWTH HABIT AND VERNALIZATION GENES

The adaptation of wheat cultivars to diverse environmental conditions is greatly influenced by flowering time (Whitechurch and Slafer, 2002), which is mainly determined by three groups of genes: vernalization response genes (*Vrn* genes), photoperiod response genes (*Ppd* genes) and developmental rate genes (earliness per se, *Eps* genes) (Snape et al., 2001). The first two groups of genes are environment-dependent, whereas the third is largely environment-independent. Vernalization genes determine growth habits which divide wheat into winter and spring types. Winter types require vernalization to promote flowering and spring types do not. Photoperiod genes determine characters of photoperiod response in wheat. Photoperiod response is described as sensitive when timely flowering occurs only in long days, and insensitive when flowering occurs in either long- or short-day environments. The different frequencies of *Vrn* alleles observed in different parts of the world suggest that these allele combinations have an adapatative value (Gotoh, 1979; Stelmakh, 1990; Goncharov, 1998; Iwski et al., 2000, 2001). Photoperiod response is closely associated with adaptability and grain yield in European and Canadian wheat cultivars (Martiníc 1975; Hunt 1979; Worland et al. 1994; 1998). In Asia, Mediterranean and North African regions, most landraces are sensitive to photoperiod, whereas all improved cultivars with high yield potential are insensitive (Ortiz Ferrara et al. 1998). Therefore, understanding of the vernalization and photoperiod genes present in wheat breeding programs is useful when developing high yielding cultivars broadly adapted to different regions.

Table 9. Frequencies (%) of dominant alleles at *Vrn-A1*, *Vrn-B1*, *Vrn-D1* and *Vrn-B3* loci in spring cultivars of various wheat zones

Genotype	Zone								Total
	I	II	III	V	VI	VII	VIII	X	
Vrn-A1	-	-	-	15.2	95.8	94.1	74.1	85.7	44.2
Vrn-B1	-	6.7	11.1	42.4	75.0	58.8	70.4	50.0	42.4
Vrn-D1	-	93.3	96.3	78.8	50.0	29.4	14.8	28.6	61.0
Vrn-B3	-	-	-	-	8.3	-	-	-	1.2

Totally, 278 leading Chinese wheat cultivars released since the 1960s, and collected from eight wheat growing zones (I, II, III, V, VI, VII, VIII and X) were characterized with molecular markers for the vernalization genes *Vrn-A1*, *-B1*, *-D*, and *-B3*, Heading time of the 266 cultivars tested was evaluated in a greenhouse under long days without vernalizaiton (Zhang et al., 2008).

The frequencies of the different *Vrn* allele combinations were very different across different wheat agro-ecological zones. Cultivars with recessive alleles at all the analyzed *Vrn* loci represent 38.1% of the cultivars and are mainly concentrated in Zones I, II and X. The other 61.9% includes cultivars with at least one dominant *Vrn* allele, which can be classified as spring. These cultivars are found mainly in Zones III, V, VI, VII, VIII and X. The dominant *Vrn-D1* allele showed the highest frequency in Chinese wheats (37.8%), followed by the dominant *Vrn-A1* (27.3%), *-B1* (26.3%), and *-B3* (0.7%) alleles.

Among the cultivars with at least one dominant *Vrn* allele, the frequencies of the different alleles varied in across regions (Table 9). The dominant *Vrn-B3* allele is present only in two cultivars from zone VI with the frequency of 1.2%. Among the *Vrn-1*, alleles *Vrn-D1* (61.0%) showed highest frequency, followed closely by dominant *Vrn-A1* (44.2%) and *Vrn-B1* (42.4%) alleles. The dominant *Vrn-A1* allele is not presented in Zones I, II, and III, and its frequency is low in Zone V (15.2%). However, high frequencies are observed in Zones VI (95.8%), VII (94.1%), VIII (74.1%) and X (85.7%). The dominant allele *Vrn-B1* is not presented in Zone I, and low frequencies are observed in Zones II (6.7%) and III (11.1%). However, high frequencies are observed in Zones V (42.4%), VI (75.0%), VII (58.8%), VIII (70.4%) and X (50.0%). The dominant allele *Vrn-D1* is not presented in Zone I, but is present at relatively high frequencies in Zones II (93.3%), III (96.3%), V (78.8%), VI (50.0%), VII (29.4%), VIII (14.8%) and X (28.6%).

Among the four autumn-sown wheat zones (I, II, III and V), the frequency of dominant *Vrn-D1* allele is the highest, followed by *Vrn-B1* and *Vrn-A1* (*Vrn-B3* is absent). In contrast, in three spring-sown wheat zones (VI, VII and VIII) the frequency of the dominant *Vrn-A1* allele is the highest, followed by *Vrn-B1* and *Vrn-D1*, respectively. *Vrn-B3* frequency (2.9%) is the lowest.

Frequencies of the different combinations of vernalization genes were also very different among the various wheat zones. In brief, nine combinations of dominant *Vrn* alleles were identified (Table 10). Among them, the *Vrn-D1* allele alone was the most frequent (72 cultivars), followed by the *Vrn-A1*/*Vrn-B1* (36 cultivars) combination. In summary, most cultivars released in the autumn-sown wheat regions of south China (Zones III and V) and north China (Zone II) possessed *Vrn-D1* as a single dominant allele. In contrast, in spring-sown wheat regions, cultivars carried the strongest dominant *Vrn-A1* alleles and the majority of them included additional dominant *Vrn* alleles at the *Vrn-B1*, *Vrn-D1* and *Vrn-B3* loci. On the basis of the vernalization alleles found in our study, the vernalization requirement can be ranked from strongest to weaker from Zone I, Zone II, Zone III, Zone V, with the weakest requirement in the spring-sown spring wheat regions (Zones VI, VII, and VIII).

Table 10. Frequencies (%) of different *Vrn* allele combinations and their heading times in spring cultivars of various wheat zones

Genotype and heading time	I	II	III	V	VI	VII	VIII	X	Total
Vrn-A1 alone = 39d· (33-46d)	-	-	-	-	12.5	23.5	26.0	35.7	11.0
Vrn-D1 alone² = 54d (36-109d)	-	93.3	88.9	51.5	-	-	3.7	14.3	41.9
Vrn-B1 alone = 47d (36-73d)	-	6.7	3.7	15.2	-	-	11.1	-	6.4
Vrn-B1 + *Vrn-D1* = 42d (36-49d)	-	-	7.4	18.1	4.2	5.9	11.1	-	7.6
Vrn-A1 + *Vrn-B1* = 38d (32 – 49d)	-	-	-	6.1	33.2	47.1	48.1	35.7	20.9
Vrn-A1 + *Vrn-D1* = 38d (36-44d)	-	-	-	6.1	12.5	17.6	-	-	4.6
Vrn-A1 + *Vrn-B1* + *Vrn-D1* = 38d (33-50d)	-	-	-	3.0	29.2	5.9	-	14.3	6.4
Vrn-A1 + *Vrn-B1* + *Vrn-B3* = 30d	-	-	-	-	4.2	-	-	-	0.6
Vrn-A1+*Vrn-B1*+*Vrn-D1*+*Vrn-B3* = 31d	-	-	-	-	4.2	-	-	-	0.6

· Average flowering time of tested genotypes with this genotype.

Heading dates showed a continuous distribution from 30 days to more than six months after planting in the greenhouse. Out of 266 cultivars tested in the greenhouse, the 92 cultivars, which failed to head within 109 days all possessed recessive vernalization alleles at the four *Vrn* loci. Most of them were classified as winter cultivars in the literature (Jin, 1986 and 1997; Zhuang, 2003). Among the 174 cultivars that headed within 109 days (early heading), 164 of them carried at least one of the tested dominant vernalization allele, and were classified as spring. The other cultivars, nine from Zones II (Jimai 36, Taishan 1, Lumai 23, Laizhou 953, Weimai 8, Xinmai 9408, Yumai 66, Yumai 70, Xuzhou 14) and one from Zone III (Emai 6) carried recessive alleles at the four vernalization loci. The most likely explanation for this discrepancy is the presence of an unknown allele at the four loci characterized in this study or the presence of a spring allele at the *Vrn4* locus not included in this survey because the gene is still unknown.

All 32 cultivars released in Zone I headed after 109 days, had all three recessive *vrn-1* alleles and were classified as winter. In Zone II, of 75 cultivars tested in the greenhouse, 44 (58.7%) headed after 109 days. Although both winter and spring types are found in all provinces of Zone II, the late heading cultivars were mainly cultivated in the provinces of Shandong (79.2%), Shaanxi (63.6%), and Anhui (100.0%), whereas the early heading cultivars were mostly present in the provinces of Henan and Jiangsu. Most of the cultivars from Zones III and V (>94%) headed before 109 days, the rests were released before 1960. The frequency of early heading genotypes increased gradually from north to south in the autumn-sown regions. In Zones VI, VII and VIII, all 68 cultivars tested in the greenhouse headed within 109 days. In Zone X, the frequency of early heading genotypes was 51.9%. Therefore, it was concluded that spring cultivars in China are more frequent in the high latitude regions (spring sowing) and in the low latitude area with warm winters (autumn sowing). Winter cultivars are frequently present in the middle latitude area with relatively cold winter (autumn sowing).

The 92 late heading (winter) cultivars all carried recessive alleles at the four vernalization loci. Different combinations and proportions of vernalizaion alleles were found in other 172 early heading (spring) cultivars (Table 10). Single dominant alleles were observed for the *Vrn-A1* (11.0%), *Vrn-B1* (6.4%) or *Vrn-D1* (41.9%). We observed two gene combinations including *Vrn-A1 / Vrn-B1* (20.9%), *Vrn-A1 / Vrn-D1* (4.6%) and *Vrn-B1 / Vrn-D1* (7.6%); and three dominant allele combinations including *Vrn-A1Vrn-B1Vrn-D1* (6.4%), and *Vrn-A1Vrn-B1Vrn-B3* (0.6%). In addition, one very early heading cultivar (Liaochun 10) carried all four dominant alleles (*Vrn-A1Vrn-B1Vrn-D1Vrn-B3*).

The relationships between vernalization genotypes and heading times in diverse regions are different (Table 10). In summary, the earliest cultivars were those carrying three to four dominant alleles, including the rare *Vrn-B3* allele (average 30 to 31 days to heading), followed by the one, two or three gene combinations, including *Vrn-A1* but not *Vrn-B3* (average 38 days to heading). Cultivars carrying the *Vrn-B1*/*Vrn-D1*allelle combination headed approximately 42 days after sowing, whereas those carrying only the *Vrn-B1* (average 47 days) or *Vrn-D1* (average 54 days) were among the latest spring cultivars. On the basis of these data, the strength of the dominant spring *Vrn-1* alleles can be ranked as *Vrn-A1* > *Vrn-B1* > *Vrn-D1*. *Vrn-B3* resulted in the earliest heading times in combination with other dominant *Vrn1* alleles.

DISTRIBUTION OF PHOTOPERIOD GENES

The *Ppd-D1a* allele for photoperiod insensitivity is generally considered the most potent, followed by *Ppd-B1a* and *Ppd-A1a* (Scarth and Law, 1984). A total of 926 Chinese wheat landraces (438) and improved cultivars (488) collected from nine wheat growing zones were tested for their genotypes at the *Ppd-D1* locus using allele-specific markers (Yang et al., 2008). The overall frequency of the dominant *Ppd-D1a* allele in Chinese wheats was 66.0%, but frequencies varied across regions (Table 11). The highest frequency was found in Zones V (87.8%) and VII (87.5%), followed by Zones I (48.5%), II (71.2%), III (69.2%), IV (54.5%) and X (78.1%), the lowest was in Zones VI (36.0%) and VIII (43.7%). Among the four autumn-sown wheat zones where both landraces and improved cultivars were tested in this study, the frequency of *Ppd-D1a* in the Northern China Plain (Zone I) was much lower than that in the middle (Zone II) and southern parts (Zones III and V).

Table 11. Frequencies (%) of photoperiod insensitive allele *Ppd-D1a* in landraces and improved cultivars from different zones

Zone	I	II	III	IV	V	VI	VII	VIII	X	Total
Landrace	18.2	32.8	58.2	54.5	59.1	na	na	5.3	na	38.6
Improved cultivar	89.5	96.1	98.0	na	98.3	36.0	87.5	87.9	78.1	90.6
Subtotal	48.5	71.2	69.2	54.5	87.8	36.0	87.5	43.7	78.1	66.0

na = not available.

The frequency of the *Ppd-D1a* in 263 improved cultivars released in Zones I and II (94.7%) was three fold that among 211 landraces (27.5%). The frequency of *Ppd-D1a* also increased markedly from landraces (58.3%) to improved cultivars (98.2%) in Zones III and V. For Zones VI, VII and VIII, the frequency of *Ppd-D1a* in improved cultivars was 71.9%. Generally, the frequency of the *Ppd-D1a* in landraces (38.6%) was much lower than in improved cultivars (90.6%), a consequence of the introduction of *Ppd-D1a* to improve cultivar adaptation to various environments.

Among landraces in the autumn-sown wheat zones, the frequencies of *Ppd-D1a* in Zones I to V were 18.2%, 32.8%, 58.2%, 54.5%, and 59.1%, respectively. The frequency of *Ppd-D1a* was comparably lower (5.3%) in landraces from Zone VIII, a spring-sown spring wheat area. This indicated that the presence of *Ppd-D1a* in landraces gradually increased from north to south in the five autumn-sown wheat zones (I, II, III, IV and V). Among improved cultivars, those in Zones II, III and V located in the middle and southern parts of China had high frequencies of *Ppd-D1a*, ranging from 96.1% to 98.3%, followed by Zones I (89.5%), VII (87.5%), VIII (87.9%), and X (78.1%), and the lowest frequency was observed in Zone VI (36.0%). This again showed the increasing frequency of *Ppd-D1a* in improved cultivars from north to south.

Detailed analysis indicated that all current cultivars with early maturity in Zones I, II, III, IV, VII, and VIII carry the *Ppd-D1a* where early maturity is needed to avoid sprouting damage and to allow optimal sowing of maize after wheat. Zone VI is a high latitude environment, thus cultivars are expected to have a certain level of photoperiod sensitivity (He et al. 2001). Nine improved cultivars from Heilongjiang had the *Ppd-D1a* as expected since strong photoperiod sensitivity is required in high latitude environment. In Zone X, seven improved cultivars identified with *Ppd-D1b* were developed and grown in Xinjiang, and were expected to have photoperiod sensitivity.

CHALLENGES FOR IMPROVING WHEAT GRAIN YIELD IN THE FUTURE

Due to utilization of dwarfing genes, the 1B/1R translocation, and dominant vernalization and photoperiod genes for early maturity, improvement of wheat grain yield in China has achieved significant since 1949. However, breeding for high yield potential still remains the first priority, since the Chinese population increase by more than 1% annually and arable land diminishes by 1% (He et al., 2001). The future challenge of wheat breeding in China is to raise grain yield, or to both maintain the genetic gain in grain yield and improve processing quality, without increasing inputs for the wheat based double cropping system in the main wheat regions.

The significant genetic improvement in grain yield in China was primarily attributed to increased kernel weight per spike, reduced plant height, and increased harvest index. Large kernel size is generally preferred in China and it is associated with fast grain filling under high temperatures, which are very common in north China after anthesis. Therefore, TKW is an important selection criterion for the Chinese wheat breeding programs. The increase in TKW could also be due to the use of the 1B/1R translocation, which could contribute to tolerance to high temperatures during grain filling stage (personal communication, Q.S. Zhuang). Considering the limited potential to further increase TKW in Chinese environments, and the trends of reducing spike number per unit area, it is generally believed that further increases of kernels per spike (even 2-3 kernels) could offer an opportunity for increasing grain yield potential. Experience in developing new cultivars such as CA 9722 and CA0175 in Beijing has supported this approach.

The significant increase in grain yield in China was mainly occurred in the early 1980s, largely due to the successful utilization of dwarfing genes and the incorporation of the 1B/1R translocation. All leading cultivars in Zones I and II now have plant heights around 80 cm. Chinese experiences also indicates that breeding for semi-dwarf was more difficult in Zone I compared to Zone II. This might be due to the short grain filling period (30-35 days), rapid temperature rise after anthesis, and poor tillering ability of semi-dwarf cultivars or lines (Zhuang, 2003). The release of Nongda 3291, CA9722, and Lunxuan 987 with plant heights around 85 cm, gives an indication of the progress in reducing plant height in Zone I during the last few years. However, it is very unlikely that further reduction in plant height in Zones I and II will benefit yield progress. It is generally believed that the optimum plant height is around 80 cm in Zone II, based on the experience of Chinese wheat breeders (Zhuang, 2003). Most current leading cultivars in Zones I and II have plant heights around 75-85 cm, suggesting that combinations of *Rht-B1b* or *Rht-D1b* with *Rht 8* confer optimal plant height for these regions based on our studies. Combination of *Rht-B1b* with *Rht 8* is suggested in Zone III. The GA-insensitivity genes *Rht-B1b* and *Rht-D1b* have pleiotropic effects on plant growth, causing reduction in coleoptile length and seedling leaf area. Other dwarfing genes such as *Rht 8* and *Rht 9* that do not confer GA insensitivity may therefore be more suitable in reducing final plant height without compromising early plant growth (Ellis et al., 2005). It is also essential that these GA insensitivity dwarfing genes are introduced in Chinese wheat cultivars in the future.

As indicated in our study, significant yield gains were achieved with the release of cultivars carrying the 1B/1R translocation in the early 1980s. In the early 1970s, resistance to yellow rust was the major breeding objective, largely due to the frequent breakdown of resistance of cultivars. Therefore, Lovrin 10, Lovrin 13, and Neuzucht with the 1B/1R translocation, were primarily used as resistance donors in breeding, although their agronomic traits were also acceptable. The excellent yield performance and tolerance to high temperature at late growth stages were observed in both the early generations and yield trial stages. Therefore, it is believed that the 1B/1R translocation played a major role in the improvement of grain yield in China in the early 1980s. At present, 1B/1R remains frequently present in cultivars and advanced lines in Zone I and Hebei Province, but its negative effects on pan bread and Chinese noodle quality need to be considered in the breeding programs (He et al., 2005).

Growth habits, vernalization genes and photoperiod response are of great importance for optimal adaptation of bread wheat cultivars to specific environment. Because there are obvious differences in the severity of the winter temperatures and the length of the growing season in the different wheat zones of China, the distributions of growth habit, vernalization alleles and photoperiod responsive *Vrn-D1* allele in cultivars released among various wheat zones are largely different. All cultivars released in Zone I should be winter types and carry recessive alleles at the four vernalization loci. In Zone II, both winter and spring cultivars may be present and the later usually need to carry a single dominant *Vrn-D1* allele. In Zones III and V, spring cultivars with the single dominant *Vrn-D1* allele are frequently released in the future. In spring-sown Zones VI, VII and VIII, all cultivars should be spring and most of them carry the strongest dominant vernalization gene *Vrn-A1* plus other dominant gene(s). Wheat cultivars released in the future should carry *Ppd-D1a* allele with photoperiod insensitivity except for spring wheats in high latitude northwestern China (especially Zone IV), and winter wheats in Gansu and Xinjiang.

In addition to powdery mildew and stripe rust, Fusarium head blight [caused by *Gibberella zeae* (Sacc.) Petch] is now endemic to main wheat regions, and sharp eye spot (caused by *Rhizoctonia cerealis* Van der Hoeven) and take-all [caused by *Gaeumannomyces graminis* (Sacc) Arx & D. Olivier var. tritici J. Walker] are also present. It is important that introductions of foreign wheat and alien genes from wild relative species into Chinese bread wheats increase their multiple-disease resistance. Introduced cultivars played an important role in Chinese wheat breeding and production in the past. Over two hundred synthetic hexaploid wheat accessions from CIMMYT were introduced into China in recent years. Elite synthetics were crossed and backcrossed with Chinese commercial wheat cultivars for improving stripe rust resistance and yield potential. Four synthetic derivatives, Chuanmai 38, Chuanmai 42, Chuanmai 43 and Chuanmai 47, have been released in Zone V in recent years. Of them, Chuanmai 42 with large kernels and resistance to stripe rust, had the highest average yield (> 6 t/ha) of any cultivar over two years in Sichuan provincial yield trials, outyielding the commercial check cultivar Chuanmai 107 by 22.7%. Chuanmai 42 increased yields by 0.45-0.75 t/ha in farmers' fields (Yang, W. Y., personal communication).

It is very useful for increasing of wheat output and decreasing of input to breed cultivars with higher water, nitrogen (N) and phosphorus (N) fertilizer use efficiencies in China. Drought tolerance for rainfed areas should be strengthened because cultivars with drought tolerance and better water use efficiency are urgently needed. Initially most wheat breeding programs in China developed cultivars for optimum environments, and few paid attention to drought tolerance before 1990s even though half of the country's wheat area is rainfed, particularly in the spring-sown spring wheat region (Zones VI, VII and VIII). At present, the national program gives priority to breed tolerance or resistance drought cultivars. Cultivars, such as Jinmai 47, Luohan 2 and Shimai 8 with high water use efficiency, have released. Some cultivars with higher N and P use efficiencies have been identified. For example, Abbondanza from Italy and Xiaoyan 6 have higher P absorption efficiency, Nanda 2419 from Italy, Chengduguangtou and Mazhamai have higher P utilization efficiency; Fengchan 3, Zhoumai 9, Chuangwu 134, Shixin 5418, Henong 341 and Ji 97-6360 have higher N utilization efficiency (Li, 2000; Cao et al., 2006; Li et al., 2006). These accessions will be used to breed cultivars combining higher P and N utilization efficiencies in the future.

Chinese high yield cultivars generally had poor bread making and noodle making quality before 1990. At present, the national program gives priority to bread making quality; each year a large number of crosses in Zones I and II are made between lines with good bread making quality characters and high yielding Chinese cultivars to develop cultivars combining early maturity, multiple disease resistance, good quality, and high yield potential, without increasing inputs for the wheat based double cropping system. It also indicated that the combination of high grain yield and excellent industrial quality is possible, as exemplified by Jimai 19 and Yumai 34, leading cultivars in Shandong and Henan, respectively.

REFERENCES

Cao, C. J., Li, B. X., Wang, B., Li, Y. M. & Xiao, K. (2006). Physiological mechanisms of absorption and use of phosphorus with high efficiency in wheat cultivars. *Acta Agronomica Sinica*, 32, 827-832 (in Chinese).

Ellis, M. H., Rebetzke, G. J., Azanza, F., Richards, R. A. & Spielmeyer, W. (2005). Molecular mapping of gibberellin-responsive dwarfing genes in bread wheat. *Theor. Appl. Genet.*, 111:423-430.

Goncharov, N. P. (1998). Genetic resources of wheat related species: The *Vrn* genes controlling growth habit (spring vs. winter). *Euphytica,* 100:371-376.

Gotoh, T. (1979). Genetic studies on growth habit of some important spring wheat cultivars in Japan, with special reference to the identification of the spring genes involved. *Jap. J. Breed.,* 29:133-145.

He, Z. H., Liu, L., Xia, X. C., Liu, J. J. & Pena, R. J. (2005). Composition of HMW and LMW glutenin subunits and their effects on dough properties, pan bread, and noodle quality of Chinese bread wheat. *Cereal Chem.*, 82: 633-638.

He, Z. H., Rajaram, S., Xin, Z. Y. & Huang, G. Z. (2001). *A History of Wheat Breeding in China.* CIMMYT, Mexico, DF.

Hunt, L. A. (1979). Photoperiodic responses of winter wheats from different climatic regions. *Z Pflanzenzüchtung,* 82:70-80.

Iwaki, K., Haruna, S., Niwa, T. & Kato, K. (2001). Adaptation and ecological differentiation in wheat with special reference to geographical variation of growth habit and *Vrn* genotype. *Plant Breed.,* 120:107-114.

Iwaki, K., Nakagawa, K., Kuno, H. & Kato., K. (2000). Ecogeographical differentiation in East Asian wheat, revealed from the geographical variation of growth habit and *Vrn* genotype. *Euphytica,* 111:137-143.

Jin, S. B. (1986). *Chinese Wheat Cultivars and Their Pedigrees* (1962-1982). Beijing, China Agriculture Press (in Chinese).

Jin, S. B. (1997). *Chinese Wheat Cultivars and Their Pedigrees* (1983-1993). Beijing, China Agriculture Press (in Chinese).

Li, S. W., Zhou, Y. Z., Wen, H. D., Li, Y. M. & Xiao, K. (2006). Nitrogen use efficiency and yield traits in different wheat varieties. *Journal of Plant Genetic Resources,* 7, 204-208 (in Chinese).

Li, Z. S. (2000). New revolution of agricultural science and technology and research on crop breeding for super high yield. *China Science foundation,* 40-42 (in Chinese).

Martiníc, Z. F. (1975). Life cycle of common wheat varieties in natural environments as related to their response to shortened photoperiod. *Z Pflanzenzüchtung,* 75:237-251.

Ortiz Ferrara, G., Mosaad, M. G., Mahalakshmi, V. & Rajaram, S. (1998). Photoperiod and vernalisation response of Mediterranean wheats, and implications for adaptation. *Euphytica,* 100:377-384.

Scarth, R., Law, C. N. (1984). The control of day-length response in wheat by the group 2 chromosomes. *Z Pflanzenzuchtung,* 92, 140-150.

Snape, J. W., Butterworth, K., Whitechurch, E. & Worland, A. J. (2001). Waiting for fine times: genetics of flowering time in wheat. *Euphytica,* 119, 185-190.

Stelmakh, A. F. (1990). Geographic distribution of *Vrn* genes in landraces and improved varieties of spring bread wheat. *Euphytica,* 45:113-118.

Whitechurch, E. M. & Slafer, G. A. (2002). Contrasting *Ppd* alleles in wheat: effects on sensitivity to photoperiod in different phases. *Field Crop Res.,* 73, 95-105.

Worland, A. J., Appendino, M. L. & Sayers, E. J. (1994). The distribution, in European winter wheats, of genes that influence ecoclimatic adaptability whist determining photoperiodic insensitivity and plant height. *Euphytica,* 80:219-28.

Worland, A. J., Börner, A., Korzun, V., Li, W. M., Petrovíc, S. & Sayers, E. J. (1998). The influence of photoperiod genes on the adaptability of European winter wheats. *Euphytica,* 100:385-394.

Yang, F. P., Zhang, X. K., Xia, X. C., Laurie, D. A., Yang, W. X. & He, Z. H. Distribution of the photoperiod insensitive *Ppd-D1a* allele in Chinese wheat cultivars. *Euphytica* (in press).

Zhang, X. K., Xiao, Y. G., Zhang, Y., Xia, X. C., Dubcovsky, J. & He, Z. H. (2008). Allelic variation at the vernalization genes *Vrn-A1*, *Vrn-B1*, *Vrn-D1* and *Vrn-B3* in Chinese wheat cultivars and their association with growth habit. *Crop Sci.*, 48, 458-470.

Zhang, X. K., Yang, S. J., Zhou, Y., He, Z. H. & Xia, X. C. (2006). Distribution of the *Rht-B1b*, *Rht-D1b* and *Rht8* reduced height genes in autumn-sown Chinese wheats detected by molecular markers. *Euphytica,* 152, 109-116.

Zhou, Y., He, Z. H., Sui, X. X., Xia, X. C., Zhang, X. K. & Zhang, G. S. (2007a). Genetic improvement of grain yield and associated traits in the northern China winter wheat region from 1960 to 2000. *Crop Sci.*, 47, 245-253.

Zhou, Y., He, Z. H., Zhang, G. S., Xia, L. Q., Chen, X. M., Gao, Y. C., Jing, Z. B. & Yu, G. J. (2004). Utilization of 1BL/1RS translocation in wheat breeding in China. *Acta Agronomica Sinica,* 30,531-535 (in Chinese).

Zhou, Y., Zhu, H. Z., Cai, S. B., He, Z. H., Zhang, X. K., Xia, X. C. & Zhang, G. S. (2007b). Genetic improvement of grain yield and associated traits in the southern China winter wheat region: 1949 to 2000. *Euphytica,* 157, 465-473.

Zhuang, Q. S. (2003). *Wheat Improvement and Pedigree Analysis in Chinese Wheat Cultivars.* Beijing, China Agriculture Press (in Chinese).

In: Soybean and Wheat Crops
Editors: S. Davies, G. Evans
ISBN: 978-1-60741-173-4

Chapter 11

WHEAT IN BANGLADESH: YIELD GROWTH, PRODUCTION PERFORMANCE AND DETERMINANTS*

Sanzidur Rahman*
School of Geography, Faculty of Social Science and Business, University of Plymouth, Drake Circus, Plymouth, PL4 8AA, United Kingdom
M. Kamrul Hasan†
Planning and Evaluation Division, Bangladesh Agricultural Research Institute (BARI), Gazipur - 1701, Bangladesh

ABSTRACT

Wheat is the second most important cereal crop in Bangladesh. A unique feature of wheat in Bangladesh is 100% adoption of modern varieties. The present chapter provides an account of the growth performance of wheat in Bangladesh over the past four decades. The chapter then examines the productivity performance of the wheat producers as well as its determinants at the farm-level using a survey data of 293 households collected from three wheat growing regions in 2004. Results reveal that the area under wheat increased six folds from only 132,000 ha in 1971 to 832,000 ha in 2000 but then declined sharply to 479,050 ha in 2006. Consequently, total production and yield grew at an annual rate of 6.9% and 1.9%, respectively. The actual yield level increased from 0.9 t/ha to 1.5 t/ha over this 36 year period. Farm-level result reveals that the environmental production conditions within which the farmers operate significantly affect productivity as well as technical efficiency of wheat production, an issue commonly ignored in the existing literature. Wheat productivity is significantly lower in low lying areas and poor soils. Productivity is also significantly affected by a delay in sowing. Technical efficiency of

---

* The present chapter draws heavily on materials from Rahman and Hasan (2008).

* Address for correspondence: Dr. Sanzidur Rahman. Senior Lecturer in Rural Development, School of Geography, University of Plymouth, Drake Circus, Plymouth, PL4 8AA. Phone: +44-1752-585911; Fax: +44-1752-585998; E-mail: srahman@plymouth.ac.uk
† E-mail: khasan41@yahoo.com

wheat production in Bangladesh is estimated at 83%, implying that production can be increased by 20% [{(100-83)/83}*100] through reallocation of resources alone. Analysis of the determinants of technical efficiency reveals that a host of managerial and socio-economic factors significantly affect performance of wheat producers. Farmers' education, access to agricultural information, training and use of mechanical power significantly improves technical efficiency, whereas a delay in sowing and fertilization, and poor sourcing of seeds (i.e., from local market and/or neighbours) significantly reduces efficiency. Large farms are more efficient relative to small and medium sized farms. Geography does matter. Productivity of wheat is significantly lower in Jamalpur region. Policy implications include, soil fertility improvement through soil conservation and crop rotation, improvement in managerial practices through extension services and adoption of modern technologies, promotion of education and training targeted to farmers, strengthening the research-extension link, and development of new varieties that have higher yield potential and are also suitable for marginal areas.

Keywords: Wheat yield, Technical efficiency, Environmental production conditions, Managerial factors, Stochastic production frontier, Bangladesh

1. INTRODUCTION

Wheat is considered as one of the main cereal crops in the world including Bangladesh. Wheat has contributed more calories and protein to the world's diet than any other food crop (Hanson et al., 1982). Also, the world trade in wheat exceeds trade in all other food grains combined. Figure 1 presents the trends in area, production, yield and export of wheat in the world over a 49 year period (1961–2008). The total harvested area of wheat increased slightly from 202.2 million ha in 1961 to 224.9 million ha over the past five decades. However, total production increased nearly three folds from 223.5 million ton in 1961 to 656.0 million ton in 2008, with corresponding increase in yield from 1.2 t/ha to 2.9 t/ha during the same period. Trade in wheat also increased nearly 2.7 times from 43.8 million ton in 1961 to 117.5 million ton in 2008. The lower panel of Figure 1 provides the average annual compound growth rate estimates of world wheat area, production, yield and exports for each decade. It is clear from the Figure 1 that, wheat area grew only during the first two decades (1961–1980) and then recorded a decline in the 1980s and remained static since then. On the other hand, total production grew faster during 1960s and 1970s and then slowed down thereafter. The growth in yield level mirrors the growth in production, implying considerable technological progress in wheat sector. Furthermore, it is interesting to see that the overall rate of growth in wheat area, production, yield and export is the same estimated at 2% per annum over the past five decades. However, there has been a sharp rise in wheat prices worldwide in recent years, with a record rise in 2007/08, which was largely blamed on the production failure due to adverse weather conditions worldwide, particularly in the southern hemisphere (i.e., Australia) (Allen, 2008). The trends presented in Figure 1, however, do not show existence of production failure of recent years and consequent record rise in prices, implying that changing trading conditions may also be responsible for the price hike instead of production failure alone. However, it is encouraging to note that record wheat production is expected in the world in year 2008/09 which will ease global supply of wheat (Allen, 2008). The reason for a surge in wheat production in 2008/09 was attributed to farmers being encouraged by a record rise in wheat prices, thus providing an incentive to plant the golden crop worldwide in 2008/09 (Allen, 2008).

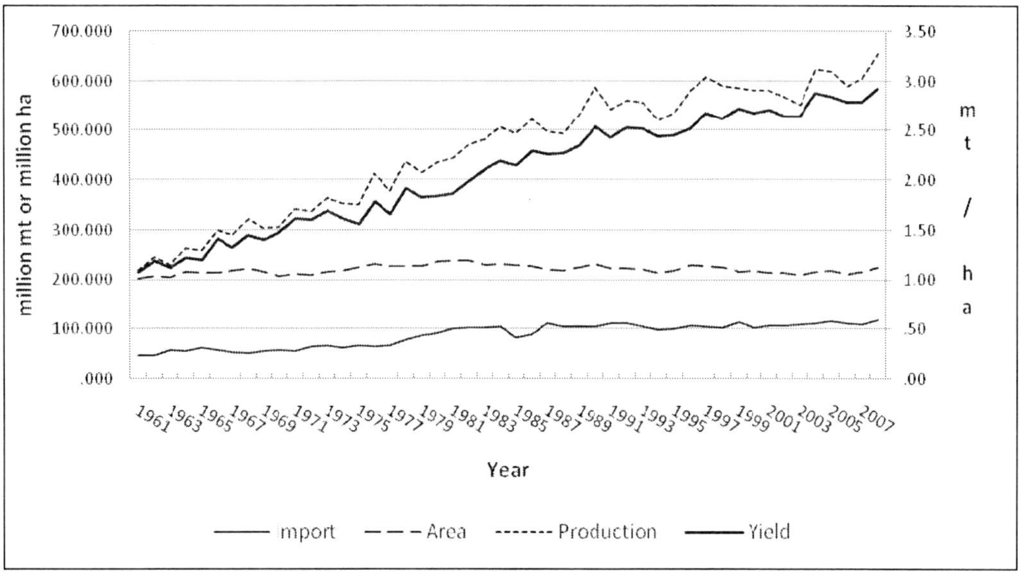

Variables	Growth rates					
	1961 – 1970	1971 – 1980	1981 – 1990	1991 – 2000	2001 – 2008	1961 – 2008
Area	0.006*	0.012***	-0.006**	-0.001	0.005	0.000
Production	0.040***	0.030***	0.021***	0.011**	0.016**	0.021***
Yield	0.034***	0.018***	0.027***	0.011***	0.011**	0.020***
Export	0.013	0.044***	0.006	-0.002	0.012**	0.019***

Note: Growth rates are estimated using semi-log trend function: $lnY = \alpha + \beta T$, where T denotes time and β is the growth rate.

*** significant at 1 % level ($p<0.01$)

** significant at 5 % level ($p<0.05$)

* significant at 10 % level ($p<0.10$)

Source: Computed from the data provided by USDA (2008): Foreign Agricultural Service, Production, Supply, and Distribution Database.

Figure 1. Trends in area, production, yield and export of wheat in the world during the period 1961 – 2008.

Bangladesh, traditionally a food deficit country dominated by rice production, also depended on wheat imports since independence in 1971 which continued well into the late 1980s. This injection of wheat through imports gradually resulted in a change in dietary habits. As a result, wheat consumption became an important supplement of rice. Also, wheat acreage now ranks second after rice area. Figure 2 presents the trends in wheat area, production, yield and imports over the past four decades (1971–2006) in Bangladesh. In general, the trends are not as smooth as observed in the world level. The wheat area increased steadily from only 125.6 thousand ha to 832.4 thousand ha in 2000 and then started to decline sharply reaching only 479.1 thousand ha in 2006. Similarly, wheat production increased almost six folds from 110 thousand ton in 1971 to 1,840 thousand ton in 2000 and then declined sharply to 735.5 thousand ton in 2006. The yield also increased from 0.9 t/ha in 1971 to 2.2 t/ha in 2000 and then fell to 1.5 t/ha in 2006. The lower panel of Figure 2 presents the average annual compound growth rates of wheat area, production, yield and import for each decade. The first decade (1971–1980) has been the period of vigorous performance with remarkable growth in wheat yield, which then stagnated in the 1980s. This increased area, production and yield of wheat in the 1970s spurred mainly because of the introduction of modern seed-water-fertilizer technologies. During the 1980s, the stagnation occurred due to natural constraints such as high temperature during the growing period of wheat and pre-monsoon rainfall before harvesting which badly affected the enthusiasm of the farmers previously observed in the 1970s (Ahmed and Meisner, 1996). Growth rates again picked up during the 1990s. However, from 2001 the fall in all three indicators has been dramatic particularly the fall in production. Overall, wheat area, production, and yield grew at a rate of 5.1%, 6.9%, and 1.9% per annum over the past 36 year period. One possible explanation of such sharp declines in area, production and yield in Bangladesh in recent years is due to adverse weather conditions characterised with short spell of cold days and day-night temperature during grain filling stage. Another important reason is the competition of limited land for high-value non-cereal crops grown during the winter months (the *Rabi* season) which are mainly destined for exports as well as urban consumer markets in the cities.

According to the Bangladesh Soil Survey report, an estimated 3.1 million hectares are suitable for wheat (Hossain, 1985). During the early 1990s, a comprehensive review of food policy in Bangladesh dismissed wheat as a competitive crop in terms of economic and social profitability (Mahmud et al., 1994). However, it was later realised that wheat provides highest returns in non-irrigated zones and in areas that are unsuitable for *Boro* rice (dry winter irrigated rice) and represents the most efficient use of domestic resources when inputs and outputs are assigned economic prices (Morris et al., 1996).

Constraints of Wheat Production in Bangladesh

Research into the constraints of wheat production in Bangladesh by wheat agronomists during 1988 to 1990 revealed that a host of natural as well as managerial factors are affecting wheat yield. The reduction in wheat yield is estimated at: (a) 23–42% due to foliar diseases; (b) 8–16% due to soil pathogens; (c) 25–46% due to farmers' fertilizer doses (which is lower than the recommended doses); (d) 2.1% and 33.7% due to lack of irrigation under high- and the low- fertility situations; and (e) late seeding at the rate of 1.3% per day of delay after November 30th (Ahmed and Meisner, 1996).

One unique feature of wheat in Bangladesh at present is 100% adoption of high yielding varieties as opposed to rice. Also, the use rate of modern inputs in wheat production is very high. Nevertheless, On-Farm Research Division (OFRD) of Bangladesh Agricultural Research Institute (BARI) reports that there is still a yield gap of 41–61% between farmers' practice and recommended package of the research station. Wheat yield with recommended package is 3.2 t/ha whereas actual production at farm level varies between 1.3 to 1.9 t/ha (OFRD, 2001). Nevertheless, best practice farmers can produce 2.8 t/ha when compared with 1.9 t/ha by the average farmers, thereby, revealing a 29% yield gap (Hasan, 2005). Such a yield gap between best practice farmers and average farmers amounts to a loss of 25% of gross margin (Tk. 9,875/ha or US$169/ha). Therefore, considerable scope exists to improve the productivity performance of the average farmers. One way to assess the extent of such scope is to empirically estimate technical efficiency in wheat production and identify its determinants, which forms the main objective of this research.

The paper proceeds as follows. The next section briefly describes the history of Research and Development of wheat crop in Bangladesh. Section 3 describes the analytical framework, study areas and the data. Section 4 presents the results. The final section concludes and draws policy implications.

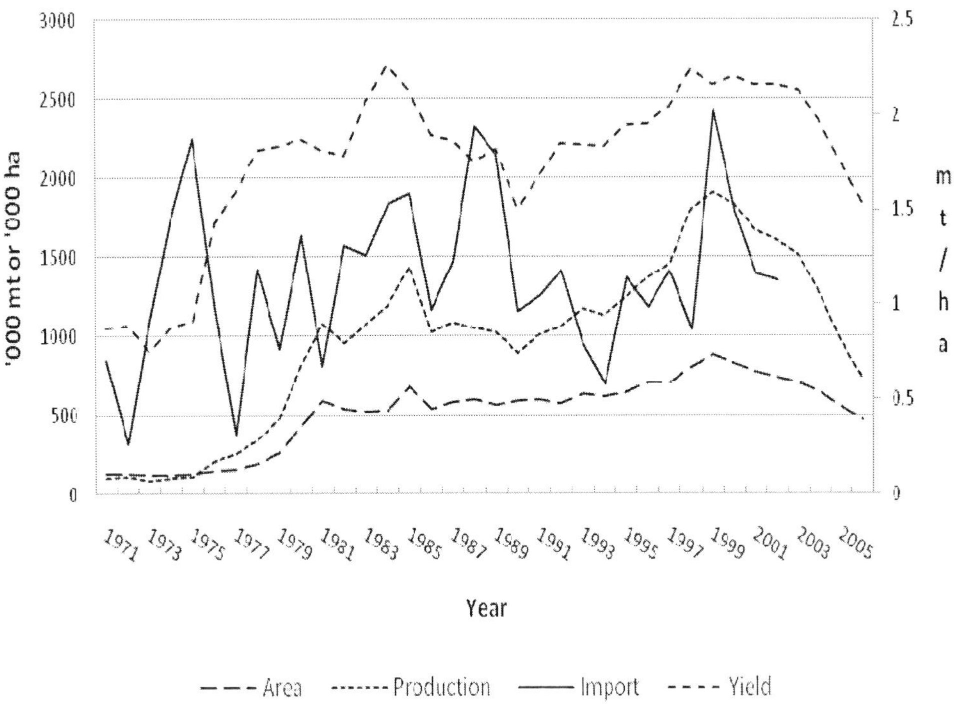

Variables	Growth rates				
	1971 – 1980	1981 – 1990	1991 – 2000	2001 – 2006	1971 – 2006
Area	0.118***	0.007	0.046***	-0.095***	0.051***
Production	0.231***	-0.012	0.076***	-0.165***	0.069***
Yield	0.113***	-0.019	0.030***	-0.070***	0.019***
Import	0.058	0.038	0.057***	--	0.014*

Note: Growth rates are estimated using semi-log trend function: $lnY = \alpha + \beta T$, where T denotes time and β is the growth rate.

*** significant at 1 % level (p<0.01).

* significant at 10 % level (p<0.10).

Source: Computed from the data provided by BBS (various issues) and Hamid (1991).

Figure 2. Trends in area, production, yield and import of wheat in Bangladesh during the period 1971 – 2006.

2. History of Wheat Development in Bangladesh

Since the independence of Bangladesh in 1971, serious food deficits necessitated food imports in large quantities, primarily wheat from the international markets. The growing dependence on wheat imports during the 1970s brought about a gradual change in dietary habits, and as a result, wheat became a cereal supplement to rice (Begum, 1998). Although wheat is a crop suited to temperate climate, the ecological conditions in Bangladesh are favourable with potential for its production during winter months, provided that disease-resistant and early maturing high yielding varieties of wheat could be developed (CIMMYT, 1982). The introduction of HYVs and the adoption of technologies for better management and seed preservation, generated through research in the mid 1970s, helped Bangladesh to become a wheat growing country. The government supported the initiative by facilitating the creation of a Wheat Research Centre (WRC) within the Bangladesh Agricultural Research Institute (BARI). Thus far, the centre has developed and disseminated 24 high yielding wheat varieties (Appendix A1). *Balaka* was the first variety selected using germplasm and technical assistance from CIMMYT and was released in 1979. However, *Kanchan*, released in 1983, has become the most popular variety which is still grown today. As mentioned earlier, wheat cultivation in Bangladesh is composed of 100% HYVs, with varieties released during the 1980s and 1990s being the most popular amongst the farmers.

3. Research Methodology

Analytical Framework

The stochastic production frontier approach, developed by Aigner et al., (1977), is utilized in this study. We extend the framework and include variables representing environmental production conditions in addition to physical inputs to explain productivity performance following Sherlund et al, (2002) and Rahman and Hasan (2008). The stochastic production frontier for the *i*th farmer is written as:

$$Y_i = f(X_i, W_i) - u_i + v_i, \quad (1)$$

where Y_i is the output, X_i is the vector of physical inputs, W_i is the vector of relevant environmental variables that control production conditions, v_i is assumed to be independently and identically distributed $N(0,\sigma^2_v)$ two sided random error, independent of the u_i; and the u_i is a non-negative random variable $(u_i \geq 0)$, associated with inefficiency in production which is assumed to be independently distributed as truncation at zero of the normal distribution with mean $-Z_i\delta$, and variance σ_u^2 $(|N(-Z_i\delta, \sigma^2_u|)$, where Z_i are the correlates of inefficiencies on farm *i*. In this formulation, output is assumed to be strictly monotonically increasing in both physical inputs as well as environmental conditions. Omissions of environmental variables biases the estimates of the parameters of the production function, overstates the level of technical inefficiency, as well as biases the correlates of inefficiency (Sherlund et al., 2002).

In determining the predictors of production efficiency, we use the single stage approach proposed by Battese and Coelli (1995) wherein the technical inefficiency parameter is related to a vector of farm-specific managerial and household characteristics subject to statistical error, such that:

$$u_i = Z_i\delta + \zeta_i \geq 0, \tag{2}$$

where, Z_i are the farm-specific managerial and household characteristics and the error ζ_i is distributed as $\zeta_i \sim N(0, \sigma_\zeta^2)$. Since $u_i \geq 0, \zeta_i \geq -Z_i\delta$, so that the distribution of ζ_i is truncated from below at the variable truncation point, $-Z_i\delta$.

The production efficiency of farm i in the context of the stochastic frontier production function is defined as:

$$EFF_i = E[\exp(-u_i) \mid \xi_i] = E[\exp(-\delta_0 - \sum Z_i\delta \mid \xi_i) \tag{3}$$

where E is the expectation operator. This is achieved by obtaining the expressions for the conditional expectation u_i upon the observed value of ξ_i, where $\xi_i = v_i - u_i$. The method of maximum likelihood is used to estimate the unknown parameters, with the stochastic frontier and the inefficiency effects functions estimated simultaneously. The likelihood function is expressed in term of the variance parameters, $\sigma^2 = \sigma_v^2 + \sigma_u^2$ and $\gamma = \sigma_u^2 / \sigma^2$ (Battese and Coelli, 1995).

Selection of the Study Area and Sample Farmers

Wheat is cultivated almost all over the country though the intensity of planted area and land suitability are not equal in all regions. Therefore, we computed a wheat area index for each greater district[1]. The wheat area index for the jth district is expressed as:

$$WI_j = (Area_j / GCA_j) * 100, \tag{4}$$

where WI is the wheat area index, $Area$ is the wheat area and GCA is the gross cropped area. Based on this index, wheat growing regions were classified into three levels of intensity: high intensity ($WI>8.0$), medium intensity ($4.01<WI<8.0$), and low intensity areas ($WI<4.0$).

A multistage sampling procedure was adopted to select the sample farmers. First, three wheat growing regions (two from high intensity areas – Dinajpur and Rajshahi, and one from medium intensity areas – Jamalpur) were selected purposively[2]. The selected three

[1] Although there are 64 districts in Bangladesh, most secondary data are still reported at the level of these 21 former greater districts.

[2] The low intensity area is excluded because it is assumed that wheat production has limited potential in these districts.

districts/regions[3] together cover 31% of the total wheat area of the country (Table 1). Also, each selected district belonged to different agro-ecological zones (AEZ) of Bangladesh, namely, AEZ-3, AEZ-11 and AEZ-9, respectively[4]. Dinajpur is located in the north-west, Rajshahi in the mid-west and Jamalpur in the mid-north of Bangladesh. In the second stage, one *upazila* (sub-district) from each district and one union from each *upazila* were selected at random. Next, three *mouzas* (one from each union) were selected at random for primary data collection from farm households. However, due to an insufficient number of households in one *mouza*, a fourth *mouza* was also selected at random to fulfil the required sample size. In the third stage, a number of steps were followed to select the households to ensure a high level of representation. At first, a sampling frame of wheat growing holdings was constructed with the assistance of village leaders, record book at the union council office and other key informants. The list included the names of household heads and their land holdings in the selected *mouzas*. These farm holdings were then stratified into three standard farm-size categories commonly adopted in Bangladesh (e.g., Hossain, 1989). Then, a total of 293 wheat producing households were selected following a stratified random sampling procedure. Two sets of structured questionnaires were administered. These questionnaires were pre-tested prior to finalization. The survey conducted in 2004 covered wheat growing period from November 2003 to March 2004.

Table 1. Selection of the study area and the sample size

Study area	Area selection criteria			Farm size categories			
	Wheat area index (WI)	Intensity Rank (Out of 21 greater districts)	% of total wheat area	Large farms (2.0 ha and above)	Medium farms (1.01 to <2.0 ha)	Small farms (up to 1.0 ha)	All categories
Dinajpur	16.94	2	16	33 (92)	29 (86)	39 (122)	101 (300)
Rajshahi	9.12	4	11	19 (32)	32 (49)	52 (228)	103 (309)
Jamalpur	7.55	6	4	8 (11)	25 (46)	56 (178)	89 (235)
All area	-	-	31	60 (135)	86 (181)	147 (528)	293 (844)

Notes: Figures in parentheses indicate sampling frame.
Source: BBS (2000), and Field survey (2004).

The Empirical Model

The empirical model is specified with a restricted Translog stochastic production frontier function.

$$\ln Y_i = \alpha_0 + \sum_{j=1}^{5} \alpha_j \ln X_{ij} + \sum_{j=1}^{5}\sum_{k=1}^{5} \beta_{jk}(\ln X_{ij} \ln X_{ik}) + \sum_{l=1}^{7} \varphi_k E_{ik} + \sum_{m=1}^{2} \tau_m R_{im} + v_i - u_i \tag{5}$$

[3] In this study the term district and region are used interchangeably to emphasize the large spatial variation between our study areas.

[4] There are a total of 29 agro-ecological zones which cut across many of the 21 greater districts/regions.

and

$$u_i = \delta_0 + \sum_{d=1}^{13} \delta_d Z_{id} + \zeta_i \qquad (6)$$

where Y_i is the wheat output (including grain equivalent of straw output); X_{ij} is jth input for the ith farmer; E_{ik} are the environmental production condition variables, R_{im} is the dummy variable for districts, v_i is the two sided random error, u_i is the one sided half-normal error, *ln* natural logarithm, Z_{id} variables representing managerial and socio-economic characteristics of the farm to explain inefficiency, ζ_i is the truncated random variable; α_0, α_j, β_j, τ_m, δ_0, φ_k and δ_d are the parameters to be estimated.

A total of five production inputs (X), seven environmental production condition variables (E), and two regional dummies (R) were used in the production function, and 13 variables representing managerial and socio-economic characteristics of the farmer (Z) were included in the inefficiency effects model as predictors of technical inefficiency. Table 2 presents the definitions, units of measurement, and summary statistics for all the variables.

Table 2. Definition, measurement and summary statistics of the variables used in the stochastic production frontier model

Variables	Measure	Mean	Standard deviation
Inputs and output			
Wheat (includes grain equivalent of straw)	Kg per farm	655.5495	346.414
Land cultivated	Hectare	0.13	0.06
Labour	Persons	13.35	6.23
Fertilizers	Kg of active nutrients (N, P, K, and S)	19.28	9.85
Draft/mechanical Power	Taka	289.12	158.32
Irrigation	Taka	22.25	8.15
Environmental production conditions			
Land type	Indexed (1 = Medium high land – most suitable; 2 = High land – suitable; 3 = Low land – not suitable)	1.49	0.74
Soil type	Indexed (1 = loamy – good; 2 = sandy loam – average; 3 = clay loam – poor)	1.86	0.71
Weed infestation[a]	Indexed (1 = 1 – 10%, 2 = 11 – 20% of crop yield)	0.73	0.53

Table 2. Continued

Variables	Measure	Mean	Standard deviation
Pest infestation[a]	Indexed (1 = 1 – 10%, 2 = 11 – 20% of crop yield)	0.80	0.49
Weather[a]	Indexed (1 = 1 – 10%, 2 = 11 – 20%, 3 = 21 – 30%, 4 = 31 – 40% of crop yield)	0.69	0.68
Late sowing[a]	Indexed (1 = 1 – 10%, 2 = 11 – 20% of crop yield)	0.68	0.80
Soil fertility[a]	Indexed (1 = 1 – 10%, 2 = 11 – 20%, 3 = 21 – 30% of crop yield)	0.50	0.81
Other variables			
Jamalpur region	Dummy (1 = Yes, 0 = No)	0.35	0.48
Rajshahi regions	Dummy (1 = Yes, 0 = No)	0.30	0.46
Managerial variables			
Age of the farmer	Years	46.99	12.08
Education of the farmer	Completed years of schooling	4.88	4.12
Experience in wheat farming	Years of growing wheat	16.14	6.23
Sowing date[b]	Indexed (1 = if sown during optimum time, i.e., November 15 – 30th, 2 = slightly late, 1 – 7th December, 3 = moderately late, 8 – 14th December, 4 = extreme late, 15 – 30th December)	2.11	0.94
Timing of first top-dressing of urea fertilizer[c]	Indexed (1 = 10 – 21 DAS [days after sowing], 2 = 22 – 28 DAS, 3 = 29 till maximum DAS)	1.22	0.44
Timing of first weeding[c]	Indexed (1 = 16 – 21 DAS, 2 = 22 – 28 DAS, 3 = 29 – 35 DAS, 4 = no weeding at all)	3.26	1.17
Timing of first irrigation[c]	Indexed (1 = 15 – 21 DAS, 2 = 22 – 28 DAS, 3 = 29 – 35 DAS, 4 = 36 – 42 DAS, 5 = 43 – 60 DAS or no irrigation)	1.57	1.08
Mechanical power	Dummy (1 = Used, 0 = No)	0.61	0.49
Source of procuring seed	Indexed (1 = Research centre or BADC – good source, 2 = Own processed – average, 3 = Local market or neighbour – poor)	2.17	0.63
Link with extension services[d]	Frequency of contact in the past year (number)	13.84	16.50

Sources of agricultural information[e]	Number of sources	2.11	0.97
Training	Dummy (1 = received training on wheat production in the past 5 years, 0 = No)	0.14	0.35
Wheat area	Amount of land under wheat	0.13	0.06
Total number of observations		293	

Note: [a] = figures are based on farmer's own account of his/her crop loss due to each specific factors.

[b] = wheat cultivation is very sensitive to sowing date. The optimum time of planting is November 15 – 30th. Failure to sow wheat by November 30th reduces crop yield by an estimated 1.3% per day.

[c] = to ensure optimum production of wheat, three operations need to be done simultaneously in the first spell. These are, applying first irrigation, first weeding and first top-dressing of urea fertilizer, all within 17 – 21 days after sowing.

[d] = information was collected on the nature of extension link: 0 = no link, 1 = weekly contact, 2 = fortnightly, 3 = monthly, 4 = quarterly. The information was then converted into frequencies in one year using: quarterly = 4 times, monthly = 12 times, fortnightly = 24 times (deducted two weeks to account for official holidays), and weekly = 42 times (deducted 10 weeks to account for official holidays and other missing days).

[e] = figures are number of sources that the farmers reported as his/her sources of information. A total of six sources were recorded. These are: electronic media (radio/television), block supervisor (lowest unit of extension service), NGO, neighbours, printed media (e.g., leaflets), and fertilizer/pesticide dealer.

Source: Adopted from Rahman and Hasan (2008).

4. Results

From the information provided in Table 2, we can see that the average farm size is very small (0.13 ha). Land type is in the suitable range whereas soil type is of average quality. Variables representing environmental production conditions are non-zero ($p<0.01$). The average age of the farmers is 47 years with 16 years of experience in growing wheat, education is less than five years, 61% of the farmers used mechanical power services, extension link is relatively high (13.4 times in a year), farmers are exposed to at least two sources of agricultural information, and only 14% of the farmers received training on wheat production in the past 5 years.

Productivity Effects of Environmental Production Conditions

The assumption underlying the inclusion of environmental production conditions in estimating the parameters of the production frontier is that they are exogenously determined. Furthermore, if these variables are asymmetrically distributed, then their omission will lead to upward bias in the estimates of firm specific technical inefficiency, which was shown by Sherlund et al., (2002) and Rahman and Hasan (2008) for Cote d'Ivoire and Bangladesh, respectively. The suite of variables chosen to control for environmental production conditions in this study includes truly exogenous (e.g., weather), quasi-fixed characteristics (e.g., soil types and land types) as well as combinations of exogenous shocks and managerial response (e.g., pest and weed infestation). Both Sherlund et al., (2002) and Rahman and Hasan (2008) showed that the production inputs were found to be strongly correlated with the variables that condition the production environment of the farmers.

Parameter estimates of the restricted Translog stochastic production frontier are reported in Table 3 using the Maximum Likelihood Estimation (MLE) procedure in STATA Version 8 (STATA Corp, 2003). A series of hypothesis tests using Likelihood Ratio (LR) test statistic were conducted regarding the model choice, inclusion of environmental variables and determinants of inefficiency, the results of which are presented in Table 4. The first test of hypothesis is the choice of the functional form, i.e., Cobb-Douglas vs Translog functional form. The result indicates that non-linearities in the production function is present and, hence, the choice of flexible Translog functional form is a better representation of the true production structure as compared to a more restricted Cobb-Douglas form. The next test of hypotheses that 'the environmental variables are jointly zero' is also strongly rejected indicating that environmental production conditions significantly affect productivity. The test of γ is also strongly rejected suggesting presence of technical inefficiency, and hence the application of stochastic production frontier model is justified (Table 4).

All five input variables significantly influence wheat productivity as expected. The input variables were mean corrected prior to estimation $(X_{ij} - \overline{X}_j)$. Therefore, the coefficients on the first order terms of the input variables can be read directly as elasticities. Land is the most dominant input followed by fertilizers, labour, animal/mechanical power services and irrigation. The hypothesis of constant returns to scale in wheat production is rejected in favour of decreasing returns to scale (Table 5) implying that farmers are not operating at the optimal scale.

Poor land types, delay in sowing and poor soil quality significantly reduces productivity, variables that are typically omitted in most studies. Since variables representing environmental production conditions were incorporated in the model, the responsiveness of the key production inputs on wheat productivity is likely to be more accurate. Both Sherlund et al., (2002) and Rahman and Hasan (2008) reported positive response of output and fall in inputs of labour and/or fertilizers when controlled for environmental production conditions. Geography does matter in wheat production performance. Wheat production is significantly lower in Jamalpur regions, although it is an intensive agricultural region but not a typical wheat growing region as compared to Rajshahi and Dinajpur.

Production Efficiency

The mean technical efficiency level in wheat production is estimated at 83% which implies that production can be increased by 20% [{(0.83–1.00)/0.83}*100] with efficiency improvements. The minimum score is 66% and the maximum is 99%. The mean estimate is comparable to estimates for other developing countries. For example, technical efficiency in wheat production varies between 57.0–78.9% in Pakistan (Battese et al., 1996), 81.0–93.4% in India (Singh, et al., 2004) and 91.0–93.0% in Iran (Bakhsoodeh and Thomson, 2001), respectively.

Table 3. Maximum likelihood estimates of production frontier

Variables	Parameters	Coefficient	t-ratio
Production function			
Constant	α_0	6.6526	95.99***
ln Land	α_1	0.6044	10.37***
ln Labour	α_2	0.0925	2.75***
ln Fertilizer nutrients	α_3	0.0979	3.81***
ln Animal/mechanical power	α_4	0.0693	4.21***
ln Irrigation	α_5	0.0506	3.22***
0.5 x (ln Land)^2	β_{11}	-0.1068	-0.31
0.5 x (ln Labour)^2	β_{22}	0.0046	0.02
0.5 x (ln Fertilizer nutrients)^2	β_{33}	-0.0714	-0.78
0.5 x (ln Animal/mechanical power)^2	β_{44}	-0.0292	-0.55
0.5 x (ln Irrigation)^2	β_{55}	0.0008	0.04
ln Land x ln Labour	β_{12}	-0.0303	-0.12
ln Land x ln Fertilizer	β_{13}	0.1576	1.00
ln Land x ln Animal power	β_{14}	-0.1068	-1.11
ln Land x ln Irrigation	β_{15}	0.0816	1.76*
ln Labour x ln Fertilizer	β_{23}	-0.0445	-0.38
ln Labour x ln Animal power	β_{24}	0.1953	2.74***
ln Labour x ln Irrigation	β_{25}	-0.1337	-3.43***
ln Fertilizer x ln Animal power	β_{34}	-0.1010	-1.91*
ln Fertilizer x ln Irrigation	β_{35}	0.0194	0.46
ln Animal power x ln Irrigation	β_{45}	0.0131	0.50
Land type	φ_1	-0.0266	-3.56***
Soil type	φ_2	0.0055	0.85
Weed infestation	φ_3	-0.0070	-0.70
Pest infestation	φ_4	0.0081	0.90
Weather	φ_5	-0.0019	-0.27
Late sowing	φ_6	-0.0179	-1.89*
Soil fertility	φ_7	-0.0538	-6.63***
Jamalpur region	τ_1	-0.0439	-2.35**
Rajshahi region	τ_2	-0.0251	-0.63
Variance parameters			
$\sigma^2 = \sigma_u^2 + \sigma_v^2$	σ^2	0.0039	10.32***
$\gamma = \sigma_u^2/(\sigma_u^2 + \sigma_v^2)$	γ	0.6968	3.77***
Log likelihood			
Inefficiency effects function			
Constant	δ_0	0.2028	1.89*
Age of the farmer	δ_1	0.0004	0.90
Education of the farmer	δ_2	-0.0033	-2.81***
Experience in wheat farming	δ_3	-0.0004	-0.35
Sowing date	δ_4	0.0413	4.98***
Timing of first top-dressing of urea fertilizer	δ_5	0.0276	2.11**

Table 3. (Continued)

Variables	Parameters	Coefficient	t-ratio
Timing of first weeding	δ_6	0.0106	0.84
Timing of first irrigation	δ_7	-0.0082	-1.05
Mechanical power	δ_8	-0.0435	-2.71***
Source of procuring seed	δ_9	0.0137	1.98**
Link with extension services	δ_{10}	-0.0001	-0.38
Sources of agricultural information	δ_{11}	-0.0149	-2.58***
Training	δ_{12}	-0.0247	-1.88*
Wheat area	δ_{13}	-0.9250	-3.20***
Total number of observations		293	

Note: *** significant at 1 % level ($p<0.01$);
** significant at 5 % level ($p<0.05$); * significant at 10 % level ($p<0.10$)

Table 4. Tests of hypotheses

Hypothesis	Critical value of χ^2(v, 0.95)	LR statistic	Decision
Choice of the functional form (H_0: $\beta_{11} = \beta_{22} = \ldots = \beta_{45} = 0$)	27.49	50.51**	Reject H_0
No effect of environmental variables on productivity (H_0: $\varphi_1 = \varphi_2 = \ldots = \varphi_7 = 0$)	14.07	81.06**	Reject H_0
Presence of inefficiency, (H_0: $\gamma = 0$)	3.84	4.27**	Reject H_0
No effect of managerial variables on inefficiency (H_0: $\delta_1 = \delta_2 = \ldots = \delta_{13} = 0$)	22.36	102.37**	Reject H_0
Constant returns to scale in production (H_0: $\alpha_1 + \alpha_2 + \ldots + \alpha_5 = 1$)	3.84	4.71**	Reject H_0

Note: In testing (H_0: $\gamma = 0$) the critical value of $\chi^2_{(1,0.95)}$ was used which is 3.84.
** significant at 5 % level ($p<0.05$)

Table 5. Technical efficiency estimates

Items	Proportion of farmers
Efficiency levels	
upto 70%	3.42
71 – 80%	32.76
81 – 90%	46.07
91% and above	17.75
Efficiency measures	
Mean efficiency score	0.83
Standard deviation	0.07
Minimum	0.66
Maximum	0.99

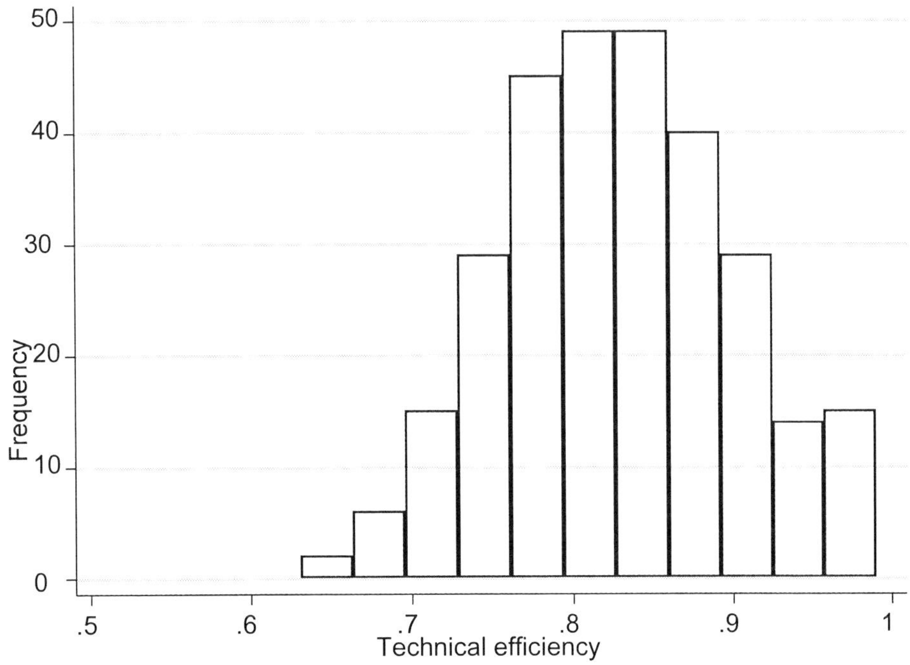

Figure 3. Technical Efficiency of Wheat Farmers in Bangladesh.

Determinants of Production Inefficiency

The lower panel of Table 3 reports the determinants of technical inefficiency in wheat production. The null hypotheses that the 'managerial variables are jointly zero' is strongly rejected (Table 4), implying that technical efficiency in wheat farming is highly sensitive to managerial factors. For example, failure to sow on time, delay in the first application of fertilizer and selection of poor quality seeds significantly decrease efficiency. On the other hand, education, agricultural information sources and training significantly increase efficiency. Use of modern technology, i.e., mechanical power services instead of animal power, also significantly improves efficiency. Efficiency increases with the size of operation.

Yield Gap Analysis and Identification of Constraints in Wheat Farming

In general, farmers' performance remains lower than their potential capacity to produce largely due to underutilization of the most recently developed production technologies. There is a scope to increase the yield level by minimizing the '*yield gap*'. The concept of yield gap came from the constraint studies carried out by the International Rice Research Institute (IRRI) which estimates a quantitative difference between experiment station yield and actual farm level yield. There are two types of yield gap under different management of the production environment: (a) *Gap-I is the difference between maximum level of yield realized at the experiment station and the potential farm level yield.* and (b) *Gap-II is the difference between the potential and actual yields in farmer's own production environment* (De Datta *et al.*, 1978). The yield Gap-I exists mainly because of the differences in the production environment between the experiment station and the average farm. Also, there may be components of experiment station technology that are not transferable to farmers' field. By definition, the yield GAP-II exists because of individual farmer's use of inputs or practices resulting in lower yields than potentially possible on his/her farm.

Table 6. Constraints affecting production performance of wheat farmers

Factors	N	Actual yield (kg/ha)	Yield loss (kg/ha)	TE
Educated farmers vs.	199	2572.46	465.12	0.85
Farmers with no education	94	2396.44	606.03	0.80
t-test for mean differences		3.29***	-5.77***	6.30***
Suitable land type vs.	193	2703.01	487.96	0.85
Unsuitable land type	100	2155.03	553.50	0.79
t-test for mean differences		12.05***	-2.65**	6.72***
No loss due to poor soil fertility vs.	202	2720.48	497.75	0.85
Loss due to poor soil fertility	91	2062.07	538.23	0.79
t-test for mean differences		15.62***	-1.61	6.48***
Timely fertilization vs.	232	2584.11	490.70	0.84
Delay in fertilization	61	2256.89	584.96	0.79
t-test for mean differences		4.53***	-3.36***	5.15***
Seed procured from Research Station vs.	37	2734.22	410.64	0.87
Seed procured from neighbours or market	256	2484.45	524.74	0.82
t-test for mean differences		5.11***	-3.00***	4.06***
Timely sowing vs.	85	2816.21	405.86	0.88
Delay in sowing	208	2393.30	553.02	0.81
t-test for mean differences		10.21***	-5.76***	8.33***
Use of mechanical power vs.	180	2576.99	424.38	0.86
No use of mechanical power	113	2418.81	647.24	0.79
t-test for mean differences		3.12***	-10.03***	9.52***
Received training vs.	42	2716.52	405.24	0.87
No training	251	2482.43	527.91	0.82
t-test for mean differences		3.26***	-3.72***	4.23***
Four or more sources of information vs.	19	2786.32	418.50	0.87
Three or less sources of information	274	2497.24	516.69	0.83
t-test for mean differences		4.46***	-2.38***	3.44***
All farms	293	2515.98	510.33	0.83

Note: Yield loss = Maximum possible yield – actual yield. The maximum yield is computed by dividing the actual yield by the technical efficiency score.
*** Significant at 1 % level ($p<0.01$).

The first step in estimating yield gap in wheat production is to determine the yield levels at different phases. In this study, experiment station yields were obtained from the WRC for the year 2000 (WRC, 2000). On the other hand, potential farm level yields were estimated from the actual farm level data and the computed technical efficiency scores (Table 4). For the HYV *Kanchan*, which presently covers 94% of the wheat area nationally, the maximum yield obtained at the research station was estimated at 6.5 t/ha (WRC, 2000). This yield level at the research station is the maximum potential yield of the *Kanchan* variety in a good production environment and under sound management practice. In fact, this level of yield was not achieved in on-farm trials and demonstrations conducted during the process of variety release in the early 1980s.

The potential farm yield (i.e., the technically efficient yield level obtainable by the sample farmers) is estimated at 3.03 t/ha [2.52 t/ha (actual mean yield)/0.83 (overall TE)]. Figure 4 presents the results of the yield gap analysis. The total yield gap is 3.98 t/ha of which yield Gap-I is 3.47 t/ha and yield Gap-II is 0.52 t/ha. The yield Gap-II can be recovered by removing technical inefficiency alone. Results from Table 3 clearly show that a total of nine factors significantly affect productivity and/or technical efficiency of the wheat farmers.

Table 6 presents the detailed analysis of these nine factors which significantly influenced productivity and/or technical efficiency of wheat farmers as evidenced from the joint parameter estimates of the stochastic production frontier and the inefficiency effects model (Table 3). It is clear from Table 6 that all these factors have significant and varied effect on the production performance of the wheat farmers. For example, educated farmers produce 7% more actual yield, incur 30% less yield loss and operate at 6% higher level of technical efficiency compared to their uneducated peers. Land suitability has the highest impact on actual yield levels. Farmers who grew wheat in medium high land (the most suitable land type for wheat production) actually produce 20% more than the others. Similarly, non-use of mechanical power incur highest level of yield loss (53%) and these farmers also operate at 8% lower level of technical efficiency when compared to those who do use mechanical power. The overall yield loss of these wheat farmers is estimated at 510 kg/ha which is 20.3% of actual yield level.

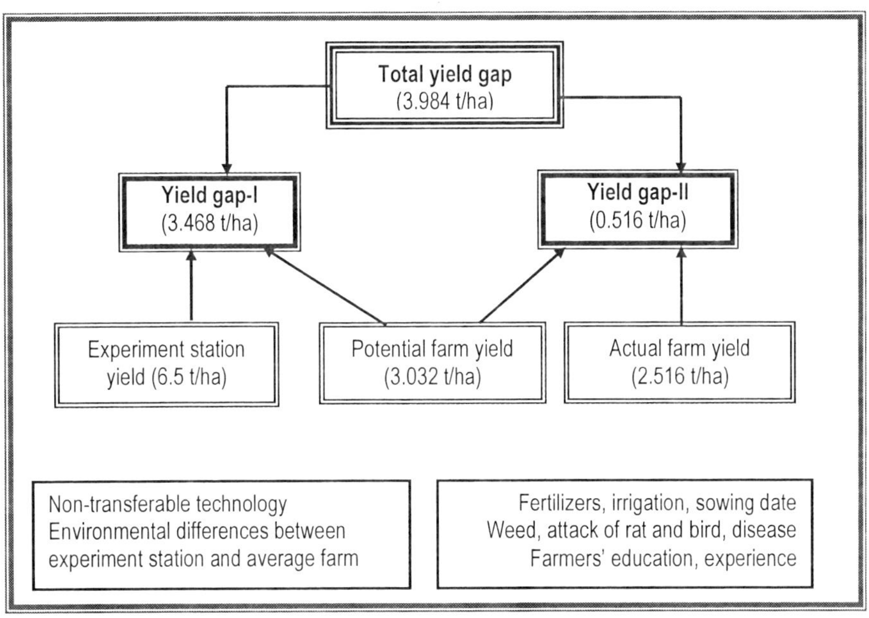

Figure 4. Yield gap analysis of the sampled farmers.

5. CONCLUSIONS

The present chapter provided an account of the growth performance of wheat in Bangladesh over the past four decades preceded by a brief overview of the world wheat outlook for the period 1961–2008. Then the study also examined productivity performance of wheat farmers as well as its determinants at the farm-level using a survey data of 293 households collected from three wheat growing regions of Bangladesh, and discussed the relevant constraints in realizing the full production potential. Results reveal that the area under wheat crops increased six folds from only 132,000 ha in 1971 to 832,000 ha in 2000 but then declined sharply to 479,050 ha in 2006. Consequently, total production grew at an annual rate of 6.9 % with corresponding growth in yield at a rate of 1.9 % per annum, increasing from 0.86 t/ha to 1.5 t/ha over this 36 year period. Farm-level result reveals that several factors significantly affect wheat productivity as well as technical efficiency. Also, significant level of yield gap exists in wheat production of which 0.52 t/ha (yield Gap-II) can be recovered by removing technical inefficiency.

In general, the environmental production conditions, within which the farmer operates, are considered vital but are often arbitrarily omitted in productivity and efficiency studies, resulting in biased estimates of the production parameters, efficiency scores and correlates of inefficiency (Sherlund et al., 2002). Our results demonstrate that environmental production has significant influence on productivity of wheat production in Bangladesh. Poor land type, poor soil fertility and delay in sowing results in significant production loss. Technical efficiency of wheat production in Bangladesh is estimated at 83%, implying that production can be increased by 20% [(100-83)/83] through reallocation of resources alone. Farmers' education, access to agricultural information, training and use of mechanical power significantly improves technical efficiency, whereas a delay in sowing and fertilization, and poor sourcing of seeds (i.e., from local market and/or neighbours) significantly reduces efficiency. Large farms are more efficient relative to small and medium sized farms. Geography does matter. Productivity of wheat is significantly lower in Jamalpur region. Nevertheless, scope to raise wheat production remains limited with the existing set of varieties and technologies because farmers have already adopted 100% of popular modern varieties and are also producing at a high level of technical efficiency (83%).

Based on the results of our study, a number of specific policy implications can be drawn. First, soil fertility improvement seems essential to raise productivity. This may be addressed through adopting soil conservation practices and/or improving crop rotation practices (e.g., including soil health enhancing crops, such as pulses and oilseeds, in the system). Of the nine total cropping patterns observed among the sample farmers, most followed rice-based cropping. Only two patterns included jute in the system and none included any pulse or oilseed crops, which is potentially highly detrimental to soil health in the long run. Second, is the improvement in managerial practices (e.g., timely sowing and fertilizer application) and the use of modern technology (e.g., mechanical power services). These can be addressed through strengthening agricultural extension services and improvements in rural infrastructure. Third, investment in education and training targeting the farming population seems crucial. Fourth, is to improve existing research-extension link. Currently, new varieties that are developed remain confined at the research stations. Dominance of only one variety at the farm-level which was released 25 years ago[5], clearly points towards the need to develop the research-extension link. Finally, our study shows that poor land type significantly reduces productivity. Therefore, research effort should be geared towards developing varieties that are suitable for marginal areas. Evidence suggests that wheat production in marginal lands accounts for 25% of global production and that research innovation has led to significant improvement in yield growth in these areas, particularly in drought and high temperature environments (Lantican et al., 2003). The challenges to realize all of these policy options are formidable. However, a boost in wheat production could significantly curb dependence on rice as the main staple in Bangladeshi diet, which is a goal worth pursuing.

Appendix Table A1. Recommended wheat varieties and their year of approval by National Seed Board with farm level expected yield during 1974 to 2005

Variety	Year of release	Farm level expected yield (t/ha)	Maximum yield (t/ha)
Sanora-64	1974	NA	NA
Mexipak-65 /Kalyansona	1974	NA	NA
Iniya-66	1974	3.5-3.8	NA
Norteno-67	1974	3.2-3.9	NA
Sonalika-68	1974	3.2-3.4	NA
Tenori-71	1975	3.4-3.5	NA
Nuri-70	1975	NA	NA
Jupatico-73	1975	3.5-3.7	NA
Pavon-76	1979	3.7-4.2	NA
Balaka	1979	3.7-4.1	NA
Doel	1979	3.5-4.0	NA
Ananda	1983	3.4-3.8	4.86
Kanchan	1983	3.5-4.8	6.60
Akbar	1983	3.5-4.2	4.63
Barkat	1983	3.4-3.8	5.39
Aghrani	1987	3.0-3.5	5.02
BARI Gam-17 (Shaugat)	1993	NA	5.13
BARI Gam-18 (Protiva)	1993	3.8-4.5	5.08
BARI Gam-19 (Sourav)	1998	3.5-4.6	5.39
BARI Gam-20 (Gaurab)	1998	3.6-4.8	5.28
BARI Gam-21 (Shatabdi)	2000	3.6-4.6	4.87
BARI Gam-22 (Sufi)	2005	3.6-4.8	NA
BARI Gam-23 (Bijoy)	2005	4.3-5.0	NA
BARI Gam-24 (Prodip)	2005	3.5-5.1	NA

Note: NA= Not available.

Source: SCA (1985 and 1992), Razzaque *et al.* (1993 and 2000), BARI (1989, 1992 and 1994), WRC (2000).

[5] Although 24 modern varieties of wheat have been released since 1974 (see Appendix A1), '*Kanchan*' released in

References

Ahmed, S.M., Meisner, C. 1996. *Wheat Research and Development in Bangladesh.* Bangladesh Australia Wheat Improvement Project/CIMMYT, Bangladesh.

Aigner, D.J., Lovell, C.A.K., Schmidt, P. 1977. Formulation and estimation of stochastic frontier production function models. *Journal of Econometrics,* 6: 21 – 37.

Allen, E.W. 2008. International wheat outlook for 2008/09. Economic Research Service Report #WHS-2008-01. *United States Department for Agriculture (USDA)*, Washington, D.C.

Bakhsoodeh, M., Thomson, K.J. 2001. Input and output technical efficiencies of wheat production in Kemran, Iran. *Agricultural Economics,* 24: 307 – 313.

BARI, 1989. *BARI Annual Report* (1989-90), Bangladesh Agricultural Research Institute, Gazipur, Bangladesh.

BARI, 1992. *BARI Annual Report* (1992-93), Bangladesh Agricultural Research Institute, Gazipur, Bangladesh.

BARI, 1994. *BARI Annual Report* (1993-94), Bangladesh Agricultural Research Institute, Gazipur, Bangladesh.

Battese, G.E., Coelli, T.J. 1995. A model for technical inefficiency effects in a stochastic frontier production function for panel data, *Empirical Economics,* 20, 325-332.

Battese, G.E., Malik, S.J., Gill, G.A. 1996. An investigation of technical inefficiencies of production of wheat farmers in four districts of Pakistan. *Journal of Agricultural Economics,* 47: 37 – 49.

BBS, (various issues). *Statistical Yearbook of Bangladesh 1999.* Bangladesh Bureau of Statistics, Dhaka, Bangladesh.

Begum, R. 1998. *An economic study of wheat production and marketing in Bangladesh.* Ph. D. dissertation, The united graduate school of agricultural sciences, Ehime University, Matsuyama, Japan. P- 194.

CIMMYT, 1982. Wheat in Bangladesh, CIMMYT Today No. 15. *International Maize and Wheat Improvement Centre,* Apartado Postal 6-641, 06600, Mexico. P-15.

De Datta, S. K., Gomez, K.A., Herdt, R.W., Barker, R. 1978. A handbook on the methodology for an integrated experiment – Survey on rice yield constraints. *The International Rice Research Institute,* Los Banos, Laguna, Philippines. Pp 60.

Hamid, M.A. 1991. *A data base on agriculture and foodgrains in Bangladesh* (1947/48 to 1989/90). Bangladesh. Maloney International, 2-Kalabagan Road, Dhaka, Bangladesh.

Hanson, H., Borlaug, N. E. Anderson, R. G. 1982. *Wheat in the third world. Westview press,* Boulder, Colorado, USA. P 174.

Hasan, M.K. 2005. Yield and benefit gaps in wheat production: comparison between two farmer management practices. *Seminar on Higher Agricultural Education and Research in Bangladesh: Prospects and Challenges.* Bangladesh Agricultural University, Mymensingh, Bangladesh.

Hossain, A.B.S. 1985. Wheat production in Bangladesh: its constraints and research priorities. In Wheat for more tropical environments. *Proceedings of the International Symposium held September 24-28,* 1984 at CIMMYT, Mexico.

1983, remains the most popular choice. In fact, 94% of our sample farmers used only 'Kanchan'.

Hossain, M. 1989. *Green Revolution in Bangladesh: Impact on Growth and Distribution of Income.* University Press Limited, Dhaka, Bangladesh.

Hossain, M.S., Rahman, M.M., Harun-ur-Rashid, M., Farid, A.T.M., Quayyum, M.A., Ahmed, M., Alam, M. S., Hussain, K.M.S.U., 2006. *Krishi Projukti Hatboi* (Handbook of Agro-technology). 4th Edition. Bangladesh Agricultural Research Institute, Gazipur, Bangladesh.

Lantican, M.A., Pingali, P.L., Rajaram, S. 2003. Is research on marginal lands catching up? The case of unfavourable wheat growing environments. *Agricultural Economics,* 29: 353 – 361.

Mahmud, W., Rahman, S.H., Zohir, S. 1994. Agricultural growth through crop diversification in Bangladesh. Food Policy in Bangladesh Working Paper # 7. *International Food Policy Research Institute,* Washington, D.C., USA.

Morris, M., Chowdhury, N., Meisner, C. 1996. Economics of wheat production in Bangladesh. *Food Policy,* 21: 541 – 560.

OFRD, 2001. Annual Report, 1999-2000. *On-Farm Research Division.* Bangladesh Agricultural Research Institute (BARI), Gazipur, Bangladesh.

Rahman, S., Hasan, M.K. 2008. Impact of environmental production conditions on productivity and efficiency: the case of wheat producers in Bangladesh. *Journal of Environmental Management*, 88: 1495-1504.

Razzaque, M.A., Samad, M.A., Hossain, A.B.S., Sufian, M. A., Haque, H., Saha, N.K., Rahimullah, Rahman, A.B.M. Hossain, A., Zaman, M.S. 1993, Technologies generated in wheat research and methods of their transfer in Bangladesh. In: Proceedings of the International conferences *Wheat in hot, dry, irrigated environments* held during 1-4 February, 1993 in Medani, Sudan and *Wheat in warm areas, rice-wheat farming systems* held during13-15 February, 1993 Dinajpur, Bangladesh.

Razzaque, M.A., Satter, M.A., Amin, M.S., Quayyum, M.A., Alam, S. 2000. *Krishi Projukti Hatboi* (Handbook on Agro-technology), 2nd edition, Bangladesh Agricultural Research Institute, Gazipur-1701, Bangladesh.

SCA, 1985. *Seed Certification Agency. Characteristics of crop varieties approved by National Seed Board (1st issue),* Ministry of Agriculture, Government of Bangladesh, Dhaka, Bangladesh.

SCA, 1992. *Seed Certification Agency. Characteristics of crop varieties approved by National Seed Board (2nd issue),* Ministry of Agriculture, Government of Bangladesh, Dhaka, Bangladesh.

Sherlund, S.M., Barrett, C.B., Adesina, A.A. 2002. Smallholder technical efficiency controlling for environmental production conditions. *Journal of Development Economics,* 69: 85 – 101.

Singh, G., Singh, S., Singh, J. 2004. Optimization of energy inputs in wheat crop in Punjab. *Energy Conservation and Management,* 45: 453 – 465.

STATA Corp, 2003. *STATA Version 8.* Stata Press Publications, College Station, Texas, USA.

USDA, 2008. *Foreign Agricultural Service Production, Supply, and Distribution Database United States Department for Agriculture,* Washington, D.C.

WRC. 2000. *Wheat production technology. Wheat Research Centre,* Bangladesh Agricultural Research Institute, Dinajpur, Bangladesh.

In: Soybean and Wheat Crops
Editors: S. Davies, G. Evans
ISBN: 978-1-60741-173-4

Chapter 12

STABILIZING PRODUCTIVITY OF DROUGHT-STRESSED CROPS BY FOLIAR APPLICATION OF ALKANOLAMINES

Hans Bergmann and Gerhard Gramss
Friedrich-Schiller-University, Institute of Geological Sciences,
Burgweg 11, D-07743 Jena, Germany

ABSTRACT

Losses in biomass production and protein caused by stress factors such as drought, waterlogging, salinity, or heavy-metal contamination are accompanied by structural damages to cellular membranes of crop plants, to alterations in enzyme activities, and to cytoplasmic dehydration. Water deficit leads to the closure of stomata, to reduced uptake of aerial CO_2, and thus to the inhibition of photosynthesis and the excessive formation of reactive oxygen species in chloroplasts and other cell compartments. Stress-activated phospholipases and esterases catalyze the separation of ethanolamine (EA), choline, and their phosphate esters, as well as fatty acids and serine from the phospholipid molecule of cell membranes. The membrane protein component is subject to proteolysis. The liberated serine is a precursor for the formation of further EA and choline. In a research project extending over 3 decades, the membrane lipid moieties, ethanolamine and choline were applied at 1.5 kg/ha as a foliar spray to potted cereal plants and to cultures in more than 150 field trials. External application of the biogenic stress metabolites was expected to initiate the plants' resistance and tolerance mechanisms by pretending a stress situation. Actually, the treatment confined yield losses in drought-stressed wheat, rye, and barley by stabilizing water household, photosynthesis, and protein production. Relative to untreated, drought-stressed plants, increases in biomass (5-20 %) and protein content (5-7 %) were recorded in 14 to 58 % of the experiments which had been conducted at 17 Experimental Agrostations of different soil conditions but little climatic extremes. Plant responses were significant under drought conditions and on poor soils. They were negligible in the absence of abiotic stress. The major short- and long-term stress responses were followed on the biochemical and ultrastructural level. Under the temperate climatic conditions of Central Europe, the ecotoxicologically unobjectionable treatment has not yet become part of daily agricultural practice. In the light of the world-wide food crisis, an application of stress control agents could help stabilize crop production in semi-arid and saline regions and during temporary rainfall deficits. The treatment should also revalorize the economic role of the multitude of traditional cereal cultivars in use which are adapted to the local soil and climate conditions.

1. Plant Response to Abiotic Stress

Adverse environmental conditions such as drought, heat, salinity, heavy-metal contamination, waterlogging, soil acidity, soil compaction, and chill exert injurious strain to plants and match therefore the definition of abiotic stress factors (Levitt 1980). Some of the short-term plant responses to stress are also incited by pathogen ingress, arbuscular-mycorrhiza formation, and even by exposure to IAA releasing soil bacteria (Bergmann et al. 1999; Kemp and Burden 1986; Leinhos and Bergmann 1995a).

Freezing, desiccation, and salinity go along with cytoplasmic dehydration (McKersie 1991) and changes in the concentration of numerous proteins (Caruso et al. 2008) and enzymatic activities. Dehydration triggers the formation of sugars, polyols, amino acids and their derivatives, including proline, N-dimethylproline, trigonelline, glycine betaine, and polyamines (Bergmann et al. 2001; Blum 1996; McNeil et al. 1999; Mullet and Whitsitt 1996; Schlee 1992) with osmoprotective properties (Blum 1996; Morgan 1991; 1995; Rajam et al. 1998). The synchronously appearing radicals and reactive oxygen species (Baker and Orlandi 1995; Lamb and Dixon 1997; Larson 1997) which are by-product of the aerobic metabolism (Bartosz 1997) and are increasingly formed in chloroplasts by inhibition of photosynthesis (Dat et al. 2000; Loggini et al. 1999; Smirnoff 1998) impair permeability and structure of cellular plasma, mitochondrial, and thylakoid membranes (Bewley 1979; Mascher et al. 2005; McKersie 1991; Santarius et al. 1979).

In short-term responses, stress-resistant plants neutralize reactive oxygen species with the antioxidative enzymes, superoxide dismutase, peroxidase, catalase, and glutathione reductase (McKersie 1991; Mascher et al. 2005) but also with radical scavengers such as glutathione, ascorbate, phenolics, amines, and terpenoids (Bartosz 1997; Elstner et al. 1994; Larson 1997; Marschner 1995).

In long-term plant responses to stress, the destructive action of enzymes and reactive oxygen species to protein and phospholipid component of cell membranes results in proteolysis and in the formation of fatty acid and amino acid derivatives, respectively (Figure 1). Activated phospholipases and esterases catalyze the release of ethanolamine (EA), choline, serine, or their phosphate esters as well as fatty acids from the membranes' phospholipid molecule (Bergmann 1996; Giddings Jr. and Hanson 1982; Keith Cowan 2006; Krishanquamurthy and Bhagwat 1990). The alkanolamines, EA and choline, or their phosphoryl- and phosphatidyl derivatives are finally transformed to the inert, osmoprotective glycine betaine (Bergmann et al. 2002; Eckert et al. 1988b; Hanson et al. 1979). Fatty acids are peroxidized by the reactive oxygen species (Bowler et al. 1992; Dat et al. 2000; Hartley-Whitaker et al. 2001) to accumulate with aldehydes and the proteinase inhibitor and phytohormone, jasmonate, which is catalyzed by lipoxygenase (Ryan 1992) from the increasing pool in linolenic acid (Figure 1; Bergmann et al. 1999; Zuniga et al. 1990). The concentration of free amino acids increases by *de-novo* synthesis and the lysis of membrane

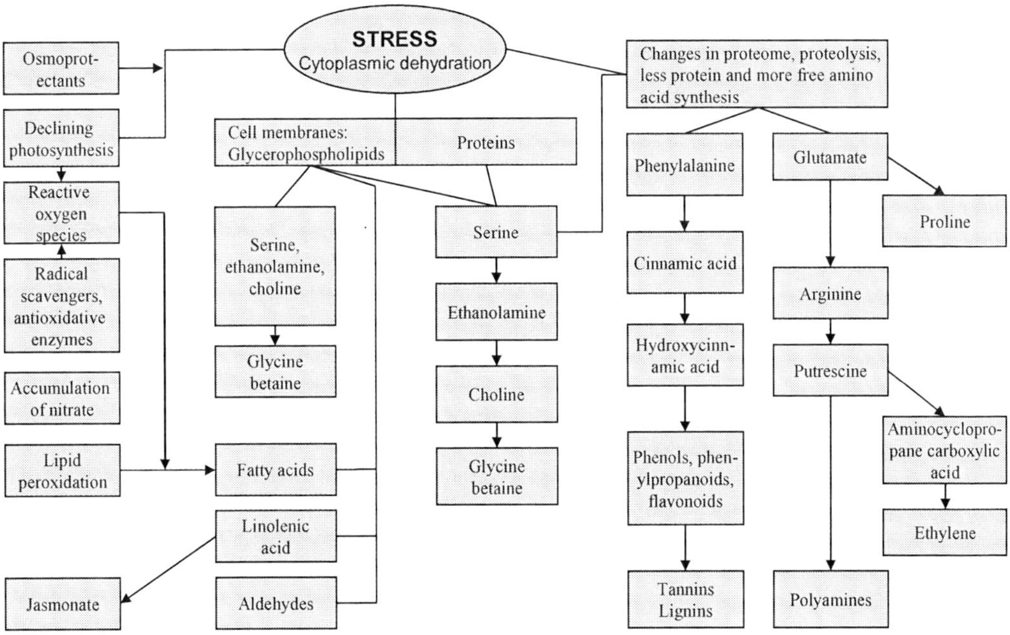

Figure 1. Some drought-stress induced events in plants (refer to Section 1).

proteins. Free serine is then decarboxylated to form the major quantities of EA (Bergmann 1996; Rontein et al. 2003). Phenylalanine is transformed to phenolic radical scavengers, whereas glutamate plays a role as precursor of proline and the polyamines, putrescine, spermidine, and spermine which accumulate in the cell. Polyamines are radical scavengers and inhibitors of lipid peroxidation (Benavides et al. 2000; Chang and Kao 1997; Borrell et al. 1997). They are thus involved in structural and functional stabilization of cell membranes and organelles under stress. Long-term responses make plants adapted, and to some degree resistant to detrimental influences of stressors (see Bartosz 1997; Bergmann et al. 1999; Chernyad'ev 2005; Rajam et al. 1998 for more details).

Among stress metabolites of interest, the alkanolamines, EA and choline and their phosphatidyl derivatives are attached to the polar phosphorus moiety which heads the glycerol backbone of the phospholipid molecule (Keith Cowan 2006). They are constituents of membrane lipids in all taxa of the plant kingdom (Fritsche 1990; Galliard and Mercer 1985; Incharoensakdi and Wutipraditkul 1999) and are also found water-soluble at 5 to 50 ppm in fresh plant tissue (Bergmann 1996). The diamine, putrescine, and the polyamines, spermidine and spermine are common to eukaryotic plants but not to several algae (Smith 1985; Tiburcio et al. 1990). The alkanolamines (Kamalyan 1971; Kamisaka 1979) and glycine betaine as their end product (Ashraf et al. 2008; Bergmann and Eckert 1984; Mäkelä et al 1996), but also the di- and polyamines (Galston 1983; Smith 1985) were found to be conducive to growth and metabolism of crop plants and protecting from damage by stress (Bergmann et al. 1983; 1991; 1999; Fujii et al. 1972; Mascher et al. 2005). Monoethanolamine (Davtyan 1981; Fujii et al. 1972; Ishigami and Suzuki 1970; Kamalyan 1971; Washio and Kiguchi 1972) and phenylethanolamine (Kamisaka 1979; Oota 1974) were recommended as the most efficient plant growth stimuli. Extensive applications of EA and choline chloride to field cultures of cereals and potatoes as a foliar spray, beginning in the late nineteen seventies, resulted then in viable results and the award of several patents (see Appendix section).

Alkanolamines are not deleterious to microbial (Moody et al. 2000; Zwart et al. 1983) or animal life (Groth et al. 1993). They are, e.g., released by vine yeasts into fermented alcoholic drinks (Caruso et al. 2002), and are regulatory to some aspects of human health.

2. Experimental

In the late nineteen seventies, the Research Center for Soil Fertility Muencheberg/Jena (Germany) initiated test trials with alkanolamines to stabilize the productivity of cereals and potato under semi-arid conditions. Sixteen to nineteen test plants each were grown in 5.5-L Mitscherlich pots containing 6.9 kg of a mineral-fertilizer amended quartz sand/loamy soil mixture (2:1 v/v). Two applications of K_2SO_4 and NH_4NO_3 followed during the shooting period (Bergmann et al. 1983; 1991; 2002). Cultures grew under outdoor conditions protected from rainfall. Drought stress was simulated by lowering soil water from 65 % WHC (Ψ-550 hPa) to 30 % (Ψ-1400 hPa) for 7 d, with daily extremes of 15 % (Ψ-3500 hPa) over 2 h. This treatment was applied two times to plants in the shooting to early flowering state and was supervised gravimetrically.

The stress control agents, ethanolamine (Merck) and choline chloride (Serva) in aqueous solution of 10^{-2} *M* were sprayed on plants at early shooting state in amounts of 0.3 to 0.5 mg of which around 10 % were resorbed by the aboveground vegetation (Eckert et al. 1988a). Samples of tillers and ears were taken at early flowering and after grain filling. For the determination of crude protein, amino acids, stress metabolites, and the uptake of $^{14}CO_2$ refer to the original papers (Bergmann et al. 2002; Eckert 1988).

Early field applications of EA (3 kg amine/ha) go back to 1977 to 1981 in the Experimental Station Straussfurt (Thuringia, Germany). Randomly distributed field plots of 25 m^2 in 4 to 6 replicates were established on a clay-chernozem soil within a period of rainfall deficit (Bergmann et al. 1983). Until 1999, results of EA and choline chloride applications (1.5 kg amine/ha) from 114 field trials and 140 large-scale experiments at 17 Experimental Agrostations in East Germany were available. They comprised the response of barley, rye, wheat, and potato crops to different soil and climate conditions but with little climatic extremes (Bergmann et al. 1999; see below for more details).

3. Productivity and Composition of Drought-Stressed Cereals in Pot Trials

In a pot trial with 5 cultivars of spring barley, moderate drought stress reduced the grain yield of Alexis, Krona, Lada, and Salome by a mean of 5 %. Treatment with EA or choline chloride compensated near-completely for these losses and improved the yield by 4.2 %. In the case of Salome and Trumpf, differences were significant (Figure 2; Bergmann et al. 2002). In Krona and Alexis, severe drought stress caused yield losses of 80/70 % (Figure 2).

Treatment with alkanolamines increased the grain yield by 38/20.4 %. Referring these values to the yield obtained under non-stress conditions, the dramatic overall losses were 7.6/6.2 % smaller. All cultivars expressed positive responses to treatment with alkanolamines.

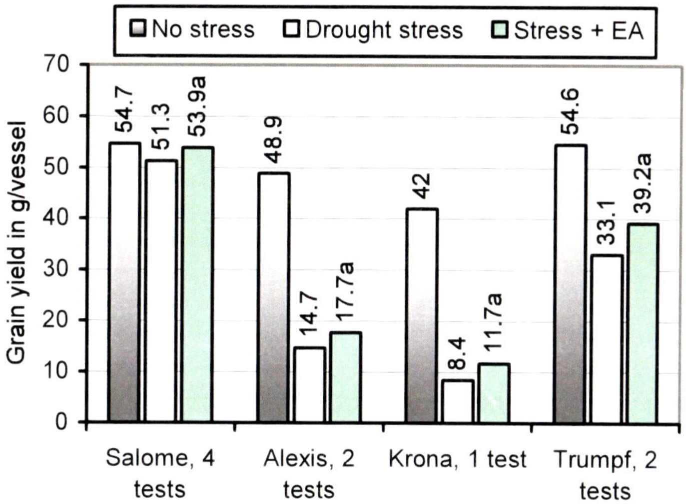

Figure 2. Grain yield (g/pot) of spring barley cultivars grown in Mitscherlich pots (6.9 kg soil, 19 plants) under drought stress. Ethanolamine (0.3 to 0.5 mg per plant) applied at early shooting state. a, Yield increases relative to drought-stressed, untreated plants significant at $p < 0.05$. Treatment of drought- stressed barley plants with choline chloride stabilized the grain yield to the same extent. Adapted from Bergmann et al. (2002).

Table 1. Association of the grain yield with first (main)- and second-order (side) tillers of cereals cultured in Mitscherlich pots (6.9 kg soil, 16 plants) under moderate drought stress

Detail	Plant organ	Spring barley (Salome)		Winter rye (Pluto)		Winter wheat (Alcedo)	
		Untreated	EA-appl.	Untreated	EA-appl.	Untreated	EA-appl.
Grain yield in g per pot	Whole plant	56.7	59.4[b]	61.3	65.7[b]	37.0	38.4
	Main tiller	16.3	16.6	34.5	34.2	29.1	29.2
	Side tillers	40.4	43.0[b]	26.8	30.5[b]	7.9	9.2[b]
Ears per pot	Whole plant	72.5	76.4[b]	38.3	41.8[b]	29.0	28.8
	Side tillers	56.5	60.3[b]	22.3	25.7[b]	13.0	12.8
Number of grains per ear	Whole plant	19.2	18.6	46.8	44.4	29.5	31.3
	Side tillers	18.3	17.7	40.2	37.2	17.0	21.5[b]
Single grain weight in mg	Whole plant	40.8	41.7	33.7	35.1	44.8	44.0
	Side tillers	39.1	40.3	29.9	31.9	35.8	33.7
Coefficient of fertile side tillers[a]		0.88	1.00[b]	0.75	0.83[b]	0.71	0.70

[a] Quotient of fertile side tillers to the number of their initials formed at the outset of shooting.
[b] Values significantly ($p < 0.05$) different from those of untreated controls.
Application of 10 mg ethanolamine per pot at early shooting. Adapted from Bergmann et al. (1991).

In drought-stressed cereals, the numbers of ears formed by second-order tillers are substantially lower (Jensen and Tophoj 1985; Kachel and Roth 1987; Weber et al. 1990). Relative grain yield increases by alkanolamine treated plants were then actually brought forth by the higher number of fertile side tillers in barley, rye, and wheat and their higher grain number in wheat, but not by the weight of the single barley and wheat grain (Table 1; Bergmann et al. 1991; Eckert 1988). This firmer position of side tillers in plants treated with alkanolamines is shown in Figure 3.

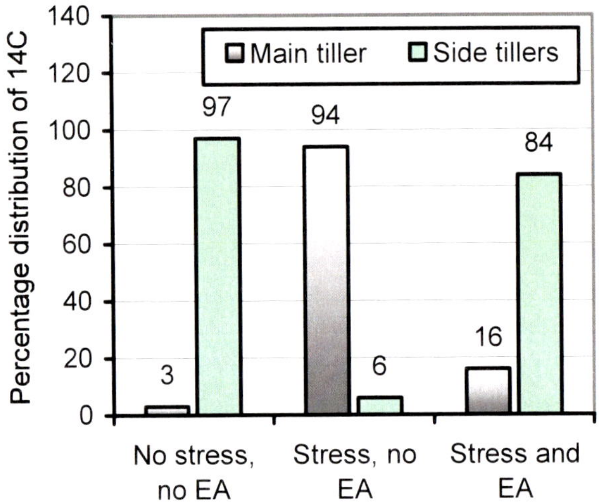

Figure 3. Distribution of ^{14}C assimilate (%) between main and side tiller ears of spring barley plants in the state of early flowering. The plants subjected to drought stress (soil water reduced from 65 to 35 % water-holding capacity) were treated with ethanolamine (10 mg per Mitscherlich pot) and exposed to $^{14}CO_2$ for 24 h. Adapted from Bergmann et al. (1991).

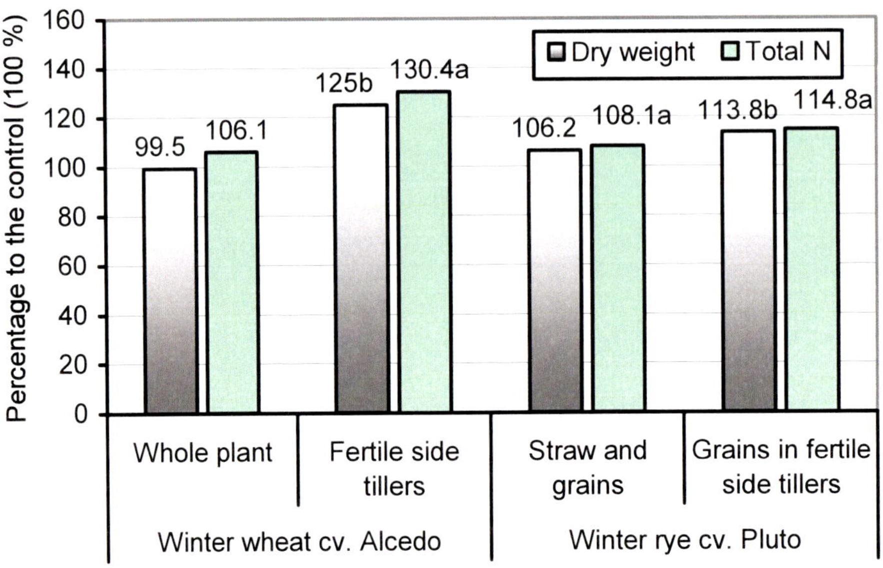

Figure 4. Influence of ethanolamine application on dry weight production and total-N accumulation of particular plant organs in drought-stressed wheat and rye cultivars grown in Mitscherlich pots (6.9 kg soil, NH_4NO_3/urea fertilization, 16 plants). Values given in percent refer to those of untreated, drought-stressed controls (100 %), and deviate significantly at $p < 0.05$ (a) or < 0.1 (b). No significant effects of EA on dry weight production and total N were detected for ears, main tiller, and root of wheat, and for tillers and main-tiller grains of rye. Adapted from Bergmann et al. (1991).

The treatment was able to reverse the preferential allocation of ^{14}C in the first-order tiller, and the near-complete neglect of the side tillers, in drought-stressed plants (Bergmann et al. 1991; Eckert 1988). The consequences are once more depicted in Figure 4. Significant gains in the dry weight were restricted to the whole fertile side tillers in wheat and to grains from these second-order tillers in rye. Both organs showed also higher concentrations in total nitrogen (Bergmann et al. 1991), although the N content in main-tiller grains did not decline. Treatment with EA improved thus the acquisition of soil nitrogen.

Table 2. Changes in the content of organic N and the stress metabolite, glycine betaine, in pre-mature grains of spring barley, resulting from drought stress and treatment with ethanolamine

Stress intensity	Cultivar	Crude protein[a]		Lysine[b]		Isoleucine[b]		Proline[c]		Glycine betaine[c]	
		Untr	EA	Untr	EA	Untr	EA	Untr	EA	Untr	EA
Low	Alexis	6.68	7.59[e]	234	239	238	255	10.2	26.5[d]	6.9	9.7[e]
	Krona	7.45	6.82	213	201	212	210	1.0	1.0	9.0	7.9
High	Alexis	2.38	3.02[d]	59	86[d]	63	92[d]	60.9[f]	20.0[d]	9.6[f]	7.5[d]
	Krona	1.80	2.35[d]	40	63[d]	46	69[d]	1.4	1.3	9.1	5.2[d]

[a] Content in g per pot; [b] content in mg per pot; [c] content in μmol/g dry matter.
[d] Values of the EA treatment significantly ($p < 0.05$) different from those of the untreated (Untr) control.
[e] Values of the EA treatment significantly ($p < 0.1$) different from those of the untreated control.
[f] High-stress values significantly different ($p < 0.05$) from the respective low-stress values.
Plants of the stress-susceptible *cv.* Alexis and the resistant *cv.* Krona cultured in Mitscherlich pots (6.9 kg rooting soil). Adapted from Bergmann et al. (2002).

In the spring barley cultivars Alexis and Krona, premature-grain concentrations in crude protein and in the amino acids, lysine and isoleucine were identical under low-stress conditions, and their losses under high-stress conditions were also of the same order (Table 2; Bergmann et al. 2002). Due to the ameliorating effect of EA, organic-N concentrations reduced by stress increased to 130 to 150 %. Synchronously, the amino acid composition of the grain protein was re-adapted to that of grains from non-stressed plants (Leinhos and Bergmann 1995a; b; Leinhos et al. 1996).

Concentrations of free proline in untreated shoot tissue of Alexis and Krona determined at early flowering increased to the 2.4/3.9-fold and reached identical levels in both cultivars when drought stress conditions changed from low to high (Bergmann et al. 2002; data not shown). In pre-mature grains of *cv.* Krona analyzed at 10 wk after treatment with EA, proline concentrations were uniformly low, irrespective of the stress conditions (Table 2). Grains of cv. Alexis retained their high stress-induced proline concentration in the untreated control but lost their proline content widely after EA treatment. Concentrations in free proline did apparently not confer the grain-yield determining stress tolerance to barley (compare Figure 2 and Table 2). This came also true for glycine betaine. It has been suggested that proline formation expresses the degree of injury to plant tissue (Hanson and Nelsen 1978). This osmoprotectant could primarily serve as C and N resource of the recovering plant (Hare and Cress 1997). Proline could also play a vital role in maintaining integrity of plasma membranes and could act as signalling molecule in response to abiotic stresses (Kishor et al. 2005; Siripornadulsil et al. 2002; Su and Wu 2004).

The set of responses to treatment with alkanolamines was bound to plants grown under (non-specific) stress conditions. Plants grown in the absence of stress factors did not respond to the treatments.

4. Productivity of Cereals under Stress Conditions in Field Trials

The grain yield of spring barley grown on field plots over 3 yr varied considerably with the water supply (Table 3, Bergmann et al. 1983). Drought stress control by treatment with EA compensated for the declining productivity with yield increases from 1 to 22.6 % (60 to 970 kg/ha) with a mean of 10.4 % (557 kg/ha) over the test period. Increases in grain yield correlated with the higher density of fertile ears/m^2 and were thus accompanied by an elevated straw output (Table 3). The gains in productivity during dry seasons went along with an improved water use efficiency of EA treated plants. Grain yield increases of 15 to 23 % per L of rainwater under semi-arid conditions (Table 3) had been confirmed by pot trials (Bergmann et al. 1983). In addition, shoot applications of EA stimulated the formation of root biomass in cereals (Bergmann et al. 1998; Lippmann et al. 1998) and tomato (Grimmer 1998) and retarded losses in chlorophyll (Bergmann et al. 1999; Horvath and van Hasselt 1985; Wejnar 1989).

Large-scale applications of alkanolamines under the variable conditions of 17 Experimental Agrostations in East Germany resulted in local yield improvements from 5 to 20 % with means up to 11 and 13 %, respectively, in barley and rye (Table 4; Bergmann et al. 1999). Comparable effects were recorded for 53 field trials of wheat with relative grain yield increases up to 20 % under unfavourable climate or soil conditions. Figure 5 relates the productivity of wheat to soil type and the intensity of drought stress at five Experimental Stations in East Germany. Significant yield improvements were bound to a pronounced drought stress and were more reproducible by the use of choline chloride instead of EA.

Table 3. Grain and straw production (kg/ha) and water use efficiency of the spring barley cultivars Trumpf and Lada on field plots 25 m^2 (6 replicates per treatment) established on clay-chernozem soil under semi-arid conditions

Precipitation deficit[a] (mm)	Grain yield (kg/ha)		Straw yield (kg/ha)		Stand density (ears/m^2)		Water use efficiency[b]	
	Untreated	EA-appl.	Untreated	EA-appl.	Untreated	EA-appl.	Untreated	EA-appl.
-32.5	5980	6040	ND	ND	ND	ND	ND	ND
-62.5	6230	6670	ND	ND	ND	ND	ND	ND
-93.2	5110	5870[c]	3440	3640	ND	ND	1.26	1.45[c]
-180.2	4290	5260[c]	3450	3960[cd]	460	640[c]	1.45	1.78[c]

ND, not determined.

[a] Difference between precipitation and evapotranspiration from April 1 to July 20.

[b] Grain yield in g per L of rainwater.

[c] Values significantly ($p < 0.05$) different from those of untreated controls.

[d] Application of EA did not alter length and diameter of sprouts. Cultures were treated with ethanolamine (3 kg/ha) at the outset of shooting. The tests were conducted by Experimental Station Straussfurt, Thuringia, Germany, in 1977, 1978, and 1981. Adapted from Bergmann et al. (1983).

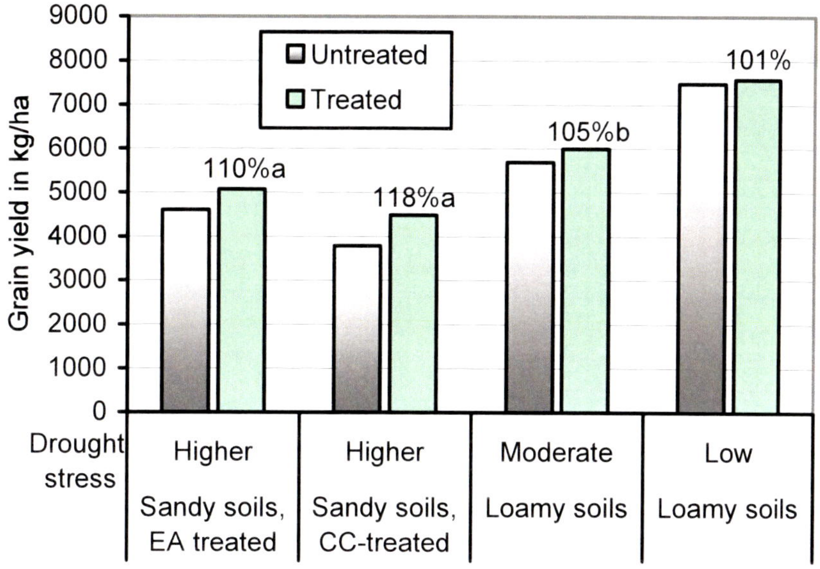

Figure 5. Effect of ethanolamine (EA, as free base or as secondary phosphate) and choline (CC, as chloride; 1.5 kg amine/ha), sprayed at shooting state, on the grain yield (kg/ha) of wheat (*T. aestivum cv.* Alcedo, Arcos, or Miras). Values given in percent refer to the yield of the untreated controls (100 %), and deviate significantly from those at $p < 0.05$ (a) or < 0.1 (b). Field experiments conducted by five Experimental Agrostations in Thuringia and Saxonia, Germany. Number of experiments on sandy soils, 6; on loamy soils at moderate drought stress, 14; and on loamy soils at low drought stress, 18.

Table 4. Increases in the grain yield (kg/ha) of spring barley and winter rye by ethanolamine or choline chloride application under the conditions of drought stress in field trials

Crop plant	Severity of drought stress	Total no. of trials	Number of trials with yield increases of					Yield of control
			<100	>100-250	>250	>250-350	>350	
Barley	Moderate to high	40	6	10	§	13	11[a]	4200
	Low	55	26	20	§	6	3	6300
Rye	High	47	9	11	27[b]	§	§	3400
	Moderate to low	30	13	13	4	§	§	4400

§ No data available for this size class.

Mean yield increases in the respective size classes amounted 472 ([a]) and 440 kg/ha ([b]).

Ethanolamine was applied at the outset of shooting as free base (1.5 kg/ha) or as phosphate. Adapted from Bergmann et al. (1999).

The detrimental influence of soil compaction on the grain yield of wheat (e.g., Miransari et al. 2008; Whalley et al. 2008) is demonstrated by Figure 6. Relative yield increases up to 32 % in field trials by EA application denote soil density to be a first-order stress factor. Its economic significance may increase upon the use of heavy agricultural machinery and no-tilling technologies.

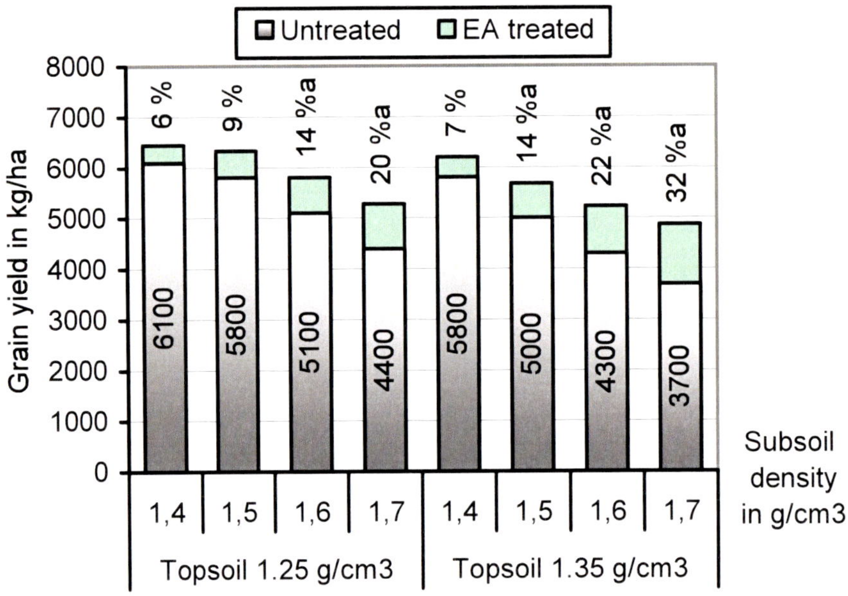

Figure 6. Effect of ethanolamine (EA; 1.5 kg/ha) on the grain yield (kg/ha) of wheat (*T. aestivum cv.* Alcedo) grown on two loamy top soils which cover clay-loam subsoils in four density states. Yield increases given in percent refer to the yield of the untreated control (100 %). Mean precipitation from shooting to grain filling 1.5 mm/day.
a, percentage increases are significant at $p < 0.05$.
Field experiments conducted by Experimental Station Grossobringen, Thuringia, Germany, over 3 yr.

5. How do Alkanolamines Act?

With the degradation of membrane phospholipids and proteins (e.g., Lin and Wu 1996) and the reduced protein synthesis in favour of the formation of free amino acids under abiotic stress, the building stones of antiproteolytic, osmoprotective, radical scavenging, cell growth promoting, and cell wall strengthening compounds are provided (Figure 1; Bergmann et al. 1999). Foliar applications of alkanolamines at < 0.5 mg to plants prior to the initiation of stress had immediate remediating effects (e.g., Bergmann et al. 2002; Mascher et al. 2005). With uptake of the applied [^{14}C]EA (MW 61.08) at a rate of 10 % (Eckert et al. 1988a), its concentration in plants of 0.25 g dry weight should increase by < 0.2 mg/g. The concentration of 3.8 to 4.8 mg/g in its transformation product, glycine betaine (MW 117.15; Bergmann et al. 2002) in drought-stressed barley tissue could then theoretically rise by < 0.38 mg/g. In practice, however, EA application lowered the glycine betaine concentration in treated barley to 3.3 to 3.8 mg/g dry weight (Bergmann et al. 2002). A key position of the cell membrane stabilizing glycine betaine (Bewley 1979; Greenway and Munns 1980) in stress control must therefore be denied. Its foliar application to maize alleviated nevertheless stress symptoms caused by water deficit (Ashraf et al. 2008; Lippmann and Bergmann 1995). This osmoprotectant is believed to stabilize photosystem II protein complexes in higher plants (Yeo 1998). A 48-h pre-treatment with EA prevented disintegration and burst of barley chloroplasts by superoxide radicals which had been formed after the exposure to the herbicide, paraquat (Figure 7; Mascher et al., 2005). Application of alkanolamines to potted cereal plants prior to simulated drought stress, or during the changing climate conditions in the field reduced convincingly the damages caused by water deficits (sections above).

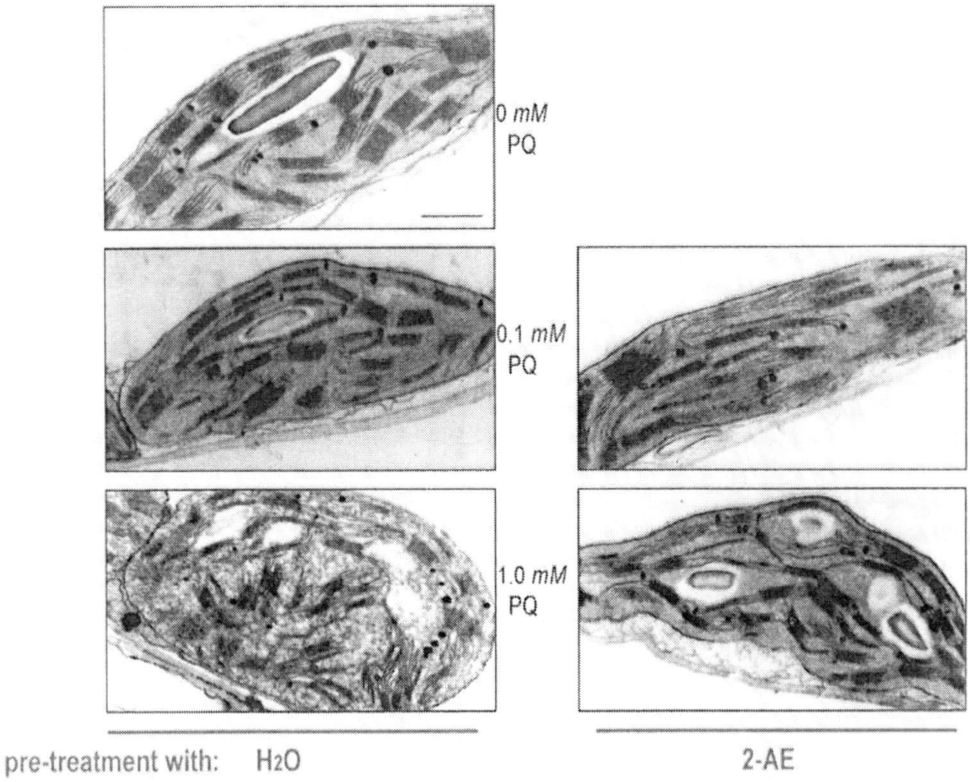

Figure 7. Protective effect of pre-treatment with EA (2-AE, 2-aminoethanol) to oxidative damage of barley chloroplasts. Electron micrographs taken at 8 h after application of 0; 0.1; and 1.0 m*M* paraquat (PQ). Scale bar, 1μm.

The working principle of alkanolamines remains nevertheless obscure. Two weeks after the treatment of barley with [^{14}C]EA, the radiotracer had been found evenly distributed among lipid, glycine betaine, choline, and the free EA fraction. The destiny of marked choline which was taken up at a rate of 5 % was quite the same (Eckert et al. 1988a; b). It is therefore surmised that exogenous supply of stress-related biogenic compounds has a signal function. It initiates the expression of the plant's stress resistance and tolerance mechanisms by pretending a stress situation (Bergmann et al. 1994). This results in immediate changes in the proteome (Bergmann et al. 1999; Caruso et al. 2008), a re-distribution of carbon assimilates in favour of second-order tillers, an increased uptake of soil nitrogen, and an improved water use eficiency (Bergmann and Eckert 1990; Bergmann et al. 1991), connected with a root stock enlarged by 20 to 40 % (Bergmann et al. 1999; Lippmann et al. 1998), a stabilized chlorophyll content (Bergmann et al. 1999; Guan et al. 1995; Horvath and van Hasselt 1985; Kogan et al. 2000; Wejnar 1989) and re-constructed cell membranes (Ilker et al. 1976; Mansour et al. 1993; Waring et al. 1976). It is left to further studies to estimate the impact of alkanolamines and polyamines (Alcázar et al. 2006) on proteome and metabolome of the plant under stress. New insight is coming from transcriptomic studies. Responses of plants, and in particular of cereals, to abiotic stress are subtly controlled by stress-related regulatory genes and their associated signalling networks (Gao et al. 2008; Landjeva et al. 2008; Rabbani et al. 2003; Walia et al. 2006; Zhou et al. 2007; Zhu 2002).

The whole of abiotic stresses are believed to cut the world-wide yield of major crops by more than 50 % (Alcázar et al. 2006). The present world food crisis should give motivation enough to stabilize the productivity of cereals during unexpected rainfall deficits, or to extend their cultivation into semi-arid and saline regions, by foliar applications of alkanolamines, further phospholipid moieties (Keith Cowan 2006), cytokinin-like acting substances (Chernyad'ev and Monakhova 2003; Monakhova and Chernyad'ev 2007; Tassi et al. 2008), or equivalent biogenic compounds (see Appendix section). The treatment should also revalorize the economic role of the multitude of traditional cereal cultivars in use which are adapted to the local soil and climate conditions. Transfer of genomic stress and salt tolerance to plants (e.g., Dita et al. 2006; Dunwell 2004; Gao et al. in press; Rajam et al. 1998; Vasil 2007) can be linked with consume of energy and assimilates, yield loss, and the accumulation of (unwelcome) secondary metabolites (Alscher and Cumming 1990; Bergmann et al. 1996; Caruso et al. 2002; Forrest 1994; Kogel et al. 1994; Schlee 1992).

6. References

Alcázar, R., Marco, F., Cuevas, J. C., Patron, M., Ferrando, A., Carrasco, P., Tiburcio, A. F. & Altabella, T. (2006). Involvement of polyamines in plant response to abiotic stress. *Biotechnol. Lett.*, 28, 1867-1876.

Alscher, R. G. & Cumming, J. (1990). *Stress Responses in Plants. Adaption and Acclimation Mechanisms*. New York: Whiley Liss Inc.

Ashraf, M., Nawaz, K., Athar, H.-ur-R. & Raza, S. H. (2008). Growth enhancement in two potential cereal crops, maize and wheat, by exogenous application of glycinebetaine. In *Biosaline Agriculture and High Salinity Tolerance* (pp. 21-35). Basel, Swiss: Birkhäuser.

Baker, C. J. & Orlandi, E. W. (1995). Active oxygen in plant pathogenesis. *Annu. Rev. Phytopathol.*, 33, 299-321.

Bartosz, G. (1997). Oxidative stress in plants. *Acta Physiol. Plant.*, 19, 47-64.

Benavides, M. P., Gallego, S. M., Comba, M. E. & Tomaro, M. L. (2000). Relationship between polyamines and paraquat toxicity in sunflower leaf discs. *J. Plant Growth Regul.*, 31, 215-224.

Bergmann, H. & Eckert, H. (1984). Einfluss von Glycinbetain auf die Wasserausnutzung und CO_2-Assimilation in Winterweizen (*Triticum aestivum* L.). *Biol. Plant.*, 26, 384-387.

Bergmann, H. & Eckert, H. (1990). Effect of monoethanolamine on growth and biomass formation of rye and barley. *Plant Growth Regul.*, 9, 1-8.

Bergmann, H. (1996). Physiologische Wirkungen von biogenen Aminen bei Pflanzen. In D. M. Beutling (Ed.), *Biogene Amine in der Ernährung* (pp. 31-58). Berlin: Springer.

Bergmann, H., Eckert, H., Kachel, K. & Roth, D. (1983). Der Einfluss von Ethanolamin auf den Kornertrag von Sommergerste bei unterschiedlichen klimatischen Wasserbilanzen. *Arch. Acker- u. Pflanzenbau u. Bodenkd.*, 27, 127-134.

Bergmann, H., Eckert, H., Weber, C. & Roth, D. (1991). Einfluss von Monoethanolamin auf den Ertrag von Kulturpflanzen. I. Untersuchungen zum Einfluss von Monoethanolamin auf den Kornertrag und den N-Haushalt von Getreide (Mitscherlichgefäss-Experimente). *J. Agron. Crop Sci.,* 166, 117-126.

Bergmann, H., Leinhos, V. & Machelett, B. (1994). Increase of stress tolerance in crop plants by using phenolic compounds. *Acta Horticult.,* 381, 390-397.

Bergmann, H., Lippmann, B., Leinhos, V., Tiroke, S. & Machelett, B. (1999). Activation of stress resistance in plants and consequences for product quality. *J. Appl. Bot.,* 73, 153-161.

Bergmann, H., Machelett, B., Leinhos, V. & Eckert, H. (1996). Effect of natural amino alcohols on stress tolerance and food quality in stressed crops. *Agri-Food Quality, Royal Soc. Chem. (UK), Special Publications,* 179, 71-74.

Bergmann, H., Machelett, B., Lippmann, B. & Friedrich, Y. (2001). Influence of heavy metals on the accumulation of trimethylglycine, putrescine and spermine in food plants. *Amino Acids,* 20, 325-329.

Bergmann, H., Rost, S. & Machelett, B. (2002). Erhöhung der Trockentoleranz sowie Veränderungen in der Glycinbetain- und Prolinakkumulation bei *Hordeum vulgare* L. durch Cholin- und Aminoethanolapplikation. *J. Appl. Bot.*, 76, 87-95.

Bergmann, H., Tiroke, S., Leinhos, V., Werner, Th. & Eckert, H. (1998). Influence of natural amines on yield and fertilizer efficiency in cereals under unfavourable environments. In *FAO Rep. Sustainable Agriculture, Food, Energy and Industry* (pp. 531-536). London: James & James Science Publishers.

Bewley, J. D. (1979). Physiological aspects of desiccation tolerance. *Annu. Rev. Plant Physiol.,* 30, 195-238.

Blum, A. (1996). Crop response to drought and the interpretation of adaptation. *Plant Growth Regul.*, 20, 135-148.

Borrell, A., Carbonell, L., Farras, R., Puig-Parellada, P. & Tiburcio, A. F. (1997). Polyamines inhibit lipid peroxidation in senescing oat leaves. *Physiol. Plant.,* 99, 385-390.

Bowler, C., Van Montagu, M. & Inzé, D. (1992). Superoxide dismutase and stress tolerance. *Annu. Rev. Plant Physiol. Mol. Biol.,* 43, 83-116.

Caruso, G., Cavaliere, C., Guarino, C., Gubbiotti, R., Foglia, P. & Laganà, A. (2008). Identification of changes in *Triticum durum* L. leaf proteome in response to salt stress by two-dimensional electrophoresis and MALDI-TOF mass spectrometry. *Anal. Bioanal. Chem.,* 391, 381-390.

Caruso, M., Fiore, C., Contursi, M., Salzano, G., Paparella, A. & Romano, P. (2002). Formation of biogenic amines as criteria for the selection of wine yeasts. *World J. Microbiol. Biotechnol.,* 18, 159-163.

Chang, C. J. & Kao, C. H. (1997). Paraquat toxicity is reduced by polyamines in rice leaves. *Plant Growth Regul.,* 22, 163-168.

Chernyad'ev, I. I. (2005). Effect of water stress on the photosynthetic apparatus of plants and the protective role of cytokinins: A review. *Appl. Biochem. Microbiol.,* 41, 115-128.

Chernyad'ev, I. I. & Monakhova, O. F. (2003). Effects of cytokinin preparations on the pools of pigments and proteins of wheat cultivars differing in their tolerance to water stress. *Appl. Biochem. Microbiol.,* 39, 524-531.

Dat, J., Vandenabeele, S., Vranová, E., Van Montagu, M., Inzé, D. & Van Breusegem, F. (2000). Dual action of the active oxygen species during plant stress responses. *Cell. Mol. Life Sci.,* 57, 779-795.

Davtyan, L. (1981). Biochemical basis of the stimulation action of ethanolamine on agriculture plants. *Tr. Erevan. Zoovet. Inst.,* 50, 15-24.

Dita, M. A., Rispail, N., Prats, E., Rubiales, D. & Singh, K. B. (2006). Biotechnology approaches to overcome biotic and abiotic stress constraints in legumes. *Euphytica,* 147, 1-24.

Dunwell, J. M. (2004). Transgenic crops. The current and next generations. In L. Pena (Ed.), *Transgenic Plants: Methods and Protocols. Methods in Molecular Biology* Vol. 286 (pp. 377-397). Berlin: Springer Humana Press.

Eckert, H. (1988). Untersuchungen über Wirkung, Wirkungsweise und Verhalten von Monoethanolamin (EA) in Getreide als Grundlage zur Auffindung neuer stressabschwächender Wirkstoffe. *Diss. B,* Berlin: Akad. Landwirtsch.-Wiss.

Eckert, H., Bergamnn, H. & Reissmann, P. (1988a). Uptake and translocation of [^{14}C]-monoethanolamine in barley plants. *Biochem. Physiol. Pflanzen,* 183, 27-36.

Eckert, H., Reissmann, B. & Bergmann, H. (1988b). Metabolism of [^{14}C]-monoethanolamine in *Hordeum vulgare* (L.). *Biochem. Biophys. Pflanzen,* 183, 15-25.

Elstner, E. F., Osswald, W. F., Volpert, H. & Schempp. H. (1994). Phenolic antioxidants. *Acta Horticult.,* 381, 304-335.

Forrest, J. M. S. (1994). Stress response in plants. In *Use of Aquatic Invertebrates as Tools for Monitoring of Environmental Hazard* (pp. 119-128). Stuttgart, Germany: Fischer.

Fritsche, W. (1990). *Mikrobiologie.* UTB-Reihe. Jena, Germany: Gustav Fischer.

Fujii, K., Kobayashi, M. & Takahashi, E. (1972). Effects of organic acids and amines on growth of rice seedlings. *Nippon Dojo Hirogeku Zasski,* 43, 211-217.

Galliard, T. & Mercer, E. J. (1985). *Recent Advances in the Chemistry and Biochemistry of Plant Lipids.* London: Acad. Press.

Galston, A. W. (1983). Polyamines as modulators of plant development. *Bioscience,* 33, 382-388.

Gao, J.-P., Chao, D.-Y. & Lin, H.-X. (2008). Toward understanding molecular mechanisms of abiotic stress responses in rice. *Rice,* 1, 36-51.

Gao, S., Zhang, H., Tian, Y., Li, F., Zhang, Z., Lu, X., Chen, X. & Huang, R. (in press). Expression of TERF1 in rice regulates expression of stress-responsive genes and enhances tolerance to drought and high-salinity. *Plant Cell Rep.,*27, 1787-1795.

Giddings Jun., T. H. & Hanson, A. D. (1982). Water stress provokes a generalized increase in phosphatidylcholine turnover in barley leaves. *Planta,* 155, 493-501.

Greenway, H. & Munns, R. (1980). Mechanism of salt tolerance in nonhalophytes. *Annu. Rev. Plant Physiol.,* 31, 149-190.

Grimmer, U. (1998). Wirkung von Preharvestbedingungen auf qualitätsbestimmende Inhaltsstoffe der Tomate. *Master Thesis,* Jena (Germany): Friedrich-Schiller University, Biological-Pharmaceutical Faculty.

Groth, G., Schreeb, K., Herdt, V. & Freundt, K. J. (1993). Toxicity studies in fertilized zebrafish eggs treated with N-methylamine, N,N-dimethylamine, 2-aminoethanol, isopropylamine, aniline, N-methylaniline, N,N-dimethylaniline, quinone, chloroacetaldehyde, or cyclohexanol. *Bull. Environ. Contam. Toxicol.,* 50, 878-882.

Guan, Y. X., Dai, J. Y., Xu, S. C., Chen, J. & Wang, Q. X. (1995). The effects of ethanolamine on physiological and yield characters of maize under soil drought. *Acta Agron. Sinica,* 21, 424-428.

Hanson, A. D., Hitz, W. D. & Scott, N. A. (1979). Biosynthesis of betaine. *Plant Research '79, Michigan State Univ., East Lansing/MI,* 120-123.

Hanson, A. D. & Nelsen, C. E. (1978). Betaine accumulation and [^{14}C]formate metabolism in water-stressed barley leaves. *Plant Physiol.,* 62, 305-312.

Hare, P. D. & Cress, W. A. (1997). Metabolic implications of stress-induced proline accumulation in plants. *Plant Growth Regul.,* 23, 79-103.

Hartley-Whitaker, J., Ainsworth, G. & Meharg, A. A. (2001). Copper- and arsenate-induced oxidative stress in *Holcus lanatus* L. clones with different sensitivity. *Plant Cell Environ.,* 24, 713-722.

Horvath, I. & van Hasselt, P. R. (1985). Inhibition of chilling induced photooxidative damage to leaves of *Cucumis sativus* L. by treatment with amino alcohols. *Planta,* 164, 83-88.

Ilker, R., Waring, A. J., Breitenbach, R. W. & Lyons, J. M. (1976). A light and electron microscopic study of chilling injury and its alleviation by ethanolamine in tomato cotyledons. *Plant Physiol.* [Suppl.], 57, 178.

Incharoensakdi, A. & Wutipraditkul, N. (1999). Accumulation of glycinebetaine and its synthesis from radioactive precursors under salt-stress in the cyanobacterium *Aphanothece halophytica. J. Appl. Physiol.,* 11, 515-523.

Ishigami, T. & Suzuki, Z. (1970). The elongation promotion of pea stem sections by indole together with ethanolamine. *Z. Pflanzenphysiol.,* 62, 460-463.

Jensen, C. R. & Tophoj, H. V. (1985). Potassium induced improvement of yield response in barley exposed to soil water stress. *Irrig. Sci.,* 6, 117-129.

Kachel, K. & Roth, D. (1987). Niederschlags-Ertragsbeziehungen bei Winterweizen und Sommergerste auf zwei Standorten mit unterschiedlichem Bodenwasserspeichervermögen. *Arch. Acker-, Pflanzenbau, Bodenkd.,* 31, 343-351.

Kamalyan, G. V. (1971). Wirkung von Ethanolamin und seiner Derivate auf Tiere und Pflanzen (in Russian). *Mat. Zak. Nauk Konf. Vop. Shiv.,* 431-434.

Kamisaka, S. (1979). Catecholamine stimulation of the gibberellin action that induces lettuce hypocotyl elongation. *Plant Cell Phyisol.,* 20, 1199-1207.

Keith Cowan, A. (2006). Phospholipids as plant growth regulators. *Plant Growth Regul.,* 48, 97-109.

Kemp, M. S. & Burden, R. S. (1986). Phytoalexins and stress metabolites in the sapwood of trees. *Phytochemistry,* 25, 1261-1269.

Kishor, P., Sangam, S., Amrutha, R., Laxmi, P., Naidu, K., Rao, K., Rao, S., Reddy, K., Theriappan, P. & Sreenivasulu, N. (2005). Regulation of proline biosynthesis, degradation, uptake and transport in higher plants: its implications in plant growth and abiotic stress tolerance. *Curr. Sci.,* 88, 424–438.

Kogan, M. J., Kristoff, G., Benavides, M. P. & Tomaro, M. L. (2000). Effect of pre-treatment with ethanolamine on the response of *Helianthus annuus* L. to salt stress. *Plant Growth Regulation,* 30, 87–94.

Kogel, K.-H., Beckhove, U., Dreschers, J., Münch, S. & Romme, Y. (1994). Acquired resistance in barley. *Plant Physiol.,* 106, 1269-1277.

Krishanquamurthy, R. & Bhagwat, K. A. (1990). Accumulation of choline and glycinebetaine in salt-stressed wheat seedlings. *Curr. Sci.,* 59, 111-112.

Lamb, C. J. & Dixon, R. A. (1997). The oxidative burst in plant disease resistance. *Annu. Rev. Plant Physiol. Plant Molec. Biol.,* 48, 251-275.

Landjeva, S., Neumann, K., Lohwasser, U. & Börner, A. (2008). Molecular mapping of genomic regions associated with wheat seedling growth under osmotic stress. *Biol. Plant.*, 52, 259-266.

Larson, R. A. (1997). *Naturally Occurring Antioxidants (Autooxidation Mechanisms / Kinetics and Mechanisms of Inhibited Autooxidation / Antioxidants).* Boca Raton: Lewis Publishers, CRC Press.

Leinhos, V. & Bergmann, H. (1995a). Effect of amino alcohol application, rhizobacteria and mycorrhiza inoculation on the growth, the content of protein and phenolics and the protein pattern of drought-stressed lettuce (*Lactuca* L. *cv.* "Amerikanischer Brauner"). *J. Appl. Bot.,* 69, 153-156.

Leinhos, V. & Bergmann, H. (1995b). Changes in yield, lignin content and protein pattern of barley (*Hordeum vulgare cv.* Alexis) induced by drought stress. *J. Appl. Bot.,* 69, 206-210.

Leinhos, V., Tiroke, S. & Bergmann, H. (1996). Influence of osmotic stress and amino alcohol treatment on protein content, protein patterns and growth of germinating barley. *J. Appl. Bot.*, 70, 199-204.

Levitt, J. (1980). In *Responses of Plants to Environmental Stresses* (Vol. 1, p. 497). New York: Academic Press.

Lin, H. & Wu, L. (1996). Effects of salt stress on root plasma membrane characteristics of salt-tolerant and salt-sensitive buffalograss clones. *Environ. Exp. Bot.,* 36, 239-254.

Lippmann, B. & Bergmann, H. (1995). Effect of a preliminary treatment of maize with aminoalcohols on the root growth and root exudation under drought stress. In W. Merbach (Ed.), *Mikrooekologische Prozesse im System Pflanze-Boden. 5th Borkheider Seminar Oekophysiologie des Wurzelraumes.* Stuttgart, Germany: Teubner.

Lippmann, B., Leinhos, V., Dautz, S. & Bergmann, H. (1998). Effect of rhizobacteria and natural amines on root formation and nutrient uptake under drought. In *FAO Rep. Sustainable Agriculture, Food, Energy and Industry* (pp. 461-465). London: James & James Science Publishers.

Loggini, B., Scartazza, A., Brugnoli, F. & Navari-Izzo, F. (1999). Antioxidative defense system, pigment composition, and photosynthetic efficiency in two wheat cultivars subjected to drought. *J. Plant Physiol.,* 119, 1091-1099.

Mäkelä, P., Mantila, J., Hinkkanen, R., Pehu, E. & Peltonen-Sainio, P. (1996). Effect of foliar application of glycinebetaine on stress tolerance, growth and yield of spring cereals and summer turnip rape in Finland. *J. Agr. Crop Sci.,* 176, 223-234.

Mansour, M. M., Stadelmann, E. J. & Lee-Stadelmann, O. Y. (1993). Salt acclimation of *Triticum aestivum* by choline chloride: Plant growth, mineral content and cell permeability. *Plant Physiol. Biochem. (Paris)*, 31, 341-348.

Marschner, H. (1995). *Mineral Nutrition of Higher Plants.* London: Academic Press.

Mascher, R., Nagy, E., Lippmann, B., Hörnlein, S., Fischer, S., Scheiding, W., Neagoe, A. & Bergmann, H. (2005). Improvement of tolerance to paraquat and drought in barley (*Hordeum vulgare* L.) by exogenous 2-aminoethanol: effects on superoxide dismutase activity and chloroplast ultrastructure. *Plant Sci.*, 168, 691-698.

McKersie, B. D. (1991). In E. J. Pell & K. L. Steffen (Eds.), *Active Oxygen/Oxidative Stress and Plant Metabolism. Current Topics in Plant Physiology* (Vol. 6, pp. 107-118). American Society of Plant Physiologists.

McNeil, S. D., Nuccio, M. L. & Hanson, A. D. (1999). Betaines and related osmoprotectants. Targets for metabolic engineering of stress resistance. *Plant Physiol.,* 120, 945-949.

Miransari, M., Bahrami, H. A., Rejali, F. & Malakouti, M. J. (2008). Using arbuscular mycorrhiza to alleviate the stress of soil compaction on wheat (*Triticum aestivum* L.) growth. *Soil Biol. Biochem.*, 40, 1197-1206.

Monakhova, O. F. & Chernyad'ev, I. I. (2007). A protector effect of cytokinin preparations on the photosynthetic apparatus of wheat plants under water deficiency conditions. *Appl. Biochem. Microbiol.,* 43, 641-649.

Moody, J. D., Heinze, T. M., Hansen Jr., E. B. & Cerniglia, C. E. (2000). Metabolism of the ethanolamine-type antihistamine diphenhydramine (Benadryl)^{TM} by the fungus *Cunninghamella elegans*. *Appl. Microbiol. Biotechnol.,* 53, 310-315.

Morgan, J. M. (1991). A gene controlling differences in osmoregulation in wheat. *Aust. J. Plant Physiol.,* 18, 249-257.

Morgan, J. M. (1995). Growth and yield of wheat lines with differing osmoregulative capacity at high soil water deficit in seasons of varying evaporative demand. *Field Crops Res.,* 40, 143-152.

Mullet, J. E. & Whitsitt, S. (1996). Plant cellular response to water deficit. *Plant Growth Regul.,* 20, 119-124.

Oota, S. (1974). Removal of the sugar inhibition of flowering in *Lemna gibba* G3 by catecholamines. *Plant Cell Physiol.,* 15, 63-68.

Rabbani, M., Maruyama, K., Abe, H., Khan, M., Katsura, K., Ito, Y., Yoshiwara, K., Seki, M., Shinozaki, K. & Yamaguchi-Shinozaki, K. (2003). Monitoring expression profiles of rice genes under cold, drought, and high-salinity stresses and abscisic acid application using cDNA microarray and RNA gel-blot analyses. *Plant Physiol.,* 133, 1755–1767.

Rajam, M. V., Dagar, S., Waie, B., Yadav, J. S., Kumar, P. A., Shoeb, F. & Kumria, R. (1998). Genetic engineering of polyamine and carbohydrate metabolism for osmotic stress tolerance in higher plants. *J. Biosci.,* 23, 473-482.

Rontein, D., Rhodes, D. & Hanson, A. D. (2003). Evidence from engineering that decarboxylation of free serine is the major source of ethanolamine moieties in plants. *Plant Cell Physiol.,* 44, 1185-1191.

Ryan, C. A. (1992). The search for the proteinase inhibitor-inducing factor, PIIF. *Plant Molec. Biol.,* 19, 123-133.

Santarius, K. A., Heber, U. & Krause, G. H. (1979). Untersuchungen über die physiologisch-biochemischen Ursachen von Empfindlichkeit und Resistenz von Biomembranen gegenüber extremen Temperaturen und hohen Salzkonzentrationen. *Ber. dt. bot. Ges., Berlin,* 209-223.

Schlee, D (1992). *Ökologische Biochemie der Pflanzen.* Jena, Germany: Gustav Fischer.

Siripornadulsil, S., Traina, S., Verma, D. & Sayre, R. (2002). Molecular mechanisms of proline-mediated tolerance to toxic heavy metals in transgenic microalgae. *Plant Cell,* 14, 2837–2847.

Smirnoff, N. (1998). Plant resistance to environmental stress. *Curr. Opin. Biotechnol.,* 9, 214-219.

Smith, T. A. (1985). Polyamines. *Annu. Rev. Plant Physiol.,* 36, 117-143.

Su, J. & Wu, R. (2004). Stress-inducible synthesis of proline in transgenic rice confers faster growth under stress conditions than that with constitutive synthesis. *Plant Sci.,* 166, 941–948.

Tassi, E., Pouget, J., Petruzzelli, G. & Barbafieri, M. (2008). The effects of exogenous plant growth regulators in the phytoextraction of heavy metals. *Chemosphere*, 71, 66-73.

Tiburcio, A. F., Kaur-Sawhney, R. & Galston, A. W. (1990). Polyamine metabolism. In B. J. Milfin & P. J. Lea (Eds.), *The Biochemistry of Plants. Intermediary Nitrogen Metabolism* (pp. 283-325). New York: Acad. Press.

Vasil, I. K. (2007). Molecular genetic improvement of cereals: transgenic wheat (*Triticum aestivum* L.). *Plant Cell Rep.,* 26, 1133-1154.

Walia, H., Wilson, C., Wahid, A., Condamine, P., Cui, X. & Close, T. J. (2006). Expression analysis of barley (*Hordeum vulgare* L.) during salinity stress. *Funct. Integr. Genomics,* 6, 143-156.

Waring, A. J., Breitenbach, R. W. & Lyons, I. H. (1976). In vivo modification of plant membrane phospholipid composition. *Biochim. Biophys. Acta, Amsterdam,* 443, 157-168.

Washio, S. & Kiguchi, Y. (1972). *Plant growth regulation containing ethanolamine and mineral acids*. Japan Kokai 7380355 and 7380356.

Weber, C., Bergmann, H., Kachel, K. & Eckert, H. (1990). Untersuchungen zum Einfluss von Monoethanolamin auf die Ertragsstruktur von Sommergerste in Abhängigkeit vom Standorttyp. *Arch. Acker- Pflanzenbau Bodenkd.* ,34, 659-666.

Wejnar, R. (1989). Untersuchungen über Photosynthese-Pigmente bei Lemnaceae. VII. Einfluss von Ethanolamin auf Pigmentbildung und Wachstum bei deetiolierten Pflanzen von *Lemna gibba* L. *Angew. Bot.,* 63, 341-346.

Whalley, W. R., Watts, C. W., Gregory, A. S., Mooney, S. J., Clark, L. J. & Whitmore, A. P. (2008). The effect of soil strength on the yield of wheat. *Plant Soil*, 306, 237-247.

Yeo, A. (1998). Molecular biology of salt tolerance in the context of whole-plant physiology. *J. Exp. Bot.,* 49, 915-929.

Zhou, J., Wang, X., Jiao, Y., Qin, Y., Liu, X., He, K., Chen, C., Ma, L., Wang, J., Xiong, L., Zhang, Q., Fan, L. & Deng, X. (2007). Global genome expression analysis of rice in response to drought and high salinity stresses in shoot, flag leaf, and panicle. *Plant Mol. Biol.*, 63, 591–608.

Zhu, J. K. (2002). Salt and drought stress signal transduction in plants. *Annu. Rev. Plant Biol.,* 53, 247-273.

Zuniga, G. E., Fernandez, J., Cristi, R., Alberdi. M. & Corcuera, L. J. (1990). Lipid changes in barley seedlings subjected to water and cold stress. *Phytochemistry,* 29, 3087-3090.

Zwart, K. B., Veenhuis, M. & Harder, W. (1983). Significance of yeast peroxisomes in the metabolism of choline and ethanolamine. *Antonie van Leeuwenhoek* , 49, 369-385.

7. APPENDIX (Applied plant biogenic compounds; from Chemical Abstracts database)

Baskakov, Y. A. et al. (1991-05-30). *Preparation of N-(hydroxyalkyl)- and N-[(aminocarbonyloxy)alkyl] carbamates as stress-reducing plant growth regulators.* Patent WO 9107381 A1.

Bech, R. et al. (1985-09-04). *Composition for increasing water assimilation and use efficiency in plants – containing mono di and tri ethanolamine(s).* Patent DD 226755 A1.

Bergmann, H. et al. (1981-10-08). *Treatments for increasing water acquisition by agricultural crops*. Patent DD 151104 A1.

Bergmann, H. et al. (1986-11-26). *Treatments for increasing the yield-determining water exploitation of agricultural crops*. Patent DD 241007 A1.

Bergmann, H. et al. (1990-04-18). *Plant stress tolerance-enhancing compositions containing ethanolamine and fatty acids from Brassica napus*. Patent DD 277831 A1.

Bergmann, H. et al. (1990-04-18). *Plant-stress tolerance-enhancing compositions containing ethanolamine and brassinolide compounds*. Patent DD 277828 A1.

Horváth, E., Szalai, G. & Janda, T. (2007). Induction of abiotic stress tolerance by salicylic acid signaling. *J. Plant Growth Regul.*, 26, 290-300.

Kochmann, W. et al. (1992-08-06). *Plant stress tolerance enhancement by synergistic mixtures of monoethanolamine with N-(2-hydroxyethyl)piperazine*. Patent DE 4103254 A1.

Schilling, G. et al. (1984-04-25). *Compositions for influencing organ proportions in cereals*. Patent DD 209340 A1.

Schmidt, A., Sammler, P. & Bergmann, H. (1985). Influence of different biologically active compounds on evapotranspiration, water-use efficiency and yield formation of *Vicia faba* L. under water stress conditions. *Arch. Acker- Pflanzenbau Bodenkd.*, 29, 539-546.

Reviewed by Dr. Heike Hahn

SKW Stickstoffwerke Piesteritz GmbH; Landwirtschaftliche Anwendungsforschung Cunnersdorf, Am Wieseneck 7, D-04451 Cunnersdorf, Germany.

In: Soybean and Wheat Crops
Editors: S. Davies, G. Evans
ISBN: 978-1-60741-173-4

Chapter 13

WHEAT: COMPOSITION AND FEEDING VALUE FOR POULTRY

Velmurugu Ravindran and Ahmed M. Amerah
Institute of Food, Nutrition and Human Health, Massey University, Private Bag 11 222, Palmerston North 4442, New Zealand

ABSTRACT

The composition and feeding value of wheat is usually more variable than that of other cereals. Protein levels in wheat, for example, can vary from 10-18%. But wheat provides 10% less metabolisable energy than corn, mainly due to the presence of soluble non-starch polysaccharides. These non-starch polysaccharides cause increased digesta viscosity in the gut, leading to reduced nutrient digestibility and metabolisable energy values, especially in young birds. Wide variation in the metabolizable energy of wheat is a major concern to the poultry industry. Currently, exogenous enzymes, containing xylanase activity, are routinely used to mitigate the adverse effects of non-starch polysaccharides and to minimize the variation in the performance of poultry fed wheat-based diets. With the use of enzyme supplementation, wheat can be used without any limitation in poultry diet. Feeding of whole wheat grains to poultry has also generated interest in recent years as a mean reducing processing and production costs, and improving gut health.

INTRODUCTION

Wheat is a major raw material in poultry feeds in many parts of the world, including European Union, Canada, Australia, New Zealand. The use of wheat as a poultry feed ingredient has increased significantly over the years, due to favorable price conditions. Compared to corn, wheat generally delivers energy into a diet at low cost. In addition, the availability of low-cost exogenous enzymes, which are effective in improving the apparent metabolizable energy (AME) content, has improved the feeding value of wheat for poultry and made it a desirable feed ingredient. Wheat can account for up to 70% of the metabolizable energy requirements of broilers and turkeys and as much as 80% of that of the laying hens. In addition, wheat may contribute up to 35% of the protein in broiler diets when included at high levels.

Wheat is highly diverse with different types of grain. The major factors used to distinguish between types of wheat are hardness and softness of the grain, whether they are spring or winter wheats, what colour bran they produce (red or white) and protein content. For example in the US, six distinct classes are grown: Durum, Hard Red Spring, Hard Red Winter, Hard White, Soft Red Winter and Soft White. In Canada, eight classes are grown and within each of these classes, there are a number of species. In terms of feed processing and feeding value, whether a grain is hard or soft is of the greatest importance.

Physical Structure of Grain

A mature, whole grain of wheat comprises 5% pericarp, 8% aleurone layer, 2% embryo and 85% endosperm. The pericarp consists mainly of cellulose and other polysaccharides, mainly hemicelluloses and pentosans (Lasztity, 1999). The aleurone layer, which is devoid of starch, is made up of equal amounts of oil, mineral matter and protein (Kent, and Evers, 1994). In comparison to other parts of the kernel, this layer is high in protein, ash, phosphorus and vitamins (especially niacin). It acts as a major store of phosphate and macronutrient mineral elements within the cereal grain. Phytate, a potassium, magnesium and calcium salt of *myo-inositol* hexaphosphoric acid, is the storage form of phosphorus. The endosperm consists mainly of starch granules that are surrounded by protein matrix.

Chemical Composition

Starch

Starch is the most abundant carbohydrate and the main energy-yielding component in wheat. The starch content varies from 60 to 75% and is inversely related to the protein content of wheat. Starch in wheat occurs as discrete particles or granules within the cells of the endosperm. These granules are relatively densely packed and are insoluble. Three types of starch granules are present in the endosperm, namely, large (10-50 μm) lenticular shape (type A), medium (8-11μm) spherical shape (type B) and small starch granules with a maximum diameter of 5.3 μm (type C). In the mature grain, type A granules form 51.6%, type B 45% and type C 3.4% of the total mass of starch.

Table 1. The types and level of NSP present in wheat (g/ 100 g dry matter)[1]

	Arabinoxylan	β-Glucan	Cellulose	Total NSP
Soluble	1.7	0.4	0	2.3
Insoluble	6.6	0.5	2.3	9.8

[1] Mean of six wheat cultivars (Ravindran, G. and Ravindran, V., unpublished data).

Non-starch Polysaccharides

Non-starch polysaccharides are the main constituents of cell wall materials, and make up the portion of plant tissue that is not digested by endogenous secretions in the digestive tract (referred to in human nutrition as 'dietary fiber').

Cell wall polysaccharides in wheat are mainly arabinoxylans and (1→3),(1→4)-β-D-glucans, but also contain small quantities of cellulose. On average, wheat contains about 12% non-starch polysaccharides, of which approximately 8% is arabinoxylans, 2% cellulose and 1% β-glucans (Table 1). Some of the most important characteristics of soluble arabinoxylans is their ability to absorb water and readily form viscous solutions or gels. It is known that arabinoxylans and other non-starch polysaccharides (NSP) create viscous contents in the gut of poultry and pigs, which impede the digestion of nutrients by reducing the mixing of digestive enzymes with their substrates. This characteristic has long been recognized as a factor limiting the use of wheat in poultry diets.

Protein

Protein is the second most abundant constituent of cereal grains, with the protein content of commercial wheat varieties ranging from 8-16%, depending on variety and growing conditions (Lasztity, 1999). The average crude protein content of wheat, however, is higher than those of other cereal grains. Glutenin (high molecular weight protein) and gliadin (low molecular weight protein) are the main protein fractions of wheat and each contribute approximately 30-50% of the total protein content in wheat, with the albumins and globulins each contributing 5-10% (Lasztity, 1999). The first two fractions are often referred to as gluten.

The concentration in amino acids in wheats with different protein levels is shown in Table 2. It can be seen that the amino acid concentrations increase linearly with increasing grain protein levels and, in view of its high protein level, that it is a better source on amino acids than corn. Therefore, wheat-based diets require less protein supplementation than corn-based diets.

Endogenous Enzymes

Wheat kernel has its own enzymes that are responsible for the hydrolysis of endospermic starch, releasing energy for growth of the developing embryo when sprouting begins (Stevens *et al.*, 1988). In addition to ∞-amylases activity, xylanases are also present in wheat (Bonnin *et al.*, 1998; Cleemput *et al.*, 1995) with a main function of degrading the walls of aleurone and endosperm cells, during the germination stage making starch and storage proteins accessible to aleurone derived amylases and proteases (Mares and Stone, 1973a,b). Wheat is also known to have high activity of endogenous phytase which increase the availability of phytate-bound phosphorus and this is responsible for the higher availability of wheat phosphorus for poultry in comparison to most other cereal grains (Selle and Ravindran, 2007).

Variation in the Nutrient Composition of Wheat

The nutrient composition of wheat of a given type and variety varies from year to year depending upon area, growing location, fertilization rate, moisture conditions and other agronomic factors. A summary of variation reported in the literature on the chemical composition of wheat is given in Table 3. Of all cereal grains, wheat is the most variable in protein content. The crude protein content (N x 5.83) of wheat may vary from 8 to 18%, but more typically between 10 and 15 depending on the type of wheat, climate and soil fertility.

Table 2. Amino acid composition of wheat as influenced by grain protein level[1,2]

	Protein level in wheat, %			Corn
	9	12	16	(8% protein)
Indispensable amino acids				
Arginine	0.47	0.57	0.75	0.39
Histidine	0.26	0.32	0.42	0.24
Isoleucine	0.37	0.46	0.64	0.33
Leucine	0.68	0.85	1.13	1.09
Lysine	0.31	0.36	0.47	0.25
Methionine	0.13	0.16	0.20	0.14
Phenylalanine	0.45	0.60	0.80	0.43
Threonine	0.32	0.38	0.52	0.37
Valine	0.45	0.55	0.72	0.44
Dispensable amino acids				
Alanine	0.38	0.46	0.56	0.67
Aspartic acid	0.53	0.61	0.85	0.57
Glycine	0.43	0.52	0.71	0.35
Glutamic acid	2.84	4.01	5.45	1.62
Serine	0.55	0.68	0.83	0.46
Tyrosine	0.24	0.33	0.48	0.44

[1] Nitrogen x 5.83.
[2] Ravindran *et al.* (1998; 2005).

Svihus and Gullord (2002) compared 16 samples of Norwegian wheats, reported variation in starch content between 61.4 and 71.2%, protein between 10.9 and 15.4%, fat between 2.2 and 3.4% and crude fiber between 2.0 and 2.6%. Choct *et al.* (1999) surveyed Australian wheat over 3 years reported large variation in AME value, starch, protein, and soluble and insoluble NSP due to year of harvest and geographical location. Carré *et al.* (2007) stated that the wheat protein content depend more on agronomic and climatic conditions, rather than on cultivars. Wiseman and McNab (1995) concluded that wheat cultivar had greater impact on AME than either production site or harvest year. In contrast, Choct (1995) showed that seasonal effects had greater influence than variety on AME and NSP contents. Wheats with high soluble-NSP contents was related to rainfall and environmental temperature patterns during the period of grain maturation. Furthermore, starch content and granular structure, composition, and distribution in the developing grain can also be influenced by the climatic conditions (Hughes and Choct, 1999).

Table 3. Summary of published data on the variability in chemical composition (g/100 g) of wheat[1]

Parameter	Average	Range
Protein	12.0	8.9 – 18.3
Starch	65.0	58.8 – 76.9
Non-starch polysaccharides		
- Total	10.5	8.1 – 15.6
- Soluble	1.1	0.9 – 1.7
Pentosans	6.0	5.4 – 8.2
Arabinoxylans		
- Total	5.6	3.7 – 7.3
- Soluble	0.45	0.40 – 0.52
ß-glucan	0.80	0.70 – 1.00
Arabinose:xylose ratio	70	61 - 82

[1] From Austin *et al.* (1999), Choct *et al.* (1999), Hughes and Choct (1999), March and Biely (1973); Mollah *et al.* (1983), Nicol *et al.* (1993), Sibbald and Price (1977) and Sibbald *et al.* (1963).

NUTRITIONAL VALUE FOR POULTRY

Apparent Metabolisable Energy

Apparent metabolizable energy is the most important measurement used in characterizing the nutritional value of wheat for poultry. However, of all cereal grains, wheat is known to be the most variable in available energy content for poultry. Studies from several parts of the world have shown that the AME of wheat for broilers varies considerably (Table 4) and, variations of up to 1400 kcal/kg have been reported. Such variation is a nutritionists' nightmare and can cause serious management problems and economic losses to poultry producers. An understanding of factors contributing to the variation in the availability of energy in wheat is therefore critical for better utilization of wheat in poultry diets.

Table 4. Variation in the AME of wheat for poultry (kcal/kg dry matter)

Reference	Country	No of samples	AME (kcal/ kg DM)
Sibbald and Slinger (1962)	Canada	25	2940-3960
Mollah *et al.* (1983)	Australia	13	2630-3800
Rogel *et al.* (1987)	Australia	38	2470-3540
Annison, (1991)	Australia	13	2690-3250
Austin *et al.* (1999)	United Kingdom	12	1990-3280
Choct *et al.* (1999)	Australia	81	2190-3580
Ravindran *et al.* (2001)	New Zealand	80	2440-3810
NRC (1994)	North America		3160

The AME value of a wheat sample depends both on the gross content and digestibility of starch, protein and lipids. Carré *et al.* (2007) estimated that the digestibility variations between broiler diets based on 50% wheat can be as high as 15, 10 and 3% for lipids, starch and protein, respectively. These digestibility variations can result in 300 kcal/kg variation in the AME of wheat-based broiler diets. Steenfeldt (2001), however, studied the effect of variability in chemical composition of 16 wheat cultivars on the performance of broilers and reported poor correlations between performance and AME. Therefore care should be taken when using AMEn values alone to predict the nutritive value of wheat and wheat-based diets.

Some wheat samples are known to suffer from very low AME values (less than 3100 kcal/kg), termed as the 'low AME wheat' syndrome. Low AME wheats have been identified with high grain pentosan contents by Australian workers (Annison, 1993; Choct *et al.* 1995). When these wheats are included at levels above 50% in broiler diets, the birds have sticky and watery droppings accompanied by poor growth and feed efficiency. Another unique feature associated with wheat is the 'new season grain' phenomenon, where the AME is low at the harvest time and improves with post-harvest storage. The use of newly harvested wheats as the main energy source in poultry diets is more problematic compared to stored grains (Choct and Hughes, 1997). However, it must be noted that not all wheat samples exhibit this phenomenon.

Starch Digestibility

The major constituent of wheat is starch, which is consequently the principal energy yielding component. Therefore any factors which influence starch digestibility will have an impact on AME.

The starches in cereal grains are generally considered to be highly digestible by the amylolytic enzymes of poultry as they contain very little resistant starch. Longstaff and McNab (1986) found that starch from wheat was almost completely digested by adult birds, with a mean value observed of 99.7% across varieties, seasons and agronomic conditions. However, in the studies of Mollah *et al.* (1983) and Rogel *et al.* (1987), when low-ME wheats were fed to chicken, considerable amount of starch was detected in excreta suggesting incomplete starch digestion. In these studies, the excreta digestibility of starch ranged from 80 to 99% and from 82 to 100%, respectively. Starch digestibility was found to be highly correlated with the AME indicating that variation in starch digestibility was responsible to the low-ME phenomenon. It was concluded that it is not that starch *per se* is poorly utilised in some samples of wheat but that other factors within the wheat are responsible for lowering the digestion of starch. The hypothesis of an access problem should be considered, but this has not been demonstrated in cereals.

Based on the strong positive relationship between starch digestibility and AME, an *in vitro* system has been recently proposed (Wiseman *et al.*, 2000) to separate low and high ME wheats. The results indicate that the rate of starch hydrolysis varies widely in different wheats, with hydrolysis of low-ME wheat being markedly slower.

Ileal Digestibility of Amino Acids

As shown in Table 5, lysine and threonine are the least digestible essential amino acids in wheat. The amino acids in high protein cultivars are more digestible than those in low protein cultivars. This protein effect was particularly marked for the digestibility of lysine and threonine, the two amino acids which are most likely to be deficient when chickens are fed wheat-based diets.

Table 5. Apparent ileal digestibility of amino acids in wheat for broilers[1]

	Grain protein level, %		
	9	12	16
Indispensable amino acids			
Arginine	0.79	0.81	0.81
Histidine	0.78	0.83	0.85
Isoleucine	0.83	0.86	0.88
Leucine	0.84	0.87	0.89
Lysine	0.76	0.77	0.80
Methionine	0.83	0.86	0.89
Phenylalanine	0.83	0.88	0.90
Threonine	0.68	0.74	0.78
Valine	0.80	0.83	0.85
Dispensable amino acids			
Alanine	0.78	0.80	0.79
Aspartic acid	0.73	0.77	0.77
Glycine	0.77	0.81	0.82
Glutamic acid	0.93	0.95	0.95
Serine	0.80	0.83	0.84
Tyrosine	0.73	0.82	0.82

[1] Ravindran *et al.* (1998; 2005).

ANTI-NUTRITIONAL EFFECTS OF NON-STARCH POLYSACCHARIDES

It is believed that the NSP fraction (consisting mainly of arabinoxylans) is the major contributing factor for the observed variability in wheat AME. Two key mechanisms have been proposed to explain the effects of NSP (Bedford and Schulze, 1998). The first mechanism is associated with the fact that starch and protein are encapsulated by the cell wall in wheat endosperm cells. Cell wall composed mainly of cellulose, hemicelluloses (primarily xyloglucans and glucuronarabinoxylans), pectins and lignin. It is noteworthy that most of the arabinoxylans in cereal grains are insoluble in water because they are anchored to the cell walls by alkali-labile ester-like cross links (Choct, 1997). Therefore, insoluble NSP can act as a physical barrier to digestive enzymes. Since the endogenous enzymes in the birds are unable to hydrolyse the cell wall, so the starch and the protein surrounded by cell wall may escape digestion.

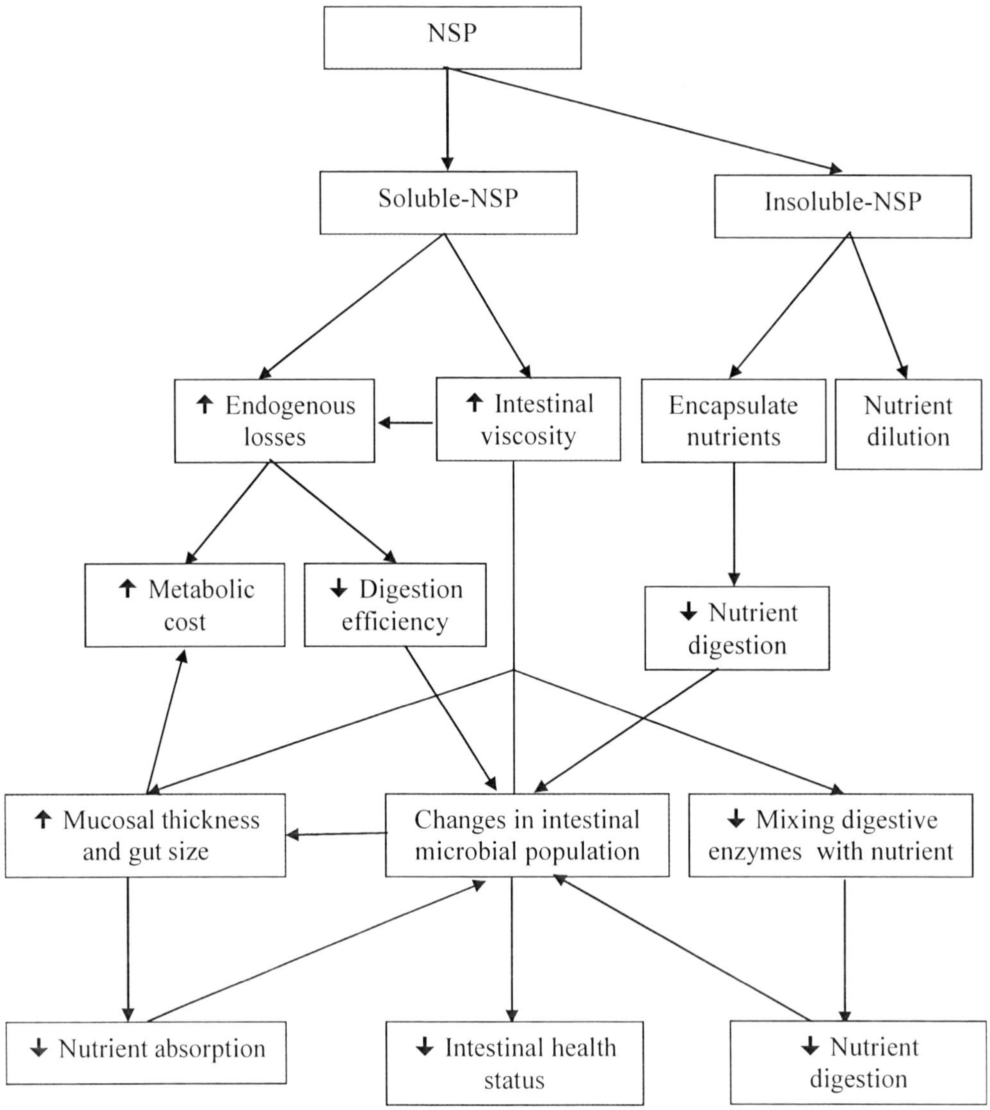

Figure 1. Proposed modes of action of wheat NSP in poultry.

The second mechanism relates to the viscous nature of digesta caused by the soluble NSP fraction. The degree of viscosity depends on the solubility of the NSP which in turn depends on the size of the molecule, whether it is branched or linear, the presence of charged groups, the surrounding structure and the concentration of the NSP. The effects of high intestinal digesta viscosity include (Smits and Annison, 1996; Choct, 1997; Bedford and Schulze, 1998) (1) reduction in passage rate; (2) reduced mixing of digestive enzymes with substrate nutrients; (3) increased secretion of endogenous enzymes and modification of the endogenous secretion of water, proteins, electrolytes and lipids which may increase the endogenous losses; (4) increased secretory response of mucus, which may increase the resistance for transport of nutrients through the unstirred water layer adjacent to the epithelial surface by increasing mucus layer thickness; and (5) interaction with gut microflora (Smits and Annison,

1996). In addition, relative weights of digestive organs increase with higher digesta viscosity (Choct, 1997) which may increase the metabolic cost of maintaining the gut. Although most of the research supports the viscosity mechanism and it has been discussed extensively in the literature, it is reasonable to suggest that both mechanisms are likely to be involved (Bedford, 2006; Cowieson *et al.*, 2006). A summary of proposed effects of soluble NSP on gut and metabolic parameters of birds fed wheat-based diets is shown in Figure 1.

The effects of insoluble NSP in wheat-based poultry diets, however, are not well understood. In general, inclusion of moderate levels of insoluble NSPs had no effect on digesta viscosity (Svihus and Hetland, 2001) but resulted in improved starch digestibility (Rogel *et al.*, 1987; Hetland and Svihus, 2001) and feed per gain (Hetland *et al.*, 2003). The suggested mechanisms by which the birds maintain normal weight gain when fed wheat-based diets diluted with insoluble NSPe are either by increasing capacity of the digestive system and/or faster passage through the digestive tract (Hetland and Svihus, 2001).

Use of Exogenous Xylanases

The effectiveness of exogenous xylanases in improving bird performance and nutrient digestibility of wheats is well documented. Currently, exogenous xylanase preparations are routinely used to mitigate the adverse effects of NSP and to minimize the variation in AME and performance of poultry fed wheat-based diets.

Xylanases have the ability to degrade either the soluble or insoluble arabinoxylans. In general, commercial xylanase preparations have broad spectrum activity on both soluble and insoluble arabinoxylans in the feed. It has been found that xylanases obtained from the same organism differ markedly in their catalytic activities on various xylans (Bedford and Schulze, 1998). Choct *et al.* (2004) compared three xylanase products, namely, xylanase A with affinity for both soluble and insoluble NSP, xylanase B is known to solubilise insoluble NSP and xylanase C with affinity for soluble NSP only. It was found that xylanase A increased the soluble NSP levels in the jejunum, but had no effect on digesta viscosity which suggested that while the enzyme attacked the insoluble cell walls to release soluble NSP into the digesta, it also hydrolysed the released soluble NSP. Xylanase B, increased the soluble NSP levels in both the jejunum and ileum and elevated digesta viscosity while xylanase C reduced digesta viscosity.

The proposed mechanisms by which enzymes improve energy and nutrient utilization of wheat based diets include degradation of NSP in the cell wall matrix and the release of encapsulated nutrients, lowering of digesta viscosity in the intestinal tract, increased accessibility of nutrients to endogenous digestive enzymes, stimulation of intestinal motility and improved feed passage rate or indirectly by enhancing cell wall breakdown through stimulation of the gizzard (Bedford and Schulze, 1998). As a result, enzyme addition has been shown to improve performance and nutrient digestibility, reduce the incidence of sticky droppings and modify the microflora of the distal gut (Cowieson *et al.*, 2006). Broiler responses to xylanase supplementation, however, have not been consistent. Factors that may cause variation in response include the type of xylanase, quality of wheat and breed and age of birds (Bedford, 1997). Some studies have shown that differences in the response of enzyme supplementation might be related to hindgut microflora (Choct *et al.*, 1996; 2006). Other factors suggested to explain variation between individual birds within the same treatment group include diurnal rhythms, water intake patterns and pancreatic serous secretion output (Bedford and Schulze, 1998).

ENDOSPERM HARDNESS AND THE NUTRITIVE VALUE

Hardness is one of the important characteristics of wheat grain that determines whether a wheat variety is classified as hard or soft and this depends on the strength of adhesion between the protein matrix and the starch granules in the endosperm. Endosperm hardness is an important aspect in determining the milling outcomes of wheat cultivars. A hard endosperm gives larger irregular shaped particles (Rose *et al.*, 2001) which are easy to handle (GreVeuille *et al.*, 2006). Due to the hard texture, the endosperm requires more energy to grind into flour, resulting in a greater level of starch damage during the milling process than experienced with soft wheats. This also results in increased water absorption during processing, for example, during the pelleting process. In the baking industry, hard wheat flours are used for bread-making, while soft wheat flours are using for biscuit and cakes.

Despite the wide use of wheat in poultry diets, studies on the effect of wheat endosperm hardness on broiler performance are limited. Harder wheat produces larger particle size flours, which may account for the better broiler performance reported using mash diets based on hard wheats (Rose *et al.*, 2001; Pirgozliev *et al.*, 2003). In pelleted diets, Salah Uddin *et al.* (1996) compared two wheat cultivars selected to be similar in nutrient composition but differing in hardness value. They found no effect of wheat hardness on broiler performance either in ground or whole grain form. Similarly, Hetland *et al.* (2007) found no relationship between wheat hardness and broiler performance.

A negative relationship between wheat hardness and the digestibility of starch in pelleted diets has been reported (Carre *et al.*, 2002; 2005; Peron *et al.*, 2006). This effect of hardness was attributed to larger particulate size reducing the surface area and accessibility for digestive enzymes (Carre *et al.*, 2005). In support of this theory, Peron *et al.* (2005) found that fine grinding of hard wheat improved starch digestibility in broilers fed pelleted diets. Similarly, Short *et al.* (2000) found that the hard wheat endosperm was associated with lower amino acid digestibility. It has been suggested that the interaction between the starch granules and the surrounding protein matrix may act to obstruct enzyme hydrolysis of starch in hard wheat (Guerrieri *et al.*, 1997; Peron *et al.*, 2006). Conversely, Salah Uddin *et al.* (1996) found that the AME of pelleted wheat diets was not affected by endosperm hardness in broilers. This lack of a relationship between grain hardness and AME or starch digestibility in wheat-based mash diets has been reported by several workers (Rogel *et al.*, 1987; Rose *et al.*, 2001; Pirgozliev *et al.*, 2003).

WHOLE WHEAT INCLUSION IN POULTRY DIETS

The need to reduce feed costs renewed the interest within the feed industry in whole grain feeding, which was a common practice at early days of commercial poultry production. Wheat is the common cereal considered in whole grain feeding and, only limited studies have examined the effect of other whole grains in broiler diets. Poultry can be offered whole grain using three techniques, namely, (1) Free choice feeding, where the birds reared in a flock are offered whole grain and another feed in separate feeders ad libitum, and they have the choice to select ingredients, (2) Mix feeding, where the whole grain is substituted for a part of ground grain in a pelleted complete diet or adding whole grain on top of the complete diet in the same feeder at the same time or (3) Sequential feeding, where the whole grain is fed in sequence with the other feed. In general, free choice feeding of whole grain with a balancer diet or a complete diet has been reported to result in poor live weight gains, but the effect on feed efficiency is variable between studies. There appears to be a general consensus that inclusion of whole grain in broiler diets using mixed feeding technique will have no adverse effect on broiler performance.

Several studies have demonstrated the beneficial effects of whole grain feeding on nutrient utilization, which may be related to the positive effect of whole grain on gizzard development. A more developed gizzard is associated with increased grinding activity, resulting in increased gut motility and greater digestion of nutrients. Higher gizzard functionality may also play a positive role in the control of bacterial populations in the gut and on gut health.

Conclusion

Amongst the cereal grains, wheat is considered as having the most variable composition. In particular, the variability in AME values is a difficult problem for the feed industry and limits the use of wheat in diets for poultry, especially the young birds. It is believed that the soluble NSP fraction (consisting mainly of arabinoxylans) is the major factor contributing to the observed variability in wheat AME. However, exogenous xylanase preparations have been effective in mitigating the adverse effects of NSP, minimizing the variation in AME and improving the performance of poultry fed wheat-based diets. With the inclusion of exogenous enzymes, wheat is an excellent energy source for poultry and can be used in poultry feed formulations without any restriction.

References

Annison, G. (1991). Relationship between the levels of soluble non-starch polysaccharides and the apparent metabolisable energy of wheats assayed in broiler nutrition. *Journal of Agricultural Food Chemistry*, *39,* 1252-1256.

Annison, G. (1993). The role of wheat non-starch polysaccharides in broiler nutrition. *Australian Journal of Agricultural Research*, *44,* 405-422.

Austin, S. C., Wiseman, J. & Chesson, A. (1999). Influence of non-starch polysaccharides structure on the metabolisable energy of UK wheat fed to poultry. *Journal of Cereal Science*, *29,* 77-88.

Bedford, M. R. (1997). Reduced viscosity of intestinal digesta and enhanced nutrient digestibility in chickens given exogenous enzymes. In R. R. Marquardt, & Z. Han (Eds.),

Enzyme in Poultry and Swine Nutrition (pp. 19–28). Ottawa, Canada: International Development Research Centre.

Bedford, M. R. & Schulze, H. (1998). Exogenous enzymes for pigs and poultry. *Nutrition Research Reviews, 11,* 91–114.

Bedford, M. R. (2006). Effect of non-starch polysaccharides on avian gastrointestinal function. In G. C. Perry (Ed.), *Avian gut function in health and disease* (pp. 159–170). Wallingford, UK: CAB International.

Bonnin, E., Le Goff, A., Saulnier, L., Chauraud, M. & Thibault, J.-F. (1998). Preliminary characterization of endogenous wheat arabinoxylandegrading enzymic extracts. *Journal of Cereal Science, 28,* 53–62.

Carré, B., Idi, A., Maisonnier, S., Melcion, J. P., Ouryt, F-X., Gomez, J. & Pluchard, P. (2002). Relationships between digestibilities of food components and characteristics of wheats (*Triticum aestivum*) introduced as the only cereal source in a broiler chicken diet. *British Poultry Science, 43,* 404-415.

Carré, B., Muley, N., Gomez, J., Ouryt, F-X., Laffitte, E., Guillou, D. & Signoret, C. (2005). Soft wheat instead of hard wheat in pelleted diets results in high starch digestibility in broiler chickens. *British Poultry Science, 46,* 66-74.

Carré, B., Mignon-Grasteau, S., Peron, A., Juin, A. H. & Bastianelli, D. (2007). Wheat value: improvements by feed technology, plant breeding and animal genetics. *World's Poultry Science Journal, 63,* 585-596.

Choct, M. (1995). *Role of soluble and insoluble fibre in broiler nutrition.* Final Report to Chicken Meat Research and Development Council on Project CSN 2CM. O'Halloran Hill, SA: CSIRO.

Choct, M. (1997). Feed non-starch polysaccharides: Chemical structures and nutritional significance. *Feed Milling International, June Issue,* 13-26.

Choct, M. & Hughes, R. J. (1997). The nutritive value of new season grains for poultry. In J. L. Corbett, M. Choct, J. V. Nolan & J. B. Rowe (Eds.), *Recent Advances in Animal Nutrition in Australia.* (pp. 146-150). Armidale, NSW: University of New England..

Choct, M., Hughes, R. J., Angkanaporn, K. & Annison, G. (1995). Non-starch polysaccharide-degrading enzymes increase the performance of broiler chickens fed wheat of low apparent metabolisable energy. *Journal of Nutrition, 125,* 485-492.

Choct, M., Hughes, R. J. & Annison, G. (1999). Apparent metabolisable energy and chemical composition of Australian wheat in relation to environmental factors. *Australian Journal of Agricultural Research, 50,* 447-451.

Choct, M., Kocher, A., Waters, D. L. E., Pettersson, D. & Ross, G. (2004). A comparison of three xylanases on the nutritive value of two wheats for broiler chickens. *British Journal of Nutrition, 92,* 53–61.

Choct, M., Sinlae, M., Al-Jassim, R. A. M. & Pettersson, D. (2006). Effects of xylanase supplementation on between-bird variation in energy metabolism and the number of Clostridium perfringens in broilers fed a wheat-based diet. *Australian Journal of Agricultural Research, 57,* 1017–1021.

Cleemput, G., Bleukx, W., Oort, M., Hessing, M. & Delcour, J. A. (1995). Evidence for the presence of arabinoxylan-hydrolysing enzymes in European wheat flours. *Journal of Cereal Science, 22,* 139–145.

Cowieson, A. J., Hruby, M. & Pierson, E. E. M. (2006). Evolving enzyme technology: impact on commercial poultry nutrition. *Nutrition Research Reviews, 19,* 90–103.

GreVeuille, V., Abecassis, J., Rousset, M., Oury, F-X., Faye, A., Bar L'Helgouac'h, C. & Lullien-Pellerin, V. (2006). Grain characterisation and milling behaviour of near-isogenic lines differing by hardness. *Theoretical and Applied Genetics, 114,* 1–12

Guerrieri, N., Aynard, L., Lavelli, V. & Cerletti, P. (1997). Interaction of protein and starch studies through Amyloglucosidase action. *Cereal Chemistry, 74,* 846-850.

Hetland, H. & Svihus, B. (2001). Effect of oat hulls on performance, gut capacity and feed passage time in broiler chickens. *British Poultry Science, 42,* 354-361.

Hetland, H., Svihus, B. & Krogdalhl A. (2003). Effects of oat hulls and wood shaving on digestion in broilers and layers fed diets based on whole or ground wheat. *British Poultry Science, 44,* 275–282.

Hetland, H., Uhlen, A. K., Viken, K. H. K., Krekling, T. & Svihus, B. (2007). Hagberg falling number and the nutritional value of wheat in broiler chicken diets. *British Poultry Science, 48,* 12 – 20.

Hughes, R. J., & Choct, M. (1999). Chemical and physical characteristics of grains related to variability in energy and amino acid availability in poultry. *Australian Journal of Agricultural Research, 50,* 689-701.

Kent, N. L. & Evers, A. D. (1994). *Kent's Technology of cereals*. Oxford, UK: Elsevier Science Ltd.

Lasztity, R. (1999). *Cereal chemistry*. Budapest, Hungary: Akademiai Kiado.

Longstaff, M. & McNab, J. M. (1986). Influence of site and variety on starch, hemicellulose and cellulose composition of wheats and their digestibilities by adult cockerels. *British Poultry Science*, *27,* 435-449.

March, B.E. & Biely, J. (1973). Chemical, physical and nutritional characteristics of different samples of wheat. *Canadian Journal of Animal Science, 53,* 569-577.

Mares, D. J. & Stone, B. A. (1973a). Studies on wheat endosperm. I. Chemical composition and ultrastructure of the cell walls. *Australian Journal of Biological Sciences, 26,* 793–812.

Mares, D. J. & Stone, B. A. (1973b). Studies on wheat endosperm. II. Properties of the cell wall components and studies on their organization in the wall. *Australian Journal of Biological Sciences, 26,* 813–830.

Mollah, Y., Bryden, W. L., Wallis, I. R., Balnave, D. & Annison, E. F. (1983). Studies of low metabolisable energy wheats for poultry using conventional and rapid assay procedures and the effects of feed processing. *British Poultry Science, 24,* 81-89.

Nicol, N. T., Wiseman, J. & Norton, G. (1993). Factors determining the nutritional value of wheat varieties for poultry. *Carbohydrate Polymers, 21,* 211-215.

Peron, A., Bastianelli, D., Oury, F-X., Gomez, J. & Carre, B. (2005). Effects of food deprivation and particle size of ground wheat on digestibility of food components in broilers fed on a pelleted diet. *British Poultry Science, 46,* 223-230.

Peron, A., Gomez, J., Mignon-Grasteau, S., Sellier, N., Besnard, J., Derouet, M., Juin, H. & Carré, B. (2006). Effects of wheat quality on digestion differ between the D+ and D- chicken lines selected for divergent digestion capacity. *Poultry Science, 85,* 462-469.

Pirgozliev, V. R., Birch, C. L., Rose, S. P., Kettlewell, P. S. & Bedford, M. R. (2003). Chemical composition and the nutritive quality of different wheat cultivars for broiler chickens. *British Poultry Science, 44,* 464-475.

Ravindran, V., Hew, L. I. & Bryden, W. L. (1998). Digestible *Amino Acids in Poultry Feedstuffs*. Rural Industries Research and Development Corporation, Canberra and Poultry Research Foundation. Sydney, NSW: The University of Sydney.

Ravindran, V., Thomas, D. V., Camden, B. J. & Voon, V. H. (2001). Apparent metabolisable energy content of New Zealand wheats. *Proceedings of the Australian Poultry Science Symposium, 13,* 240.

Ravindran, V., Hew, L. I., Ravindran, G. & Bryden, W. L. (2005). Apparent ileal digestibility of amino acids in feed ingredients for broiler chickens. *Animal Science, 81,* 85-97.

Rogel, A. M., Annison, E. F., Bryden, W. L. & Balnave, D. (1987). The digestion of wheat starch in broiler chickens. *Australian Journal of Agricultural Research, 38,* 639–649.

Rose, S. P., Tucker, L. A., Kettlewell, P. S. & Collier, J. D. A. (2001). Rapid tests of wheat nutritive value for growing chickens. *Journal of Cereal Science, 34,* 181-190.

Salah Uddin, M. S., Rose, S. P., Hiscock, T. A. & Bonnet, S. (1996). A comparison of the energy availability for chickens of ground and whole grain samples of two wheat varieties. *British Poultry Science, 37,* 347-357.

Selle, P. H. & Ravindran, V. (2007). Microbial phytase in poultry nutrition. *Animal Feed Science and Technology, 135,* 1-41.

Sibbald, I. R. & Slinger, S. J. (1962). The metabolizable energy of material fed to growing chicks. *Poultry Science, 41,* 1612-1613.

Sibbald, I. R. & Price, K. (1977). True and apparent metabolizable energy values for poultry of Canadian wheats and oats measured by bioassay and predicted from physical and chemical data. *Canadian Journal of Animal Science, 57,* 365-374.

Sibbald, I. R., Czarnocki, J., Slinger, S. J. & Ashton, G. C. (1963). The prediction of the metabolizable energy content of poultry feedingstuffs from a knowledge of their chemical composition. *Poultry Science, 42,* 486-492.

Short, F. J., Wiseman, J. & K. N. Boorman. (2000). The effect of the 1B/1R translocation and endosperm texture on amino acid digestibility in near-isogenic lines of wheat for broilers. *The Journal of Agricultural Science, 134,* 69–76.

Smits, C. H. M. & Annison, G. (1996). Non-starch plant polysaccharides in broiler nutrition-towards a physiologically valid approach to their determination. *World's Poultry Science Journal, 52,* 203–221.

Steenfeldt, S. (2001). The dietary effect of different wheat cultivars for broiler chickens. *British Poultry Science, 42,* 595–609.

Stevens, B. J .H., Selvendran, R. R., Bayliss, C. E. & Turner, R. (1988). Degradation of cell wall material of apple and wheat bran by human faecal bacteria in vitro. *Journal of the Science of Food and Agriculture, 44,* 151-166.

Svihus, B. & Gullord, M. (2002). Effect of chemical content and physical characteristics on nutritional value of wheat, barley and oats for poultry. *Animal Feed Science and Technology, 102,* 71-92.

Svihus, B. & Hetland, H. (2001) Ileal starch digestibility in growing broiler chickens fed on a wheat-based diet is improved by mash feeding, dilution with cellulose or whole wheat inclusion. British Poultry Science 42: 633-637.

Wiseman, J. & McNab, J. M. (1995). Nutritive value of wheat varieties fed to non-ruminants. HGCA Project Report No. 111. Home Grown Cereals Authority, London.

Wiseman, J., Nicol, N. T. & Norton, G. (2000). Relationship between apparent metabolisable (AME) values and *in vivo/in vitro* starch digestibility of wheat for broilers. *World's Poultry Science Journal, 56,* 305-318.

INDEX

A

B

C

D

E

G

H

I

J

K

L

M

N

Q

R

S

T

U

V

W

X

Y

Z